Apoptosis Handbook

Edited by **Max Cowen**

New York

Published by Callisto Reference,
106 Park Avenue, Suite 200,
New York, NY 10016, USA
www.callistoreference.com

Apoptosis Handbook
Edited by Max Cowen

International Standard Book Number: 978-1-63239-073-8 (Hardback)

Printed in the United States of America.

Contents

Preface

This book aims to highlight the current researches and provides a platform to further the scope of innovations in this area. This book is a product of the combined efforts of many researchers and scientists, after going through thorough studies and analysis from different parts of the world. The objective of this book is to provide the readers with the latest information of the field.

Apoptosis refers to the procedure of programmed cell death (PCD). This book reflects the recent research studies on apoptosis in medicine. It provides an introduction of apoptosis to the uninitiated. It also discusses its applications in normal physiology during bone resorption under mechanical stress and deals with the analysis of apoptosis in a number of pathological conditions with special emphasis on cancer.

I would like to express my sincere thanks to the authors for their dedicated efforts in the completion of this book. I acknowledge the efforts of the publisher for providing constant support. Lastly, I would like to thank my family for their support in all academic endeavors.

Editor

Introduction to Apoptosis

Extrinsic and Intrinsic Apoptosis Signal Pathway Review

Zhao Hongmei

Additional information is available at the end of the chapter

1. Introduction

Apoptosis, as a programmed cell death (PCD) is essential for normal cell mechanism. The word "Apoptosis" derives from Greek language "απόπτωσις" and means trees shedding their leaves in autumn, which describes the "dropping off" or "falling off" of petals from flowers, or leaves from trees. This language imaginarily described the cell death triggered by physiological and pathological stimulation. The apoptosis phenomena were first described by German Scientist Carl Vogt in 1842 year, while until 1972 year, apoptosis term was first used by John Foxton Ross Kerr group. They use apoptosis word to describe the tissue cell death. The above is the beginning of apoptosis researches and this period is the apoptosis formation; the second period about apoptosis is the biochemical level and apoptosis cell morphological changes research. In this age, people know that apoptosis accompany with cell membrane wrinkled, DNA fragmentation, cytosol calcium increased and form the apoptosis body which contain its own content so on. In this time the electron microscope play a vital role in the research. Following the apoptosis research came in the third period in recent years; scientists began to research the molecular mechanism of apoptosis and to use this cell death for clinical treatment. Some key proteins in the procession of the apoptosis have been found, such as Bcl-2 family protein, caspase-3, caspase-8, caspase-9, Bid, Bax and so on.

In the past, the apoptosis was focused on the caspase, a family of cysteine protease. While, using the caspase inhibitor to block the apoptosis pathway, the researchers found that the apoptosis still happen. So another pathway that is caspase-independent was found. Now, apoptosis is classified to type I, Type II, Type III PCD: type I PCD is the classic apoptosis, the well know caspase denpendent apoptosis; type II PCD's morphology characters are the appearance of the autophagic and double membrane of vacuole; type III PCD occurs without the condensate chromatin and has not been well-known. Type IIand type III PCD are the caspase-independent apoptosis. For example, the apoptosis induce factor(AIF), a mitochondria intermembrane flavoprotein, that can be released from mitochondria to translocate into the nuclear and cause many high molecular weight DNA fragmentation and chromatin condensation in cells, this type of apoptosis is in the caspase-independent manner. [65] No mater the typeI, Type II or type III PCD, the apoptosis will assist the host to defense the outer or inner aliens and toxic compounds and help the organism survive.

In this chapter, we will key on the apoptosis signal pathway and some ligands have the clue with apoptosis and mainly on concluded the apoptosis on two aspects, from in vivo and in vitro cells, and we can clarify the network of this cell death and give the conclusion of the latest research about this.

2. Apoptosis research time line

Cell death is essential for body homeostasis, currently years research, cell death can be classified as necrosis, apoptosis, pyroptosis. Necrosis is the first to be found, and then in the 1972 year, the apoptosis term began to be used; several years ago, pyroptosis was been known by the symbol of formation hole in cell membrane and released inflammatory factors. Apoptosis research started in the nineteen century. According to the past research, we collect and analyze the research of apoptosis and give the following summary of history and highlights about apoptosis in Table1

Year	Scientist	Research
1842	Carl Vogt	First describe the principle of apoptosis
1885	Walter Flemming	Give more precise description of PCD
1965	John Foxton Ross Kerr	Distinguish the apoptosis used by electronic microscopy
1972	John Foxton Ross Kerr	Initially used the apoptosis term
2002	Sydney Brenner, Horvitz and John E suston	Awarded to Nobel prize in medicine according they contribute in apoptosis research area.

Table 1. History and highlights about apoptosis research

In 2002 year, the Nobel Prize in medicine was awarded to the Sydney Brenner, Horvitz and John E suston, according they contribute to the apoptosis research. They make a stone research work in organ development research and genetic systemic regulation programme cell death. Sydney Brenner's contribution is the construction of nematode of C.elegan model. And subsequently, suston found the cell lineage of C.elegan and found the first gene(Nuc-1) related with apoptosis. Now the C.elegan is still the classic model to research apoptosis and organ development research, Now, apoptosis has been found in the tumor development, and it has been clear that alteration of apoptotic pathways is a common feature of tumors, enabling cancer cells to survive chemotherapeutic interventions, so apoptosis signal pathway molecular become attractive targets in cancer therapy. Beside involving in the tumor development, apoptosis have been found in many other diseases, such as neurodegenerative diseases, diabetes, stroke and so on. Here, we review the fundamental apoptotic pathways and summarize many ligands that can trigger apoptosis and give some insights in the designing some drugs, plus, we also give the opinions that changes the dietary arable may be help to improve the people's healthy.

3. Apoptosis signal pathway

Apoptosis is triggered by multi-signal pathways and regulated by multi-complicated extrinsic and intrinsic ligands. The process of apoptosis is controlled by diversity cell signals pathway and involved in regulation of cell fate death or survival. There are two major apoptosis pathways distinguished according to whether caspases are involved or not. The mitochondria, as the cross-talk organelles, can connect the different apoptosis pathway.

3.1. Caspase dependent pathway

Caspase-dependent apoptosis is the classic programmed cell death pathway, the capase-8, caspase-9, caspase-12, caspase-7, caspase-3 cascade usually participate in this type of apoptosis pathway, Variety of receptors take part in this type of apoptosis pathway, such as the TNF-alpha receptor, FasL receptor, TLR, Death receptor and so on. Some iron channels may also be involved in apoptosis pathway. The typical iron channel is the calcium channel, Since calcium's concentration in the cytosol plays an important role in the signal transduction regulation and participates in the cell proliferation and cell death, the cell fate can be controlled by the calcium channel opening or closing. In this part, we will discuss the caspase-dependent apoptosis transduction and review the complex signal crosstalk in the cells.

TNF-alpha induced caspase-8-dependent pathway relies on the TNF-alpha receptor and activates the caspase-8 through the death complex, and then the Bcl-2 protein is activated, Bcl-2 family protein activation may induced the mitochondria membrane changed and stimulates the cytochrome c released. Cytochrome c is the proapoptosis signal molecular which can activates the caspase cascade reaction and induced the apoptosis in the end. Some radiation UV or X ray can make mitochondria depolarization and membrane permeabilization, subsequently, the ROS increased; cytochorme c released, and then trigger caspase-9, caspase-3 activation,, In the end, the substracts will be cleavaged by the activation caspases and the fate of cells will be apoptosis; Some pathogen infection induced the apoptosis may be also have the caspase-8 dependent pathway, The alien pathogens can be recognized by FasL receptor and recruit FADD and caspase-8, for example, the intracellular pathogen herpervirus infection can induced the caspase-8 dependent apoptosis. Beside caspase-8 dependent apoptosis, some pathogens can trigger others caspases dependent apoptosis pathway. For example, Mycobacterium tuberculosis can induce programmed cell death on macrophage, and this apoptosis pathway is the caspase-12 dependent. NO and ROS production, stimulated by ER stress, also take part in apoptosis triggered by Mycobacterium tuberculosis; [1] Exclude bacteria, virus also can induce the apoptosis. The latest research found that an alternative Kaposi's sarcoma-associated herpesvirus replication can trigger host cell apoptosis in caspase dependent manner. [2]. Apart from the bacteria and virus, we found that many researchers' data show that RNA fragments and DNA can also trigger caspase dependent apoptosis, such as RNA fragment produced by mycobacterium tuberculosis which in the early log-phase growth can trigger caspase-8 dependent apoptosis[3]; In vivo, DNA damage can trigger apoptosis through enhancing ROS level and changing the mitochondria membrane permeability; Many proteins or peptides will also make cell apoptosis, amyloid β peptide cytotoxicity can induce the intracellular calcium disturbance, and then the calpain will be activated by imbalanced calcium storage, While the calpain can activate caspase-12 which can located in ER to inactivate the Bcl-Xl, this is a novel caspase-12 dependent apoptosis pathway[4]. This way can be used by the mitochondria to connect with other cell death pathway.

Above all, we can give the conclusion that pathogens, RNA or DNA, proteins or peptides, some chemical compounds or native compounds can all trigger caspase dependent apoptosis. They may have the different receptors and induce cell death through different caspase as the transducer to make the downstream signals transduction. The host used this way to defense the harmful factors and maintain the healthy physiological state. In the figure 1, we summarized the caspase dependent pathway transduction; we clearly known that the different ligands and receptors involved in this type of apoptosis, this picture will give us the direct impression about this type of apoptosis.

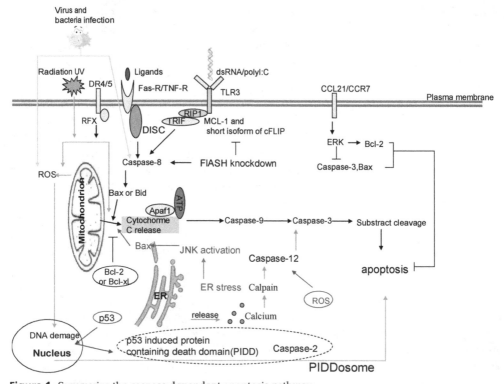

Figure 1. Summarize the caspase dependent apoptosis pathway

In this figure, we can clearly know that mitochondrion and nucleus organelles play the pivotal role in this type cell death and these organelles can connect different signals to induce the caspase activation, in this process period, ROS; Cytochorme C release; mitochondrion membrane potential change; Apart the mitochondrion pathway, some ligands can trigger MAPK pathway, such as activate ERK and then activate the caspase family which can be confluenced with apoptosis pathway.

3.2. Caspase independent pathway

Apoptosis have many mechanisms. It can be triggered by in vitro and in vivo cell ligands, and have different signal transduction pathways. Now, Apoptosis pathways are classified as caspase dependent pathway and caspase-independent pathway. In the above, we describe and conclude the caspase dependent pathway, which is characterized by the involvement of caspase in this type cell death process. In this paragraph, we will deep research the caspase independent apoptosis pathway. We will know that caspase does not participate in this type of process from the literally meaning. Up to now, this type of apoptosis has been founded by many researchers, and the research data give much information to us, this information can help us to understand the apoptosis complex mechanism, beside the mechanisms, the complicated ligands, which can induce this type of cell death, can also be found by many researchers. In the following, we will give some detailed contents about caspase independent apoptosis.

In the cell, a lot of ligands can induce mitochondria membrane potential change, the mitochondria damage will be the first step of the apoptosis, then ROS production increase, and ROS may the mainly factor to induce caspase independent apoptosis. For example, Denis Martinvalet found that granzyme A can directly induce the ROS increase and caspase independent mitochondria damage. Then the target of granzyme A, ER associated complex (SET complex), will translocated to nuclear and contributed to apoptosis [5]; AIF has been found the major important caspase-indenpendent pro-apoptosis factor, which can release from the mitochondria and translocate in the nuclear to cleavage the DNA, in the end, if the DNA damage have not been repaired by cells, the apoptosis will happen. Recently, many researchers found some compounds which can accompany with AIF production and induce cell death, such as simvastatin, staurosporine, cadmium and so on. These factors triggered caspase-independent PCD and fitted for the organism requirement; Beside AIF triggered caspase-independent PCD, ROS also participate in this type cell death. ROS can mediate poly (ADP-ribose) polymerase-1 (PARP-1) activation, and PARP-1 activation is necessary for AIF release from mitochondria. So ROS also involved in this type of cell death networks.[6].However, ROS participated in the caspase dependent apoptosis pathway also, Consequently, ROS might be the important bridge to connect two types of apoptosis in vivo. ROS mainly come from mitochondria, so mitochondria play important role in apoptosis pathways crosstalk. And the ligands usually trigger complicated reactions, including that AIF nuclear translocation, ROS increase and mitochondrial dysfunctions, these changes can cause to the caspase independent apoptosis pathway. For example, Cyclohexyl analogues of

Ethylenediamine Dipropanoic Acid, the compound that can induce peripheral blood mononuclear cells apoptosis of both healthy controls and leukemic patients through stimulating many apoptogenic factors activation(such as: AIF nuclear translocation, ROS increase and mitochondrial dysfunction). [7] Referring the ROS, we propose that the GSH , NO, or other free radical groups may also take part in this type cell death, By this way, the GSH or NO modification proteins may also take part in the apoptosis pathway, GSH and NO can block some active thiol group and make the protein S-glutathionylation or S-nitrosylation modification. These types modification may affect the protein's functions and make the cell to apoptosis result. Beside the AIF and ROS, there are many other ligands and signal molecular from the vitro or vivo cells as apoptogenic factors involving in caspase independent apoptosis pathway, such as lysosomal membrane permeabilization; some virus's protein; drugs; p53 suppression tumor factors or many other unknown compounds.

Up to now, the caspase independent apoptosis mechanism is still unknown clearly. Although some researchers have found AIF; ROS and other ligands can stimulate this type of PCD, the signal pathway still stay the phenomena stage and the deep mechanism should be dig out by us. No matter either apoptosis form, this type cell death has very important functions and deserves to be researched deeply. We collected a variety of information about this cell death and found that the caspase independent apoptosis existed in a mount of species and played indispensable role in cell growth, proliferation and death. In the figure 2, we give the outline about this type apoptosis pathway and the crosstalk manner between the different pathways, this picture will help us to know this type apoptosis well in the whole level.

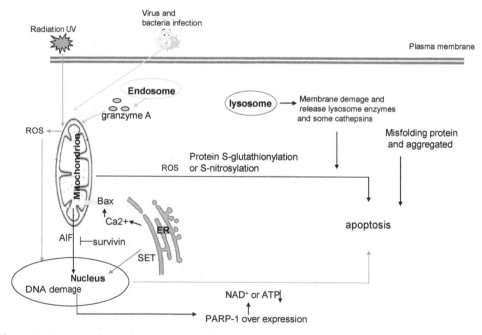

Figure 2. Caspase independent apoptosis pathway

In this type apoptosis, caspase family members did not involve in this cell death, and can not be inhibited by the caspase inhibitor {such as: the z-VAD-FMK; quinolyl valyl-o methylaspartyl[-2,6-difluorophenoxy]-methyl ketone(Q-VD-OPH);Ac-DEVD-FMK and so on}. In the cells, some components and events, such as the AIF; ROS; Ca^{2+}; NAD^+ and ATP; protein misfolding and modification, can trigger the caspase independent apoptosis.

3.3. Mitochondria dynamics and apoptosis

Mitochondria, as a semiautonomous organelle in cells, apart from containing their own genetic material, play an important role in the energy metabolism. It can produce ATP to maintain the cell life activity and be known as an energy company in cells. Beside this major role, mitochondria are the places that the lot of biological reaction processes happened, such as ROS production, apoptosis, and regulation of aging [8] and so on. Mitochondria's dysfunction has the relation with many diseases (Alzheimer's disease; Parkinson's disease, cancer, diabetes) [9, 10]. These diseases have been identified to have some relation with the apoptosis; ROS produced by mitochondria have been regarded as one of important factors for apoptosis. Currently, many researches found the ROS produced increased when some pathogen infected, ROS can trigger apoptosis, through apoptosis, the pathogens may lost the perfect living environment, the host may defense the pathogen's diffusion by this manner. Due to these roles, mitochondria may be used as a proper therapeutic target to cure diseases related with this type cell death [11].

As a dynamic organelle, mitochondria can change their shape and structure constantly to respond to the different stimuli and metabolic demands of cells. According to the latest researches about biochemistry and cell biology, the mitochondrial shape changes between fusion and fission play a very important role in the regulation of apoptosis [12, 13]. There were some debates about the opinion that apoptosis occurred relating with the mitochondria fission. These debates focused on which process happened firstly, either apoptosis is the result of the mitochondria fission and fragmentation, or reversely, as a following up affair, the mitochondria's fission and fragmentation happened at the downstream of apoptosis. While, we can sure that mitochondria shape dynamic changes must be connected with apoptosis according to the follow latest researches:

Calcium irons act as the upstream stimulus which can activate the cellular mitochondrial fission, the mitochondrial became fragmentation rapidly depended by the increased calcium level in intracellular. If the calcium level increased protractedly, mitochondria's fragmentation will be non-reversible and lead to apoptosis. So Jennifer R. Hom group regarded that the calcium involved in mitochondria morphology and participated in the apoptosis processing;[14] some mitochondria membrane proteins also have been found to control mitochondrial morphology, for example, the Bcl-2 protein resides in the outer mitochondrial membrane, and this protein acts as a central regulator of the intrinsic apoptotic cascade; while some other toxins or proteins can also regulate the mitochondrial fission/fusion, and these variation shapes of mitochondria was found to have relation with some diseases, For instance, the PD(Parkinson's disease) have been found with the mitochondrial morphology changes, there were two toxin proteins parkin and PINK1 which were detected to play a role in maintaining mitochondrial homeostasis through targeting the mitochondria and regulating the mitochondrial dynamics[15]; Fusion and fission often occur swiftly and are found to have relation with the mitochondria membrane potential changes .there are DRP1-dependent division and FZO1-dependent fusion reaction in mammalian cells, and division of mitochondria accompany with apoptosis., If mitochondrial fission/fusion dynamics losses balance, it will lead to some neurodegeneration diseases. So in brief, mitochondria act as the important role to keep cell healthy.

Mitochondrial iron transporter cytochrome C plays an classic role in apoptosis. As a bridge, it connect with caspase cascade reaction. When cytochrome C is released to the cytoplasm from the mitochondria as a result of response to some intrinsic or extrinsic ligands, it can trigger the downstream caspases and induce the intrinsic apoptosis. So mitochondria are not only the energy company of cell, but also have the ability to control the intrinsic apoptosis pathway through either cytochrome C, calcium, morphology changes(fission/fusion) or some membrane proteins expression imbalance(Bcl-2).

4. Trigger apoptosis ligands and cell environment materials

Apoptosis, as a major cell death procession for healthy, play an important role in homeostasis of whole life. In the period of fetation, apoptosis happened, as the essential affair for the finger and toe formation. due the cells between fingers occured apoptosis, we can form five fingers per hand. Apoptosis also can enhance embryonic stem cell survival during stress by increased expression Bcl-2 protein which significantly weaken the apoptosis and help colony formation [16];Apoptosis also take part in some tissues regeneration and homeostasis [17,18]; In addition to the function that we have described above, apoptosis can still anti-inflammation and hamper pathogens persistence infection, especially interfering the intracellular parasites diffusion; In conclusion, apoptosis are good for people's healthy in variety aspects, and apoptosis become research areas hotspots at present. Summarizing the currently researcher's data, we found that many ligands can trigger or inhibit apoptosis. We classified these ligands to two parts by its derivation: 1. extrinsic signal ligands. 2. Intrinsic ligands. The details about the ligands which can trigger the apoptosis will be described below:

4.1. Extrinsic cell materials

4.1.1. Cytokines

TNF-alpha plus z-VAD can trigger cell apoptosis, and this method is well-known in creating cell apoptosis model. TNF-alpha can bind to extracellular domain of TNF-alpha receptor , and the cytoplasm domain can aggregate FADD and FLICE which can initiated the apoptosis; Another famous cytokine, IFN$-\gamma$, which can induce the macrophage apoptosis, plays a key role in clearance of the mycobacterium tuberculosis by inducing host cell apoptosis depended by the nitric oxide(NO).[19] TGF-β1 acts as a chemo-attractant and is very important for the immune response, this cytokine also play a predominant suppressive role in inhibiting the cell proliferation and stimulating B cells to apoptosis. [20]

Above all, the cytokines are the positive inducer of apoptosis. The investigators also found another mechanism of cytokines related apoptosis which is negative induction by loss of suppressive signal. Many cells viability required supplying the constant or intermittent cytokines or growth factors, Lack of cytokine or growth factors, the cell will go to apoptosis. In this circumstance, these cytokines or growth factors act as the suppressors of apoptosis. For example, when we culture hematopoietic cells, we should add colony-stimulating factors and IL-3; In addition, IL-12 is required to maintain the viability of activated T cells in in vitro culture systerm; the neurorophic cells required the nerve growth factor (NGF) for survival. Moreover, there are many other cytokines and growth factors as the suppressors of apoptosis, such as: epidermal growth factor, platelet-derived growth factor and insulin like growth factor-I and so on. They are also act as the survival factors and inhibit the apoptosis, while if we block or remove these factors; the related cells will be in the procession of apoptosis. So in above all, we have known that cytokines may as inducer or suppressor to participate in apoptosis pathway.

4.1.2. Drugs

Some cytotoxic drugs (Cispkatin, Gemcitabin, Tooitecan, and paclitaxel) can trigger apoptosis; Didymin induce apoptosis by preventing N-Myc protein expression and make the cell G2/M arrest, which may be a novel mechanism to anti-neuroblastoma.[21]; apart from this anti-neuroblastoma properties, didymin have an anti-no small cell lung cancer ability by induced the Fas-mediated apoptosis, it may be a novel new chemotherapeutic agent to treat the lung cancer. [22] Gomisin N have anti-hepatotoxic, anti-oxidative and anti-inflammation abilities, while it also have anti-cancer activity through triggered the TRAIL-induced apoptosis.[23] Andrographolide as an anti-bacteria drug have been found having anti-cancer activity adrographolide treated cancer cell can activate the p53 by increasing p53 phosphorylation, p53 activation can make DR4 protein expression increased and then trigger the TRAIL-induced apoptosis.[24] Ursolic acid can stimulate the ROS production and trigger JNK activation, ROS and activated JNK can make the DR up expression, and in the end, TRAIL-induced apoptosis happened in p53 independent manner.[25] Ciglitazone, as an anti-diabetic drug, has been found by Plissonnier ML group that ciglitazone have antineoplastic effectiveness in a lot of cancer cell lines through up-regulation of soluble and membrane bound TRAIL, make caspase activation, death receptor signal pathway activation and induce Bid to be cleaved as response to TRAIL. These data gave the evidence that ciglitazone can down regulate the c-FLIP and surviving protein, and then triggered the TRAIL-induced apoptosis to kill the cancer cell.[26] These data provided the powerful evidence that triggering apoptosis may be a feasible method for clinically cure.

4.1.3. Hormone

Apoptosis occurs in the embryonic development and during the formation of organs, Hormones is usually a peptide or steroid, produced by one tissue and conveyed by the bloodstream to another place to affect physiological activity, such as growth, proliferation, metabolism. Some researchers also found that hormones can regulate the apoptosis, and through this way hormone can control the metabolism of tissues or organs. For example, the leptin is one hormone produced by adipocytes, and it has been found as a inhibitor of apoptosis, leptin can down-regulate cleaved caspase-3 and Bcl-2 associated X protein and up-regulate Bcl-2 protein, if this hormone level is normal, there will be no apoptosis, however, if the hormone level decreased, the apoptosis will happen[27], this is one mechanism by which the hormone act as inhibitor of apoptosis, the other mechanism is that hormone act as an inducer of apoptosis. The majority relations between the hormone and apoptosis are concluded as below the table 2

Hormones	Apoptosis cell (Target)	Reference
Known inhibitors of apoptosis		
Testosterone	Prostate	51
Oestradiol	ovarian cells	53,52
Growth hormone	Human monocytes or human promyelocytic leukaemia	54
Leptin	myometrial cells	27
Dihydrotestosterone	Prostate	51
progeterone	cardiomyocytes	55
Act as inducers of apoptosis		
Glucocorticoids	Human small cell lung cancer	56
progeterone	human endometrial cell	57
Thyroid hormones	Play an important role in Amphibian organ remodeling during metamorphosis through inducing apoptosis	58
Estrogen	Breast cancer cell	59
Phytoestrogens	Breast cancer cell	60

Table 2. Involvement of hormones in apoptosis

4.1.4. Pathogen effectors

During the fight between host and pathogen, there are many roads that benefit the cell survival. There is the proverb "Survival of the fittest". If failing to defense the pathogen, host will be ill, and give the phenomenon of inflammation or cell death (apoptosis, necrosis, auto-phage, pyroptosis). In this part, we will give the conclusion about host cell apoptosis which can be triggered by some pathogens, If cell occured apoptosis, the pathogen can not survival either, so through this way, host cells can clear the pathogen with little bad effects. The pathogen effectors, which possess the ability to trigger apoptosis, have been found as following below:

Mycobacterium can be cleared by macrophage's apoptosis which was induced by the NO and IFN- γ [28]; Chlamydia pneumoniae infection can induce the human T lymphocyte cell apoptosis, through this way, Chlamydia pneumonia could induce immunologic tolerance and would make pathogen persistence infection and inflammation[29]; Dendritic cells(DC) were well known immune cells, as an antigen-process cell, DC can inhibit pathogen replication and diffusion by caspase-3 dependent apoptosis in early infection stage, For instance, *Legionella pneumophila* was unable replicated in DC, because DC go to apoptosis when *Legionella pneumophila* infected these cells in the early stage.[30] Beside the bacteria infection induced the apoptosis, some virus also be found to involve in the apoptosis. For example, transmissible gastroenteritis virus infection can up-regulate the FasL; Subsequently, the Bid protein can be cleaved and cytochrome c release; in the end, Caspase-8 can be activated and the host cell happen apoptosis. [31] Through this way, the pathogens, especially the obligate intracellular parasites, will lost the suitable environment for their life and will die, however, the host will survival and suffer lowest bad effect.

From this part, we give the conclusion that apoptosis play the key role in host-pathogen confrontation. There are about two mechanisms in this confrontation, one mechanism is that pathogen's target cells can clear the pathogen and inhibit it persistence diffusion by itself apoptosis. Another is that pathogens infection didn't induce the targeted cell apoptosis but cause the related immune cells apoptosis. By this way, the host cells will fail to clear the pathogen. The loss function of host immune cells will benefit the pathogen's replication and help pathogen's diffusion persistence. There are always two sides to every thing, the apoptosis also have no exception..

4.1.5. Native activities compounds

Although apoptosis is the programmed cell death and can be recognized as the normal cell death by the immune system; and apoptosis have many important functions in the tissue development. While everything have two sides, in some cases, apoptosis also have the damage and negative effect to the life healthy. Recently, food scientists and biologist found that some native compounds from the daily life dietary can block or hamper apoptosis, and through this way, these native compounds can help to keep the body healthy.

Vitamin E, another name is tocopherol. As an antioxidant, Vitamin E has an important role in redox balance. Recently, apart it's major role in antioxidant ability, vitamin E can block the reduction of the mitochondrial membrane potential and inhibit the activation of caspase-3, in a brief, vitamin E is conducive to cell viability through blocking the caspase-3 triggered apoptosis. [32] Moreover, Purple Sweet Potato Pigments can scavenge ROS and protect the murine thymocyte by inhibiting caspase-dependent pathway apoptosis also [33]. Lycopene, another name is rhodopurpurin, Lycopene can be taken from the tomatoes, fruits and vegetables easily. As an antioxidant, Lycopene was been found that it contribute to body's heath. Nearly, researchers found that lycopene have anti-prostate cancer activity; Apart from the anti-tumor properties, it have anti-infection ability. For example, Jang SH' group gave the proof that lycopene can inhibit ROS increased, DNA damage and apoptosis in gastric epithelial AGS cells induced by helicobactera pylori infection. [34]. Soy isoflavone also have been documented as dietary nutrients broadly, it was been classified as "natural agents" which play the important role in reducing the incidence of hormone-related cancers in Asian countries, and have shown inhibitory effects on cancer's development and progression in vitro and in vivo. [35] Beside this soy isoflavone, in recent years, dietary compounds which was from the bounties of nature have been paid any attentions, the latest researches data have shown that there are some associations between the consumption of some native dietary compounds and the reduced risk of several types of cancers, [36, 37] because there are lots of active compounds containing in the food which have some associations with the apoptosis, and due by apoptosis, these native compounds will inhibit cell abnormal proliferation.

Apart from these food derived natural compounds, there are many plant components which can be used by Chinese traditional medicine (TCM), these plant active components can also have trigger apoptosis, such as fisetin, wongonin, emodin and so on. Fisetin is a natural flavonoid which can induce several type cancer cells to apoptosis by dose and time dependent manner. Fisetin can activate caspase-8/caspase-3 dependent apoptosis pathway, and these pathway transducer molecular will be the candidates for cancer therapy; [38] Wongonin, as an O-methylated flavonoind, was detected to have anxiolytic activity, and also have the ability to trigger some cancer cells apoptosis. For instance, wongonin and fisetin can make PARP to be cleaved, then pro-caspase-3 will be activated in HL-60 cells, while they can induce the ROS decreased, so it is not the ROS dependent apoptosis pathway. Several compounds have the similar structural with flaconoids, including the luteolin, nobiletin, wogonin, baicalein, apigenin, myricetin and fisetin, they all have biological activities. Up to now, wogonin and fisetin have been found that they have the potential ability to trigger apoptosis through caspase-8 dependent manner[40,41]. Rheum palmatum is traditional Chinese medicines which have anti-bacteria; anti-inflammation; and improving microcirculation, emodin is an active component of Rheum palmatum and have apoptosis-inducing activity, Chen YC group found that the emodin can activate caspase-3 cascade and trigger HL-60 cell apoptosis independent of ROS production also. [41]

Some researches gave us many inspirations in the apoptosis research. We known that many tradition Chinese medicines (TCM) have some active constituents, these active components have anti-tumor activity through apoptosis, so combination the molecular biology to dig out the mechanism of drug functions will be effective and scientifical, Now, the functions of some tradition Chinese medicines are still stay the phenomena, we did not know the real mechanism of activity materials. Although TCM have not curative effect faster than western medicine, but the TCM have the lowest side effect and help for prognosis, so many patients choose TCM to defense the diseases in china. More and more investigators are focusing on elucidating the molecular mechanisms of these natural products and identifying its targets. Moreover, I think the researches of active components from TCM will be meaningful and broadly potential in the future. In the table 3, we summarized some other native compounds that can trigger apoptosis, I hope that these informations will guild us to research some active components in the drug.

Native compound	Caspase dependent apoptosis	Reference
luteolin	Trigger mitochondria- dependent apoptosis. And activate Bax, Bcl-xl, Bcl-2, Mcl-1, caspase-9, caspase-3, and PARP	61
Apigenin	Induce cytochrome C release and ROS enhance	62
phytosphingosine	Leading caspase-8 activation and mitochondria-dependent cell death	63
β-Lapachone	Leading ER stress and JNK activation and mitochondria mediate apoptosis	64

Table 3. Another native compounds which can trigger apoptosis apart from above paragraph's related

4.2. Intrinsic cell apoptosis signal materials

4.2.1. Oxidative stress (ROS; NO; GSH).

Keratin is a cytoskeleton protein which have some abilities to maintain the cell shape, Guo-Zhong Tao group found that keratin can modulate the shape of mitochondria and contribute to hepatocyte predisposition to apoptosis and oxidative injury [42].Depletion the mitochondria GSH in the human B lymphoma cell line by treatment with L-buthionine sulfoximine can induce caspase-3 activation and apoptosis, and indicating that GSH may be the potential early activator of apoptotic signal [43]. ROS is a type of toxic compound and usually detoxified by cells GSH, when the oxidative stress occur, the ROS detoxify will be failed, and ROS will participate in apoptosis through redox-sensitive death pathway.

4.2.2. Cytochrome C

Cytochrome C, as a proapoptotic protein, plays an important role in triggering programmed cell death, The activation of cytochrome C is related with the changes of Bak/Bax ratio. The latest researches shown that the interactions of heterotypic mitochondrial membrane will change the lipid milieu, in the end, mitochondrial membrane will be permeatable and cytochrome c will release. [44]; Apart the changes of lipid milieu, arachidonic acid, triiodothyronine (T3), or 6-hydroxdopamine can also effect the permeability of mitochondrial membrane and release Ca^{2+} and cytochrome c. [45] Cytochrome c can thrigger caspase activation via oligomerization of APAF1 protein. Caspase activation can catalyze the PARP-1, Finally, the apoptosis will happen. In short, cytochrome C is the one of major intrinsic cell apoptosis signal molecules.

4.2.3. Calcium iron

The concentration of calcium in vivo is the key role in maintain the permeatability of mitochondrial membrane. The increased intra-mitochondrial calcium can result in enhanced ROS, Furthermore, cytochrome c will be stimulated to release. [46] And calcium also trigger the ER stress, and then activate JNK pathway, afterwards, JNK activation can stimulate Bax activation; Moreover, calcuim can regulate the cysteine protease calpain, It's well known that calpain participate in the cell proliferation, cell cycle, and apoptosis. Calpain can cleave the N terminal of Bax and generate a proapoptotic fragment, and in the same time, the cells will enter the apoptosis. In a brief, calcium can trigger Bcl-2 independent cytochrome c release, and through regulating the activity of the calpain, calcium iron can play its roles in modulating the apoptosis. [47]. Beside involving in the apoptosis, Cacium iron can take part in many other signal pathways by controlling the iron channel's open or close.

4.2.4. Endoplasmic reticulum (ER) stress

As the apoptogenic factor, the permeabilization of lysosomal membrane can induce apoptosis by both caspase-dependent and caspase-independent pathway [48]; Tackled with the unfolded proteins is the one of the important ER functions, cell can regulate the unfolded proteins in ER according to metabolically needed, while if numerous unfolded proteins stimulate the ER and make the ER overload stress, the cells which have lots of unfolded proteins will apoptosis[49]; Differing from the ER stress, chaperons will protect this cell death. For example, the HSP72 protein can hamper the apoptosis through down regulating the unfolded protein signal response sensor IRE1alpha. Some neurodegeneration diseases usually accompany with unfolded protein accelerated and ER stress. So it is meaningful to research the relationship between the ER stress and neurodegenration diseases, and by this way, we will well known the function of apoptosis in the some neurodegeration diseases.

The intrinsic components, which can trigger apoptosis pathway, can connect with each other by the vivo organelles. By this way, lots of materials in the cells will consisted in the network of apoptosis to feedback the vitro stimulus, This type cell death will mostly help for the body health.

5. Apoptosis and diseases

During the last decade, exceptional for the basic research, the apoptosis have attracted many attentions due to its potential application in therapying the various human diseases. In order to maintain the function of whole organism, millions of cells will die and proliferate every day, cell death like apoptosis is the essential for the regulation of organism growth and maintenance the tissue homeostasis. If the cell death and proliferation go to imbalance, many diseases will happen. Such as: some acute pathologies (stroke, heart attack, liver failure); cancer; neurodegenerative syndromes; diabetes and so on. Due to its no lethal effect to the body, Apoptosis play a fundamental role in organism development and tissue homeostasis, while if apoptosis was not under controlled, a variety of diseases will occur.

Some neurodegenerative diseases (Alzheimer's, Parkinson's and Huntington's diseases) usually have the phenomena of the ER dyshomeostasis, mitochondrial dysfunction, and unfolded protein accelerated. In these diseases, some neural cells occur to apoptosis and brain tissue damage. In the ischemia/reperfusion injury tissue, the intracellular calcium and ROS level will increased, these factors all contribute to induce AIF to release and translocate to the nucleus, in the end, the caspase-independent apoptosis will happen, if we prevent this cell death, the prognosis of ischemia/reperfusion will be favorable. However, if we promote this cell death in cancer cells , the cancer cells will form several apoptotic body containing the cell content, and did not release the cytosol and will not lead to the fatal inflammation, in this way, the patient will have the least bad effect and help them to fight against the disease; Apoptosis occur accompanying with inflammation, inflammasome in the cells may be related with the caspase family, the tissues or organs may be existed the apoptosis and deduct the immune reaction which induced by the inflammation, we proposed that if the apoptosis do not happen in this case, the persistent proliferation and serious inflammation will be occurred in cancer cells, this is bad for the organism.

5.1. Drug design

Apoptosis is the normal cell death in order to maintain the balance of homeostasis. Unlikely the necrosis can induce inflammation. apoptosis can give little side effects. So as the therapy targets, apoptosis will be the reasonable way to cure some diseases. such as obesity [50], cancer, neuron-degeneration diseases and so on. Due to the genetic changes often existed in the human tumor cells and apoptosis have the little side effect in curing some diseases, ,it is not surprised that the antitumor drug have direct or indirect to target the apoptosis pathway molecular. The identification of the apoptosis signal pathways, and together with the increased knowledge about the apoptosis mechanisms, have given the great lots of evidences for the discovery of new drugs which can target to the apoptosis.

Above all, from the life beginning and during the whole life, apoptosis always existed to make our life healthy. Apoptosis can not release the intracellular content in the end, this cell process will not lead to inflammation, and therefore, apoptosis is the injury-limiting mode of cell disposal. Above all, it is necessary and meaningful to research the apoptosis mechanism, it will give the deep inside to guild the clinical treatment and drug design. Beside, in the nutrition research area, there are some native compounds which can be from the fruits, vegetables and some marine products can help for healthy through inducing or inhibiting apoptosis signal pathway, so knowing the mechanism of apoptosis deeply not only good for guiding people to use proper drug or balance dietary in daily life, but also far reaching impacts in design some new drugs.

Author details

Zhao Hongmei
Tianjin Key Laboratory of Food Biotechnology,
College of Biotechnology and Food Science, Tianjin University of Commerce, Tianjin, China
Chinese Academy of Medical Science & Peking Union Medical School, China

Acknowledgement

This work was supported by National Natural Science Foundation of China (81101220); Tianjin Municipal Science and Technology Commission (12JCQNJC08100); Key (Key grant) Project of Chinese Ministry of Education (210010). In the end, I would like to extend my sincere gratitude to the Dr Ruan Haihua for her instructive advice and useful suggestions on this thesis.

6. References

[1] Yun-Ji Lim,Ji-Ae Choi, Hong-Hee Choi, etal. Endoplasmic Reticulum Stress Pathway-Mediated Apoptosis in Macrophages Contributes to the Survival of Mycobacterium tuberculosis. PLoS One. 2011, 6(12): e28531.

[2] Prasad A, Lu M, Lukac DM, etal. An Alternative Kaposi's Sarcoma-Associated Herpesvirus Replication Program Triggered by Host Cell Apoptosis. J Virol. 2012,86(8):4404-4419.

[3] Obregón-Henao A, Duque-Correa MA, Rojas M et al. Stable extracellular RNA fragments of Mycobacterium tuberculosis induce early apoptosis in human monocytes via a caspase-8 dependent mechanism. PLoS One. 2012, 7(1):e29970.

[4] Toshiyuki Nakagawaa, and Junying Yuana. Activation of Caspase-12 by Calpain in Apoptosis J. Cell Biol.2000, 150 (4): 887-894.

[5] Denis   M, PengchengZhu and Judy   L.GranzymeA,Induces   Caspase-Independent Mitochondrial Damage, a Required First Step for Apoptosis Immunity, 2005,22(3):355-370.

[6] Kang YH, Yi MJ, Kim MJ,et al. Caspase-independent cell death by arsenic trioxide in human cervical cancer cells: reactive oxygen species-mediated poly(ADP-ribose) polymerase-1 activation signals apoptosis-inducing factor release from mitochondria. Cancer Res.2004, 64(24):8960-8967.

[7] Misirlic Dencic S, Poljarevic J, Vilimanovich U, et al. Cyclohexyl Analogues of Ethylenediamine Dipropanoic Acid Induce Caspase-Independent Mitochondrial Apoptosis in Human Leukemic Cells. Chem Res Toxicol. 2012 Mar 22 online.

[8] Marchi S, Giorgi C, Suski JM, et al Mitochondria-Ros crosstalk in the control of cell death and aging, J signal Transduct. 2012, 1-17.

[9] Reddy RH. Role of mitochondria in neurodegenerative diseases: mitochondria as a therapeutic target in Alzheimer's disease. CNS spectra 2009, 14(8):8-13.

[10] Kwong JQ, Beal MF, Manfredi G. The role of mitochondria in inherited neurodegenerative diseases. J Neurochem. 2006, 97(6):1659-75.

[11] Paula l. Moreira, Xiongwei Zhu, Xinglong Wang, et al. Mitochondria: A Therapeutic Target in Neurodegeneration. Biochim Biophys Acta 2010, 1802(1):212-220.

[12] Mariusz Karbowski. Mitochondria on guard: role of mitochondrial fusion and fission in the regulation of apoptosis. Adv Exp Med Biol. 2010, 687:131-142.

[13] Clare Sheridan, Petrina Delivani, Sean P.Cullen.etal. Bax- or Bak-Induced Mitochondrial Fission Can Be Uncoupled from Cytochrome *c* Release. Mol cell. 2008, 31(4):570-585.

[14] Jennifer R. Hom, Jennifer S. Gewandter, Limor Michael, et al. Thapsigargin induces biphasic fragmentation of mitochondria through calcium-mediated mitochondrial fission and apoptosis. Journal of cellular physiology, 2007, 212(2):498-508.

[15] Van Laar VS, Berman SB. Mitochondrial dynamics in Parkinson's disease. Exp Neurol. 2009, 218(2):247-256.

[16] Ardehali R, Inlay MA, Ali SR, et al. Overexpression of BCL2 enhances survival of human embryonic stem cells during stress and obviates the requirement for serum factors. Proc Natl Acad Sci U S A. 2011, 22; 108(8):3282-3287.

[17] Luo H, Zhang Y, Zhang Z, et al. The protection of MSCs from apoptosis in nerve regeneration by TGFβ1 through reducing inflammation and promoting VEGF-dependent angiogenesis. Biomaterials.2012,33(17):4277-4287.

[18] Bergmann A, Steller H.Apoptosis, stem cells, and tissue regeneration. Sci Signal. 2010, 3(145):1-16.

[19] Susanne Herbst, Ulrich E.Schaibel, Bianca E.Schneider. Interferon Gamma Activated Macrophage by Nitric Oxide Induced Apoptosis. PLoS one, 2011, 6(5):1-8.e19105.

[20] Lebman D, Edmiston J. The Role of TGF-beta in growth, differentiation, and maturation of B lymphocytes. Microbes Infect. 1999, 1(15):1297-1304.

[21] Singhal J, Dalasanur Nagaprashantha L, Vitsyayan R. et al. Didymin Induced Apoptosis by Inhibiting N-Myc and Up-regulating RKIP in Neuroblastoma. Cancer Prev Res(Phila) 2012,5(3):473-483.

[22] Hung JY, Hsu YL, Ko YC.et al. Didymin, a dietary flavonoid glycoside from citrus fruits, induced Fas-mediated apoptotic pathway in human non-small-cell lung cancer cells in vitro and in vivo. Lung cancer. 2010, 63(3):366-374.

[23] Inoue H, Waiwut P, Saiki I,et al. Gomisin N enhances TRAIL-induced apoptosis via reactive oxygen species-mediated up-regulation of death receptors 4 and 5. Int J Oncol.2012, 40(4: 1058-1065.

[24] Zhou J, Lu GD, Ong CS, et al. Andrographolide sensitizes cancer cells to TRAIL-induced apoptosis via p53-mediated death receptor 4 up-regulation. Mol Cancer Ther. 2008, 7(7):2170-2180.

[25] Prasad S, Yadav VR, Kannappan R, et al. Ursolic acid, a pentacyclin triterpene, potentiates TRAIL-induced apoptosis through p53-independent up-regulation of death receptors: evidence for the role of reactive oxygen species and JNK. J Biol Chem. 2011, 286(7):5546-5557.

[26] Plissonnier ML, Fauconnet S, Bittard H. The Antidiabetic Drug Ciglitazone Induces High Grade Bladder Cancer Cells Apoptosis through the Up-Regulation of TRAIL. PLoS One. 2011, 6(12):1-12.

[27] Wendrenmaire M, Bardou M, Peyronel C. Effects of leptin on lipopolysaccharide-induced myometrial apoptosis in an in vitro human model of chorioamnionitis. Am J Obstet Gynecol. 2011, 205(4):1-363.

[28] Susanne Herbst, Ulrich E.Schaibel, Bianca E.Schneider. Interferon Gamma Activated Macrophage by Nitric Oxide Induced Apoptosis. PLoS one, 2011, 6(5):1-8.e19105.

[29] Olivares-Zavaleta N, Carmody A, Messer R, et al. Chlamydia pneumoniae inhibits activated human T lymphocyte proliferation by the induction of apoptotic and pyroptotic pathways.J Immunol. 2011, 186(12):7120-7126.

[30] Catarina V. Nogueira,Tullia Lindsten,Amanda M. Jamieson, et al. Rapid Pathogen-Induced Apoptosis: A Mechanism Used by Dendritic Cells to Limit Intracellular Replication of Legionella pneumophila. PLoS Pathog. 2009, 5(6): e1000478.

[31] Ding L, Xu X, Huang Y.Transmissible gastroenteritis virus infection induces apoptosis through FasL- and mitochondria-mediated pathways. Vet Microbiol. 2012 Jan 28 online.

[32] Wang J, Sun P, Bao Y. et al. Vitamin E renders protection to PC12 cells against oxidative damage and apoptosis induced by single-walled carbon nanotubes. Toxicology in Vitro.2012,26(1):32-41.

[33] Han YT, Chen XH, Xie J, Zhan SM,et al. Purple Sweet Potato Pigments Scavenge ROS, Reduce p53 and Modulate Bcl-2/Bax to Inhibit Irradiation-induced Apoptosis in Murine Thymocytes. Cell Physiol Biochem. 2011, 28(5):865-872.

[34] Jang SH, Lim JW, Morio T, etal. Lycopene inhibits Helicobacter pylori-induced ATM/ATR-dependent DNA damage response in gastric epithelial AGS cells. Free Radic Biol Med. 2012, 52(3):607-615.

[35] Li Y, Kong D, Bao B, et al. Induction of cancer cell death by isoflavone:The Role of Multiple Signaling Pathways. Nutrients, 2011,3(10)：877-896.

[36] Smith-Warner SA, Spiegelman D, Yaun SS, et al. Fruits, vegetables and lung cancer: a pooled analysis of cohort studies. Int J cancer. 2003, 107(6):1001-1011.

[37] Lee MM, Gomez SL,Chang JS, etal. Soy and isoflavone consumption in relation to prostate cancer risk in China. Cancer Epidemiol Biomarkers Prev. 2003,12(7):665-668

[38] Tsung-Ho Ying, Shun-Fa Yang, Su-Ju, et al. Fisetin induces apoptosis in human cervical cancer HeLa cells through ERK1/2-mediated activation of caspase-8-/caspase-3-dependent pathway. Arch toxicol. 2012,86(2):263-273.

[39] Woan-Rouh Lee, Shing-Chuan Shen, Hui-Yi Lin, etal. Wogonin and fisetin induce apoptosis in human promyeloleukemic cells, accompanied by a decrease of reactive oxygen species, and activation of caspase-3 and Ca^{2+}-dependent endnuclease. Biochem pharmacol, 2003,63:225-236.

[40] Chen YC,Shen SC,Lee WR, etal. Wogonin and fisetin induction of apoptosis through activation of caspase 3 cascade and alternative expression of p21 protein in hepatocellular carcinoma cells SK-HEP-1 Arch Toxicol. 2002, 76(5-6):351-359.

[41] Chen YC,Shen SC,Lee WR, etal. Emodin induces apoptosis in human promyeloleukemic HL-60 cells accompanied by activation of caspase 3 cascades but independent of reactive oxygen species production. Biochem Pharmacol. 2002, 64(12):1713-1724.

[42] Guo-Zhong Tao, Kok Sun Looi, Diana M. Toivola, etal. Keratins modulate the shape and function of hepatocyte mitochondria: a mechanism for protection from apoptosis. J Cell Sci. 2009, 122(21):3851-3855.

[43] Armstrong J S, Steinauer K K, Hornung B, et al. Role of glutathione depletion and reactive oxygen species generation in apoptotic signaling in a human B lymphoma cell line. Cell Death Differ, 2002, 9:252–263.

[44] Chipuk JE, McStay GP, Bharti A,et al. Sphingolipid Metabolism Cooperates with BAK and BAX to Promote the Mitochondrial Pathway of Apoptosis. Cell. 2012, 148(5):988-1000.

[45] Kanno T, Fujita H, Muranaka S,et al. Mitochondrial swelling and cytochrome c release: sensitivity to cyclosporin A and calcium. Physiol Chem Phys Med NMR. 2002, 34(2):91-102.

[46] Kumar S, Kain V, Sitasawad SL.High glucose-induced Ca (2+) overload and oxidative stress contribute to apoptosis of cardiac cells through mitochondrial dependent and independent pathways. Biochim Biophys Acta. 2012 Feb 28 online.

[47] Gao G, Dou QP.N-terminal cleavage of Bax by calpain generates a potent proapoptotic 18-kDa fragment that promotes bcl-2-independent cytochrome C release and apoptotic cell death.J Cell Biochem. 2000, 80(1):53-72.

[48] Pupyshev AB. Lysosomal membrane permeabilization as apoptogenic factor. Tsitologiia. 2011, 53(4):313-24.

[49] Walter P, Ron D.The unfolded protein response: from stress pathway to homeostatic regulation. Science. 2011, 334(6059):1081-1086.

[50] Zhang Y, Huang C. Targeting adipocyte apoptosis: A novel strategy for obesity therapy. Biochem Biophy Res Commu. 2012,17(1):1-4.

[51] AS Wright, LN Thomas, RC Douglas, et al. Relative potency of testosterone and dihydrotestosterone in preventing atrophy and apoptosis in the prostate of the castrated rat. *J Clin Invest*. 1996,98(11):2558–2563.

[52] Murdoch W.J.; Van Kirk E.A. Oestradiol inhibits spontaneous and cisplatin-induced apoptosis in epithelial ovarian cancer cells: relationship to DNA repair capacity. Apoptosis, 1997, 2(5):478-484.

[53] Seilicovich A. Cell Life and Death in the Anterior Pituitary Gland: Role of Oestrogens. J Neuroendocrinol, 2010, 22(7): 758–764.

[54] C Cherbonnier, O Déas, G Carvalho et al. Potentiation of tumour apoptosis by human growth hormone via glutathione production and decreased NF-ΚB activity. British Journal of Cancer 2003,89:1108–1115.

[55] Morrissy S, Xu B, Aguilar D, et al. Inhibition of apoptosis by progesterone in cardiomyocytes. Aging Cell. 2010, 9(5):799-809.

[56] Paul Kay, George Schlossmacher, Laura Matthews, et al. Loss of glucocorticoid receptor expression by DNA methylation prevents glucocorticoid induced apoptosis in human small cell lung cancer cells. PLoS one, 2011,6(10):e24839

[57] Tang L, Zhang Y, Pan H et al. Involvement of cyclin B1 in progesterone-mediated cell growth inhibition, G2/M cell cycle arrest, and apoptosis in human endometrial cell. Reprod Biol Endocrinol. 2009, 7:144-152.

[58] Ishizuya-Oka A . Amphibian organ remodeling during metamorphosis: insight into thyroid hormone-induced apoptosis. Dev Growth Differ. 2011, 53(2):202-212.

[59] Lewis-Wambi JS, Jordan VC. Estrogen regulation of apoptosis: how can one hormone stimulate and inhibit. Breast Cancer Res. 2009; 11(3):206-218.

[60] Seo HS, Choi HS, Choi HS, et al. Phytoestrogens induce apoptosis via extrinsic pathway, inhibiting nuclear factor-kappaB signaling in HER2-overexpressing breast cancer cells. Anticancer Res. 2011,31(10):3301-3313.

[61] Chen Q, Liu S, Chen J, Luteolin induces mitochondria-dependent apoptosis in human lung adenocarcinoma cell. Nat Prod Commun. 2012,7(1):29-32.

[62] Lu HF, Chie YJ, Yang MS, Apigenin induces caspase-dependent apoptosis in human lung cancer A549 cells through Bax- and Bcl-2-triggered mitochondrial pathway. Int J Oncol. 2010 ,36(6):1477-1484.

[63] Park MT, Choi JA, Kim MJ. Suppression of extracellular signal-related kinase and activation of p38 MAPK are two critical events leading to caspase-8- and mitochondria-mediated cell death in phytosphingosine-treated human cancer cells. J Biol Chem. 2003, 278(50):50624-50634.

[64] Lee H, Park MT, Choi BH, et al. Endoplasmic reticulum stress-induced JNK activation is a critical event leading to mitochondria-mediated cell death caused by β-lapachone treatment. PLoS One. 2011,6(6):e21533.

[65] Daugas E, SA susin, N Zamzami, et al. Mitochondria-nuclear translocation of AIF in apoptosis and necrosis. FASEB J. 2000,14(5):729-739.

Apoptosis and Physiology

Osteocyte Apoptosis-Induced Bone Resorption in Mechanical Remodeling Simulation – Computational Model for Trabecular Bone Structure

Ji Yean Kwon, Hisashi Naito, Takeshi Matsumoto and Masao Tanaka

Additional information is available at the end of the chapter

1. Introduction

Biomechanical function is one of primal concern in musculo-skeletal system supporting our body as a whole. The functional adaptation of bone to the mechanical environment has been most interesting topic in bone mechanics. We have proposed the generalized model of bone remodeling simulation considering osteocyte apoptosis and targeted remodeling. These are important factors of bone metabolism under physiological and pathological conditions. This remodeling model has the capability to speculate the changes in the trabecular bone structures under various loading conditions. In this chapter, we describe how the osteocyte apoptosis-induced bone resorption has importance in computer simulations for disuse-meditated trabecular bone remodeling resulting in the osteoporotic bone structure in the human femur.

1.1. Cells activity in bone

Bone tissue is the component of the skeletal system supporting human body. Bone structure is continuously renewed by remodeling, that is, the alternately-repeated events of local bone resorption and formation. Three mature cell types of osteoblasts, osteoclasts and osteocytes play crucial roles in bone remodeling process. Bone metabolism is regulated by bone cells which respond to various environmental signals coming from chemical, magnetic, electrical and mechanical stimuli (Klein-Nulend et al., 2005).

Osteoblast are bone forming cells that synthesize and secrete bone matrix, participate in the calcification and formation of bone, and regulate the flux of calcium and phosphate in and out of bone. Osteoclasts are multinucleated giant cells; their role is to resorb bone. Actively resorbing osteoclasts are usually found in cavities on bone surfaces, called resorption cavities. Osteocytes are the most abundant cell type in mature bone. During bone formation some osteoblasts are left behind in the newly formed osteoid as osteocytes when the bone formation moves on. The embedded osteoblast in lacunae differentiate into osteocytes by losing much of their organelles but acquiring long, slender processes encased in the lacunar-canulicular network that allow contact with earlier incorporated osteocytes and with osteoblast and bone lining and periosteal cells lining the bone surface and the vasculature. The osteocytes are the cells best placed to sense the magnitude and distribution of strains. They are strategically placed both to respond to changes in mechanical strain and to disseminate fluid flow to transduce information to surface cells of the osteoblastic lineage via their network of canalicular processes and communicating gap junctions (Cowin, 2001).

1.2. Mechanical stimuli in bone

Mechanical stimuli are one of the important regulation factors of bone remodeling. According to recent findings, osteocytes might play a role in the mechanical regulation of bone, receiving mechanical input signals and transmitting these stimuli to other cells in bone. Osteocyte is believed to comprise a sensory network that monitors mechanical load and tissue damage, and triggers appropriate adaptive responses, either formation or resorption. Mechanisms by which osteocytes could sense mechanical load have been understood by means of fluid movements throughout the lacunar-canalicular system, with some combination of shear stress and streaming potentials providing the proximate stimuli. The relations between mechanical stimuli and mature bone have been examined by experimental studies. There are suggested that higher rates of mechanical loading would evoke grater adaptive responses than lower rates of loading in mature bone (LaMonte et al., 2005). Exercise can substantially alter the physical states of a bone and generate adaptive responses. In a randomized controlled clinical trial, Fuchs et al.(2001) showed that, in jumpers, jumping significantly increases bone mineral content in femoral neck and lumbar skeleton relative to controls. McKay et al. (2000) found that jumping three times weekly for 8 months significantly augmented a real bone mineral density in the femoral trochanteric region. These studies have hypothesized that loading induced stimuli is the main signal-generating factor and play a key role in mechanobiology. However, disuse uncouples bone formation from resorption, leading to increased porosity, decreased bone geometrical properties, and decreased bone mineral content which compromises bone mechanical properties and increases fracture risk. The removal of routine bone stresses (e.g. from immobilization, inactivity, or reduced gravity) has deleterious consequences on bone integrity. Reduced skeletal loading causes net bone loss by unbalancing bone formation and bone resorption (Takata and Yasui, 2001; Caillot-Augusseau et al., 1998). Rat hindlimb immobilization, human spaceflight, and human bedrest can all cause increased bone resorption and decreased bone formation (Caillot-Augusseau et al., 1998; Weinreb et al., 1989; Li et al., 2005; McGee et al., 2008). Therefore, the process of mechanotransduction of bone, the conversion of a mechanical stimulus into a biochemical response, is known to occur in osteoblast, osteoclast and osteocytes in response to it.

1.3. Mechanical stimuli and osteocyte apoptosis

It is well recognized that the origin of osteocyte is osteoblasts embedded in its own matrix and some osteocytes die eventually (apoptosis). Aging, loss of estrogen, loading, and chronic glucocorticoid administration is known to increase osteocyte apoptosis and loading decreases the proportion of apoptotic osteocyte in cortical bone in rats (Noble et al., 2003; Gu et al., 2005; Delgado-Calle et al., 2011; Kennedy et al., 2012). These studies investigated the role of osteocytes in the control of loading related remodeling and they exhibited the effects on osteocyte viability of mechanical loading applied to the bone. Osteocyte apoptosis triggers osteoclast precursor recruitment to initiate bone resorption, whether it is induced by weightlessness or fatigue loading and precedes osteoclast resorption. An osteoclast/osteoblast-dependent process related on the osteocytes population may participate in making the decision as to where and when a new remodeling cycle will be initiated. That is, osteocyte network could be considered the main sensor of loads. To take these observations into account, Frost proposed the concept of mechanostat (Duncan & Turner, 1995; Frost, 2003; Hughes et al., 2010). According to this concept, the authors have proposed the mathematical model for bone remodeling described in the following section.

2. Bone remodeling simulation

To develop the computer simulation model of bone remodeling considering bone metabolism macroscopically, this section describes the bone remodeling model referring to the strain-based mechanostat theory submitted by Harold Frost and refined more than 5 decades. Physiological, low, and high strain ranges were defined in mechanostat theory, and different mechanisms were considered to describe the mechanical surface remodeling of trabecular architecture resulting in the structural change (Kwon et al., 2010a).

The following was assumed as the result of a remodeling turnover for each strain range: 1) in the physiological strain range, the occurrence of bone resorption or formation depends on the degree of the local stress non-uniformity; 2) in the low strain range, only bone resorption occurs owing to osteocyte apoptosis and its frequency increases with decreasing strain; 3) in the high strain range, only bone formation occurs by targeted remodeling and its frequency increases with strain. Each window is distinguished by threshold values of equivalent strain ε_c in the context of the mechanostat theory, and low, physiological, and high strain range correspond to the strain ranges of ε_c smaller than ε_{du}, between ε_{pl} - ε_{pu} and larger than ε_{ol} (Fig 1).

Trabecular architecture is discretized using voxel finite elements, as schematically shown in Fig. 2. The trabecular surface movement by remodeling is expressed by adding and removing voxel elements on the surface according to the mechanical stress/strain conditions determined by a finite element analysis. Each trabecular element was classified by equivalent strain into low, physiological, and high strain range. Bone surface remodeling was assumed to occur according to the mathematical models of each window described in the following sections in detail.

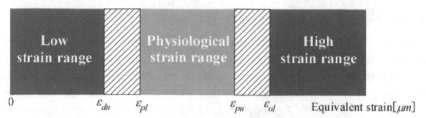

Figure 1. Strain ranges categorized by the equivalent strain and in-between transition ranges.

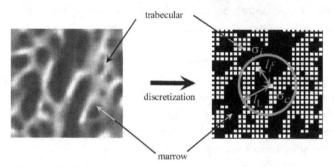

Figure 2. Discretized trabecular bone.

2.1. Physiological strain range: Mechanical adaptation by remodeling

In the physiological strain range ($\varepsilon_{pl} < \varepsilon_c < \varepsilon_{pu}$), the non-uniformity of the stress/strain distribution was taken as driving stimuli (Adachi et al., 1997). For the voxel representing a point c on bone surface, the magnitude of local stress non-uniformity Γ_c was quantified as

$$\Gamma_c = \ln\left(\sigma_c \sum_i^N w\left(l_i^c\right) \Big/ \sum_i^N w\left(l_i^c\right)\sigma_i \right) \tag{1}$$

where σ_c denotes the stress at the point c and σ_i denotes the representative stress at a point i with the distance l_i^c from c. The function $w(l_i^c)$ is described as

$$w\left(l_i^c\right) = \begin{cases} 1 - l_i^c/l_L & \left(0 \leq l_i^c < l_L\right) \\ 0 & \left(l_L \leq l_i^c\right) \end{cases} \tag{2}$$

defining a positive weight for the neighboring point i within the limit of sensing distance l_L from the point c under consideration. This represents the sensory integration by osteocyte network.

The probability $f(\Gamma_c)$ of bone resorption or formation at the point c is described by

$$f\left(\Gamma_c\right) = \begin{cases} P_f\left(\Gamma_c\right) & \left(0 < \Gamma_c\right) \\ 0 & \left(\Gamma_c = 0\right) \\ -P_r\left(\Gamma_c\right) & \left(\Gamma_c < 0\right) \end{cases} \tag{3}$$

$$P_f(\Gamma_c) = \begin{cases} \dfrac{1}{2}\left\{ \sin\pi\left(\dfrac{\Gamma_c}{\Gamma_u} - \dfrac{1}{2}\right) + 1\right\} & (0 < \Gamma_c \le \Gamma_u) \\ 1 & (\Gamma_u < \Gamma_c) \end{cases} \qquad (4)$$

$$P_r(\Gamma_c) = \begin{cases} \dfrac{1}{2}\left\{ \sin\pi\left(\dfrac{\Gamma_c}{\Gamma_l} - \dfrac{1}{2}\right) + 1\right\} & (\Gamma_l \le \Gamma_c < 0) \\ 1 & (\Gamma_c < \Gamma_l) \end{cases} \qquad (5)$$

where Γ_u and Γ_l denote the lower and upper threshold values, respectively. More specifically, voxel elements are added around the surface point c when $f(\Gamma_c) = 1$, and the voxel c is subtracted when $f(\Gamma_c) = -1$. The case of $f(\Gamma_c) = 0$ represents local remodeling equilibrium and no change occurs at and around c. Moreover, in the cases of $0 < \Gamma_c \le \Gamma_u$ and $\Gamma_l \le \Gamma_c < 0$, bone formation and resorption occurs stochastically according to probability $f(\Gamma_c)$. In this physiological strain range, the activation frequency AF_f or AF_r, defined as a relative rate for bone formation or resorption, is set to unity, i.e., $AF_f = AF_r = 1$.

2.2. Low strain range: Osteocyte apoptosis- induced bone resorption

In the low strain range ($\varepsilon_c < \varepsilon_{du}$), only osteocyte apoptosis-accelerated bone resorption is assumed to occur (Noble et al. 2003; Gu et al., 2005; Li et al. 2005). The probability of bone resorption is unity, i.e.,

$$P_{LS} = 1 \quad \text{and} \quad f(\Gamma_c) = -1. \qquad (6)$$

The degree of osteocyte apoptosis increases with the decrease of strain. Thus, the activation frequency for bone resorption AF_r is also elevated with the decrease of strain and is described by

$$AF_r(\varepsilon_c) = \begin{cases} AF_{r\max} & (0 < \varepsilon_c < \varepsilon_{\min}) \\ \left(\dfrac{\varepsilon_{du}}{\varepsilon_c}\right)^a & (\varepsilon_{\min} < \varepsilon_c < \varepsilon_{du}), \quad a = \dfrac{\ln AF_{r\max}}{\ln(\varepsilon_{du}/\varepsilon_{\min})} \end{cases} \qquad (7)$$

where AF_r reaches the maximum $AF_{r\max}$ at $\varepsilon_c = \varepsilon_{\min}$.

2.3. High strain range: Targeted remodeling by over use

In the high strain range ($\varepsilon_{ol} < \varepsilon_c$), only bone formation is assumed to occur. The probability of bone formation as the result of remodeling turnover is unity, i.e.,

$$P_{HS} = 1 \quad \text{and} \quad f(\Gamma_c) = 1. \qquad (8)$$

Bone formation in high strain range undergoes the targeted remodeling mechanism (Burr, 2002; Da Costa Gómez et al., 2005), and thus the activation frequency for bone formation AF_f is also elevated with the increase of strain and is described by

$$AF_f\left(\varepsilon_c\right)=\begin{cases}\dfrac{AF_{fmax}-1}{2}\left\{sin\pi\left(\dfrac{\varepsilon_c-\varepsilon_{ol}}{\varepsilon_{max}-\varepsilon_{ol}}-\dfrac{1}{2}\right)+1\right\} & \left(\varepsilon_{ol}\le\varepsilon_c<\varepsilon_{max}\right)\\[2ex] AF_{fmax} & \left(\varepsilon_c\ge\varepsilon_{max}\right)\end{cases}\tag{9}$$

where AF_f reaches the maximum AF_{fmax} at $\varepsilon_c=\varepsilon_{max}$. It is noted that the pathological overuse is not considered here and the upper limit ε_{max} is enforced for ε_c in high strain range.

2.4. Integrated mathematical model and simulation procedures for bone remodeling

The mathematical models of bone remodeling established in low, physiological, and high strain ranges were integrated by introducing transition regions. In the transition regions between low and physiological range ($\varepsilon_{du}<\varepsilon_c<\varepsilon_{pl}$) or between physiological and high range ($\varepsilon_{pu}<\varepsilon_c<\varepsilon_{ol}$), the probabilities of bone resorption or formation in each window were combined as a linearly weighed sum as given by Eqs. (10)-(12) resulting the resultant probability f^* of bone resorption/formation.

$$f^*\left(\Gamma_c,\varepsilon_c\right)=\begin{cases}-P_{LS} & \left(\varepsilon_c<\varepsilon_{du}\right)\\[2ex] -\left(1-\dfrac{\varepsilon_c-\varepsilon_{du}}{\varepsilon_{pl}-\varepsilon_{du}}\right)P_{LS}+\dfrac{\varepsilon_c-\varepsilon_{du}}{\varepsilon_{pl}-\varepsilon_{du}}f\left(\Gamma_c\right) & \left(\varepsilon_{du}<\varepsilon_c<\varepsilon_{pl}\right)\\[2ex] f\left(\Gamma_c\right) & \left(\varepsilon_{pl}<\varepsilon_c<\varepsilon_{pu}\right)\\[2ex] \left(1-\dfrac{\varepsilon_c-\varepsilon_{pu}}{\varepsilon_{ol}-\varepsilon_{du}}\right)f\left(\Gamma_c\right)+\dfrac{\varepsilon_c-\varepsilon_{pu}}{\varepsilon_{ol}-\varepsilon_{pu}}P_{HS} & \left(\varepsilon_{pu}<\varepsilon_c<\varepsilon_{ol}\right)\\[2ex] P_{HS} & \left(\varepsilon_{ol}<\varepsilon_c\right)\end{cases}\tag{10}$$

$$P_f^*=f^*(\Gamma_c,\varepsilon_c)\,(f^*(\Gamma_c,\varepsilon_c)\ge0)\tag{11}$$

$$P_r^*=f^*(\Gamma_c,\varepsilon_c)\,(f^*(\Gamma_c,\varepsilon_c)<0)\tag{12}$$

Using this integrated model, trabecular bone structure was simulated by the following procedures.

1. The initial shape of the trabecular bone is discretized by voxel finite elements and Young's modulus and Poisson's ratio are given.
2. The equivalent stress and strain distributions are determined under a given mechanical boundary condition.
3. The probabilities P_f^* and P_r^* of bone formation or resorption are calculated for every voxel on bone surface by Eqs. (10) - (12).
4. The activation frequency AF_f and AF_r of bone formation or resorption are calculated for every voxel on bone surface by Eqs. (7) and (9). These are rounded to the nearest integer number and truncated by AF_{fmax} or AF_{rmax}.
5. Bone surface movement (voxel removal or addition) is determined stochastically based on P_f^* and P_r^*.

6. For voxels with activation frequency AF_r or AF_f larger than one, the same surface movements in procedure (4) are repeated AF_r or AF_f times.
7. Procedures (2)-(6) are repeated until equilibrium is achieved.

3. Simulation of trabecular bone structures for human proximal femur

3.1. Voxel FE model of human proximal femur

The three-dimensional profile of human proximal femur is reconstructed using computed tomography (female, 70 years, hip osteoarthritis), the resolution of the images was 0.7 mm so that the entire volume could be captured with slice image of 512×512 cubic voxels. As the random initial trabecular structure in proximal femur, an artificial trabecular structure was given by the random-placement of hollow spheres with inside and outside diameters of 2100 and 4200 μm, respectively (Kwon et al., 2010a). One voxel was transferred to one finite element of eight node brick element. The isotropic elastic body with Young's modulus of 20.0 GPa and Poisson's ratio of 0.3 was assumed for trabecular bone material. The cortical outer surface was fixed throughout the remodeling simulation, and the intra-structure, from the head to mid-shaft, was remodeled in accordance with the same remodeling procedure. The fixed boundary conditions are imposed on the distal end. In the daily loading condition, we considered the one-legged stance, abduction, and adduction, which accounted for 50, 25, and 25% of this condition as illustrated in Fig. 3. The loading magnitudes and frequencies of these three stances were shown in Table 1 (Beaupré et al., 1990; Adachi et al., 1997). The extreme ranges of motion of abduction and adduction were assumed. The trabecular structure obtained under this daily loading condition was referred to as the normal structure and also used for the initial structure for the remodeling simulation under reduced weight-bearing conditions (Kwon et al., 2010b).

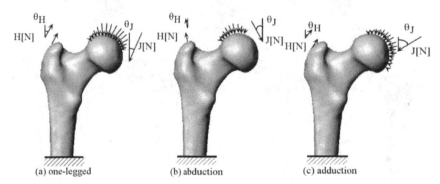

(a) one-legged   (b) abduction   (c) adduction

Figure 3. Standard weight-bearing cases at each stance (daily-loading condition).

3.2. Healthy trabecular structures in proximal femur

The simulation results from initial random structure to an equilibrium state are shown in Fig. 4. In the diaphyseal region, random inner structures disappeared whereas the thickness of cortical bone increased with simulation step to the similar level observed in CT image (Fig.

4(g)). On the other hand, trabecular bone in the metaphyseal region, showing an isotropical random pattern at the initial (Fig. 4(a)), changed to an anisotropic, non-uniform structure. Bone volume was maintained in femoral head region A but decreased in neck region B reproducing Ward's triangle characteristic of trabecular structure in the proximal femur.

	One-legged (n=ol)	Abduction (n=ab)	Adduction (n=ad)
Frequency f_n [cycle/day]	6000	2000	2000
θ_J	24°	-15°	56°
J [N]	2317	1158	1548
θ_H	28°	-8°	35°
H [N]	703	351	468

Table 1. Standard magnitude of force and frequency at each stance (daily-loading condition).

Regions A and B were subdivided to blocks of 10×10×10 voxels to compare local bone volume fraction in each block with local bone mineral density(Duchemin et al., 2008) i.e., averaged bone mineral density in the corresponding CT voxels referring to the phantoms of trabecular bone (B-mas 200). Trabecular bone volume fraction was correlated significantly with bone mineral density as shown in Fig. 5.

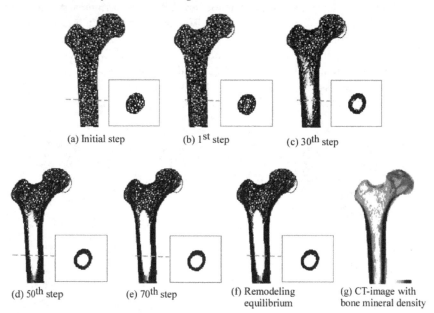

(a) Initial step   (b) 1st step   (c) 30th step

(d) 50th step   (e) 70th step   (f) Remodeling equilibrium   (g) CT-image with bone mineral density

Figure 4. Simulated trabecular structure in human proximal femur and CT-image (Subject #1).

3.3. Disuse mediated change in proximal femur

To estimate the effect of weightlessness-induced osteocyte apoptosis in trabecular bone in proximal femur, two reduced weight-bearing conditions were considered here: the infrequent

Osteocyte Apoptosis-Induced Bone Resorption in Mechanical Remodeling Simulation – Computational
Model for Trabecular Bone Structure

33

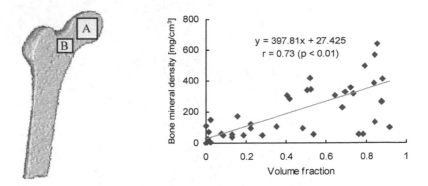

Figure 5. Bone mineral density vs. simulated volume fraction.

walking and the cane-assisted walking. In the infrequent walking condition, we wedged the interval with no loading into the daily loading condition; that is, the infrequent walking condition consisted of unloading of zero load and a part of the daily loading conditions. The proportion of unloading interval changed from 10 to 70% while the ratio of one-legged stance, abduction, and adduction were remained unchanged. In the simulation, the parameter of bone resorption and formation were represented with loading frequencies f_{ol}, f_{ab}, f_{ad}, and $f_{unloading}$ as Eqs. (13) ~ (15). In the cane-assisted walking, we referred to the study on relative changes in muscle activity and kinetics during cane-assisted walking (Neumann et al., 1998). The cane use reduced the demand on the hip abductors and decreased joint compression forces related to muscle contraction in the contralateral side. We calculated joint force and hip abductor force at various rates of leaning force when using a cane (Fig. 6). That is, in the infrequent walking, the number of loading cycles per day decreased from the daily-loading condition (Table 1), while the same loading forces were applied to the femur; in the cane-assisted walking, the numbers of loading cycles per day remained unchanged but the loading forces were altered from the daily-loading condition (Kwon et al., 2010b).

$$\Gamma_c = \frac{f_{ol}\Gamma_{c\_ol} + f_{ab}\Gamma_{c\_ab} + f_{ad}\Gamma_{c\_ad}}{f_{ol} + f_{ab} + f_{ad} + f_{unloading}} \tag{13}$$

$$AF_r = \frac{f_{ol}AF_{r\_ol} + f_{ab}AF_{r\_ab} + f_{ad}AF_{r\_ad}}{f_{ol} + f_{ab} + f_{ad} + f_{unloading}} \tag{14}$$

$$AF_f = \frac{f_{ol}AF_{f\_ol} + f_{ab}AF_{f\_ab} + f_{ad}AF_{f\_ad}}{f_{ol} + f_{ab} + f_{ad} + f_{unloading}} \tag{15}$$

The trabecular structure of the human proximal femur under the infrequent walking condition was shown for each rate of unloading interval and the initial, i.e., normal trabecular structure was also shown in Fig. 7. In the initial structure of Fig. 7(a), visible are all the normal groups of trabecular (Fig. 8), i.e. the compressive and tensile trabecular bones cross each other and the upper end of femur is completely occupied by cancellous tissue. Even Ward's triangle shows some thin trabecular bone (Singh et al., 1970).

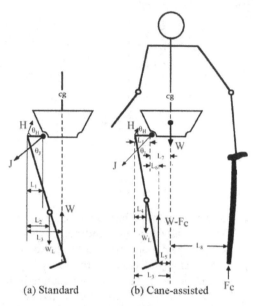

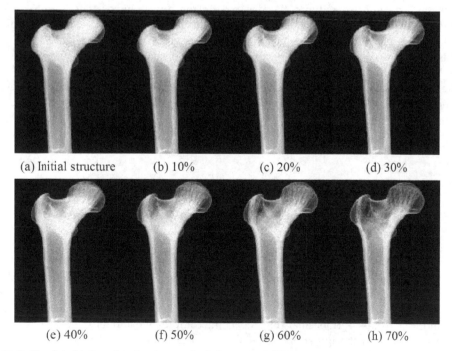

Figure 6. Diagram for boundary conditions of cane-assisted walk.

(a) Initial structure   (b) 10%   (c) 20%   (d) 30%

(e) 40%   (f) 50%   (g) 60%   (h) 70%

Figure 7. Simulated trabecular structure under infrequent walking.

Osteocyte Apoptosis-Induced Bone Resorption in Mechanical Remodeling Simulation – Computational Model for Trabecular Bone Structure

35

Increasing the rate of unloading interval led to the increased bone loss, first in the femoral head and the greater trochanter. Ward's triangle was emptier in the case of higher rate of unloading frequency. Subsequently, the secondary compressive group disappeared and the principal tensile group became thinner with further increasing of the rate of unloading. When unloading level increased up to 70%, the principal compressive group little remained although the greater trochanter group was still observed clearly.

Figure 9 shows the trabecular structure changes under cane-assisted walking conditions. Using cane decreased joint compression forces related to muscle contraction in the contralateral side. Bone loss attracts attention on the greater trochanter group and Ward's triangle became clear when reducing hip abductor forces. Subsequent disappearing of the secondary compressive group and the thinning of the principal tensile group were observed with further increasing. This is similar to those observed for the infrequent walking. However, bone loss in the trabecular femoral head differed from that under the infrequent walking condition. The hip joint load will be involved in this difference.

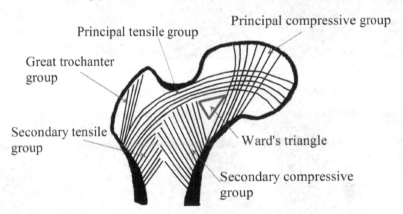

Figure 8. Trabecular structure pattern of human femur.

Clinically, the correlation of bed rest (reducing loading frequency) and muscle activity has been reported. The disuse due to bed rest with diseases accompanies with reduced muscle activity, and then leads to the cane use for walking. Thus, we simulated the trabecular remodeling by combining the two reduced weight-bearing conditions. Figure 10 shows that trabecular loss increased with increasing in the rate of unloading interval. Combination of reductions of loading frequency and loading forces relatively accentuated the structure of principal compressive and principal tensile groups and enhanced the thinning of the secondary compressive trabecular bone. Therefore, empty Ward's triangle is clearly observed. Further increase in the rate of unloading interval led to a marked reduction of tensile trabecular bone and discontinued the principal tensile group. Thus, the tensile trabecular bone is observed only in the upper part of the femoral head, where trabeculae are still comparable in density to the principal compressive trabeculae. Finally, even the principal compressive group became less obvious.

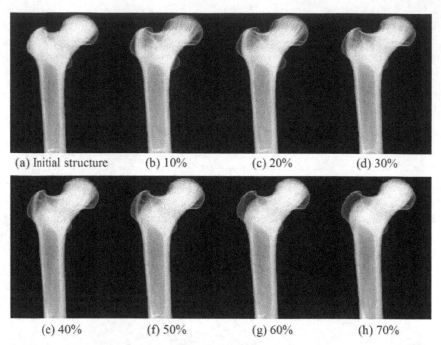

Figure 9. Simulated trabecular structures under cane-assisted walking

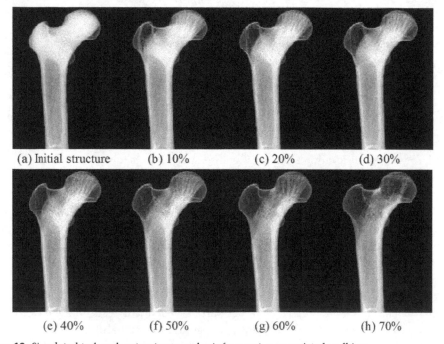

Figure 10. Simulated trabecular structures under infrequent cane-assisted walking.

Osteocyte Apoptosis-Induced Bone Resorption in Mechanical Remodeling Simulation – Computational
Model for Trabecular Bone Structure

37

Trabecular structure of the proximal femur is remodeled differently in response to various forms of mechanical stimulation (Wolff, 1986; Carter, 1987). In this study, we investigated that the structure changes of human trabecular bone under the reduced weight-bearing conditions by a surface remodeling model (Kwon et al., 2010a), in which osteocyte apoptosis plays a crucial role below a physiological strain range (Gu et al., 2005). The reduced weight-bearing conditions were imposed by assuming infrequent and/or cane-assisted walking. Depending on the reduced weight-bearing conditions, the trabecular structure reached different equilibrium structures even if the same initial structure and model parameters are used. In all cases, trabecular bone loss occurred in relation to the mechanical stimuli, although there was a regional difference in the pattern of bone loss between the two conditions. In the infrequent and cane-assisted walking conditions, significant bone loss occurred in the great trochanter and in the femoral head, respectively.

In the results of imposing the condition combining two reduced weight-bearing conditions, we found out the clinically observed trabecular structure in osteoporotic human proximal femur. The description for degree of osteoporotic trabecular structure (Singh Index) is shown in Table 2. The present stimulations showed that decreasing mechanical stimuli enhanced the degree of osteoporosis along with the grade defined by Singh et al. (1970). Figure 10 shows trabecular loss at various degrees of unloading. In the initial structure (Fig. 10 (a)), all groups of trabecular bones are visible, which corresponds to Grade 6 in Singh index. The compressive and tensile trabeculae intersect each other and the upper end of the femur is completely occupied by cancellous rich structure. Ward's triangle is not clearly delineated and there are some thin trabeculae in it. With increasing the proportion of unloading, there occurs an apparent accentuation of the structure of the principal compressive and principal tensile groups, while the secondary compressive trabeculae became thinner (Grade 5, Fig. 10 (b)~(c)). As a result, an empty region appears in Ward's triangle. Further increase in the proportion of unloading leads to the marked reduction in the tensile trabecular bones (Grade 4, Fig. 10 (d)~(e)); in due course, discontinuity occurs in the principal tensile group. At the stages of Fig. 10 (e) and (f), the tensile trabeculae are seen

Grade 6	All the normal trabecular groups are visible and the upper end of the femur seems to be completely occupied by trabecular bone.
Grade 5	The structure of principal tensile and principal compressive trabeculae is accentuated. Ward's triangle appears prominent.
Grade 4	Principal tensile trabeculae are markedly reduced in number but can still be observed in lateral cortex to the upper part of the femoral neck.
Grade 3	There is a break in the continuity of the principal tensile trabeculae opposite the greater trochanter. This grade indicates definite osteoporosis.
Grade 2	Only the principal compressive trabeculae stand out prominently; the others have been resorbed more or less completely.
Grade 1	Even the principal compressive trabeculae are markedly reduced in number and are no longer prominent.

Table 2. Singh Index

only in the upper part of the femoral head, where the trabeculae are still comparable in density to the principal compressive trabeculae (Grade 3). Finally, even the principal compressive group ceases to stand out (Grade 2, Fig. 10 (g)) and decreases markedly in number (Grade 1, Fig. 10 (h)). Therefore, the present remodeling model has the competence to study the trabecular bone loss in disuse-mediated osteoporosis, where bone resorption-dominant remodeling due to less mechanical stimuli should be characterized reasonably and sufficiently. That is introduced into the present model by considering the effect of osteocyte apoptosis under low strain stimuli, and it is essential in predicting the osteoporotic change of trabecular bone structure.

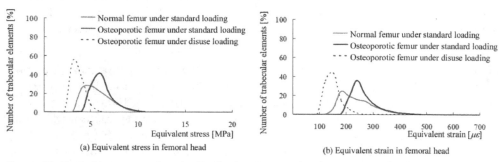

Figure 11. Equivalent stress and strain distributions in normal and osteoporotic femurs.

Finite element analyses were conducted for normal and osteoporotic human femurs obtained by remodeling simulation, and trabecular bone stress and strain were examined for a spherical volume of interest in the femoral head (Rietbergen et al., 2003). Figure 11 shows the distribution of voxel at different stress/strain for the normal femur of Fig. 10(a) and an osteoporotic femur of Fig. 10 (b) that is obtained under the disuse loading condition of reduced loading frequency and force. Solid and broken blue lines distinguish the stress/strain distribution for osteoporotic femur under standard and disuse loading conditions, respectively. Trabecular bone stress and strain in the osteoporotic femur were distributed more uniformly than those in the normal femur. In the osteoporotic femur, trabecular bones orthogonal to the alignment of the principal compressive group (non-principal trabeculae) are relatively-scarce because osteocytes in those trabeculae, especially in non-principal group originally exposed to relatively small loading, are susceptible to apoptosis when subjected to the disuse condition, resulting in scarcely distributed non-principal trabeculae. Consequently, trabeculae undergoing relatively-low stress/strain decrease and the peaks of stress and strain distributions increased in the osteoporotic femur. Trabeculae in principal compressive group are likely to preserve their role of relatively high load-bearing even under the disuse condition.

3.4. Importance of osteocyte apoptosis-induced bone resorption in remodeling simulation

To evaluate the advantage of including low and high strain ranges, we performed the simulation assuming the response model defined for physiological strain range to the entire strain range including low and high strain ranges, referred to physiological response-only model hereafter. Results by physiological range-only model were compared with that obtained by the present integrated model considering individual response model for low and high strain ranges. This physiological response-only model determines the trabecular structure only by relative non-uniformity of strain distribution over all the strain range with no local acceleration of bone resorption and formation. Figure 12(a) shows the distribution of initial equivalent strain. Diaphyseal structure was formed with the loss of trabecular bone both integrated and physiological response-only models (Fig. 12(b)). However, bone resorption occurred to the higher degree in the integrated model, especially in low strain regions as clearly shown in dashed circles of Fig. 12(b). Figure 13 shows the simulation result for change of volume fraction in ward's triangle region at each model. The integrated model was reached to equilibrium state, faster than physiological response-only model. Ward's triangle region has a low strain range, as seen in Fig. 12(a). Figure 14 shows the histogram of number of trabecular bone elements against initial equivalent strain distribution for both simulation models. In the low strain range below $100\mu\varepsilon$, as much as 43, 37, and 46% of trabecular bone preserved in the physiological response-only model were resorbed by the integrated model in different subjects of #1, #2, and #3, respectively. In low strain regions at the initial, the integrated model could reproduce more suitable trabecular structure by taking the characteristic response in low strain range into account. Furthermore, the correlation between bone volume fraction simulated and bone mineral density by CT image was higher in the integrated model than that in the physiological response-only model (Fig. 15).

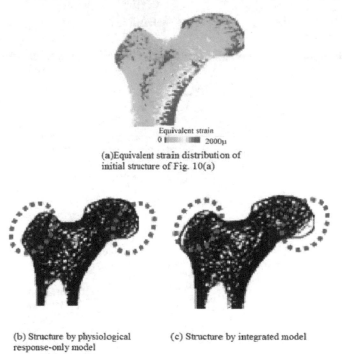

Equivalent strain
0 ▮▮▮▮   ▮▮▮▮ 2000μ

(a)Equivalent strain distribution of
initial structure of Fig. 10(a)

(b) Structure by physiological   (c) Structure by integrated model
response-only model

Figure 12. Initial stress distribution and simulated trabecular structures.

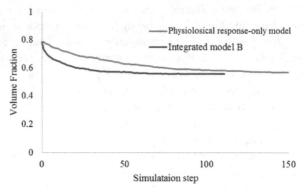

Figure 13. Volume fraction in Ward's triangle during remodeling simulation.

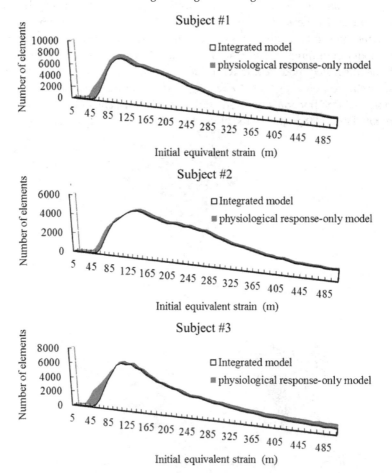

Figure 14. Number of trabecular bone elements in simulated proximal femur against initial equivalent strain levels before disuse-mediated remodeling.

Osteocyte Apoptosis-Induced Bone Resorption in Mechanical Remodeling Simulation – Computational Model for Trabecular Bone Structure

41

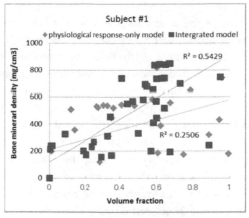

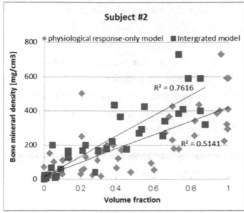

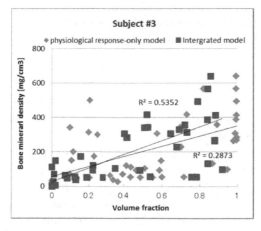

Figure 15. Bone mineral density at proximal femur by CT and simulated volume fraction by integrated and physiological response-only models.

4. Conclusions

In this chapter, described was the mathematical model of bone remodeling extending the established previously (Adachi et al., 1997) based on the mechanostat theory (Duncan & Turner, 1995; Frost, 2003; Hughes et al., 2010). Extension was the bone resorption-dominant response in low strain range by disuse and formation-dominant response in overuse windows. The osteocyte apoptosis in the low strain range due to weightlessness was a key aspect in the extended model, and the targeted remodeling was in the high strain range for bone formation. Trabecular structure in human proximal femur was simulated using the model extended and the reproduced trabecular structure exhibited good agreement with that in the actual femur more reasonably by taking the osteocyte apoptosis in low strain range and the targeted remodeling in high strain range. Especially, the consideration of osteocyte apoptosis was crucial in the mathematical model of bone remodeling model in reproducing the healthy normal bone structure and in simulation the disuse-mediated osteoporotic bone structure. The former was demonstrated by comparison of simulated volume fraction and the X-ray CT bone mineral density. For the latter, disuse-accelerated bone loss was examined for infrequent and cane-assisted walking conditions. The osteoporotic trabecular structure was examined in terms of Singh Index, a diagnostic criterion for the stage of progression of osteoporosis. This competence of the model was coming from the effect of osteoporosis due to less mechanical stimulus and its result of bone loss acceleration, and the present remodeling model is effective in bone remodeling simulation under reduced loading condition.

In conclusion, the effect of osteocyte apoptosis on bone resorption is inevitable aspect in the computer simulation of bone remodeling.

Author details

Ji Yean Kwon, Hisashi Naito, Takeshi Matsumoto and Masao Tanaka
Graduate School of Engineering Science, Osaka University, Japan

5. References

Adachi, T., Tomita, Y., Sakaue, H. & Tanaka, M., (1997), Simulation of trabecular surface remodeling based on local stress nonuniformity, *JSME International Journal, Series C*, Vol. 40, No. 4, pp. 782-792

Adachi, T., Tsubota, K., Tomita, Y. & Hollister, S.J., (2001), Trabecular surface remodeling simulation for cancellous bone using microstructural voxel finite element models, *Journal of Biomechanical Engineering*, Vol. 123, pp. 403-409

Baiotto, S., Labat B., Vico, L. & Zidi, M., (2009), Bone remodeling regulation under unloading condition: Numerical investigations, *Computer in Biology and Medicine*, Vol. 39, pp. 46-52

Barrera, G., Bunout, D., Gattás, V., Maza, M.P., Leiva, L. & Hirsch, S., (2004), A high body mass index protects against femoral neck osteoporosis in healthy elderly subjects, *Nutrition*, Vol. 20, No. 9, pp. 769-771

Beaupré, G.S., Orr, T.E. & Carter, D.R., (1990), An approach for time-dependent bone modeling-application: a preliminary remodeling simulation, *Journal of Orthopedic Research*, Vol. 8, pp. 662-670

Burr, D.B., (2002), Targeted and nontargeted remodeling, *Bone*, Vol. 30, No. 1, pp. 2-4

Caillot-Augusseau, A., Lafage-Proust. M.H., Soler, C., Pernod, J., Dubois, F. & Alexandre, C., (1998), Bone formation and resorption biological markers in cosmonauts during and after a 180-day space flight (Euromir 95), *Clinical Chemistry*, Vol. 44, pp.578–585

Carter, D.R., (1987), Mechanical loading history and skeletal biology, *Journal of Biomechanics*, Vol. 20, pp. 1095-1109

Cowin, S.C., (2001), *Bone mechanics handbook* (second edition), CRC press LLC, ISBN 0-8493-9117-2, USA

Da Costa Gómez, T.M., Barrett, J.G., Sample, S.J., Radtke, C.L., Kalscheur, V.L., Lu, Y., Markel, M.D., Santschi, E.M., Scollay, M.C. & Muir, P., (2005), Up-regulation of site-specific remodeling without accumulation of microcracking and loss of osteocytes, *Bone*, Vol. 37, No. 1, pp. 16-24

Delgado-Calle, J., Arozamena, J., García-Renedo, R., García-Ibarbia, C., Pascual-Carra, M.A., González-Macías, J. & Riancho, J.A., (2011), Osteocyte deficiency in hip fractures, *Calcified Tissue International*, Vol. 89, pp.327-334

Duncan, R.L. & Turner, C.H., (1995), Mechanotransduction and the functional response of bone to mechanical strain, *Calcified Tissue International*, Vol. 57, No.5, pp. 344-358

Frost, H.M., (2003), Bone's mechanostat: a 2003 update, *The Anatomical Record, Part A, Discoveries in molecular, cellular, and evolutionary biology*, Vol. 275, No. 2, pp. 1081-1101

Fuchs, R.K,, Bauer, J.J. & Snow, C.M.,(2001), Jumping improves hip and lumbar spine bone mass in prepubescent children: a randomized controlled trial, *Journal of Bone Mineral Research*, Vol. 16, No. 1, pp.148–156

Garber, M.A., McDowell, D.L. & Hutton, W.C., (2000), Bone loss during simulated weightlessness: a biomechanical and mineralization study in the rat model, *Aviation Space Environmental Medicine*, Vol. 71, pp. 586–592

Gu, G., Mulari, M., Peng, Z., Hentuen, T.A. & Väänänen, H.K., (2005), Death of osteocytes turns off the inhabitation of osteoclasts and triggers local bone resorption, *Biomechanical and Biophysical Research Communications*, Vol. 335, No. 4, pp. 1095-1101

Hughes,J.M. & Petit, M.A.,(2012), Biological underpinnings of Frost's mechanostat thresholds:The important role of osteocytes, *Journal of Musuloskelet Neuronal Interact*, Vol. 10, No. 2, pp. 128-135

Kaneps A.J., Stover S.M. & Lane N.E., (1997), Changes in canine cortical and cancellous bone mechanical properties following immobilization and remobilization with exercise, *Bone*, Vol. 21, pp. 419–423

Kennedy, O.D., Herman, B.C., Laudier, D.M., Majeska, R.J., Sun, H.B. & Schaffler, M.B., (2012), Activation of resorption in fatigue-loaded bone involves both apoptosis and active pro-osteoclastogenic signaling by distint osteocyte popuations, *Bone*, Vol. 50, pp. 1115-1122

Klein-Nulend, J., Bacabac, R.G. & Mullender M.G., (2005), Mechanobiology of bone tissue, *Pathologie Biologie*, Vol. 53, pp. 576-580

Kwon, J.Y., Naito, H., Matsumoto, T. & Tanaka, M., (2010a), Simulation model of trabecular bone remodeling considering effects of osteocytes apoptosis and targeted remodeling, *Journal of Biomechanical Science and Engineering*, Vol. 5, No. 5, pp. 539-551

Kwon, J.Y., Naito, H., Matsumoto, T. & Tanaka, M., (2010b), Computational study on trabecular bone remodeling in human femur under reduced weight-bearing conditions, *Journal of Biomechanical Science and Engineering*, Vol. 5, No. 5, pp. 552-564

Lang, T., LeBlanc, A., Evans, H., Lu Y., Genant, H. & Yu, A., (2004), Cortical and trabecular bone mineral loss from the spine and hip in long-duration spaceflight, *Journal of Bone Mineral Research*, Vol. 19, pp. 1006– 1012

LaMonte, J.M., Hamilton, N.H. & Zernicke, R.F., (2005), Strain rate influences periosteal adaptation in mature bone, *Medical Engineering & Physics*, Vol. 27, pp.277-284

LeBlanc, A.D., Schneider, V.S., Evans, H.J., Engelbretson, D.A. & Krebs, J.M., (1990), Bone mineral loss and recovery after 17 weeks of bed rest, *Journal of Bone and Mineral Research*, Vol. 5, pp. 843–850

Li, C.Y., Majeska, R.J., Laudier, D.M., Mann, R. & Schaffler, M.B., (2005), High-dose risedronate treatment partially preserves cancellous bone mass and microarchitecture during long-term disuse, *Bone*, Vol. 37, pp. 287–295

Luo, G., Cowin, S.C., Sadegh, A.M. & Arramon, Y.P., (1995), Implementation of strain rate
as a bone remodeling stimulus, *Journal of Biomechanical Engineering*, Vol. 117, pp. 329-338

McGee, M.E., Maki, A.J., Johnson, S.E., Nelson, O.L., Robbins, C.T. & Donahue, S.W., (2008), Decreased bone turnover with balanced resorption and formation prevent cortical bone loss during disuse(hibernation) in Grizzly bears(Ursus arctos horribilis), *Bone*, Vol. 42, pp. 396-404

McKay, H.A., Petit, M.A., Schutz, R.W., Prior, J.C., Barr, S.I. & Khan, K.M., (2000), Augmented trochanteric bone mineral density after modified physical education classes: a randomized school-based exercise intervention study in prepubescent and early pubescent children, *Journal of Pediatrics*, Vol. 136, No. 2, pp.156–162

Noble, B.S., Peet, N., Stevens, H.Y., Brabbs, A., Mosley, J.R., Reilly, G.C., Reeve, J., Skerry, T.M. & Lanyon, L.E., (2003), Mechanical loading: biphasic osteocyte survival and targeting of osteoclasts for bone destruction in rat cortical bone, *American Physiological Society*, Vol. 284, No. 4, pp. C934-C943

Neumann, D.A., (1998), Hip abductor muscle activity as subjects with hip prostheses walk with different methods of using a cane, *Physical Therapy*, Vol. 78, pp. 490-501

Osteocyte Apoptosis-Induced Bone Resorption in Mechanical Remodeling Simulation – Computational
Model for Trabecular Bone Structure

45

Parfitt, A.M., (1994), Osteonal and hemi-osteonal remodeling: the spatial and temporal framework for signal traffic in adult human bone. *Journal of Cellular Biochemistry*, Vol. 55, pp. 273–286

Rietbergen, B., Huiskes, R., Eckstein, F. & Rüegsegger, P., (2003), Trabecular bone tissue strains in the healthy and osteoporotic human femur, *Journal of Bone and Mineral Research*, Vol. 18, No. 10, pp. 1781-1788

Ruimerman, R., Huiskes, R., Van Lenthe, G.H. & Janssen, J.D., (2001), A computer simulation model relating bone-cell metabolism to mechanical adaptation of trabecular bone, *Computational Methods in Biomechanical and Biomedical Engineering*, Vol. 4, pp. 433-448

Shackelford, L.C., LeBlanc, A.D., Driscoll, T.B., Evans, H.J., Rianon, N.J., Smith, S.M., Spector, E., Feeback, D.L. & Lai, D., (2004), Resistance exercise as a countermeasure to disuse-induced bone loss, *Journal of Applied Physiology*, Vol. 97, pp. 119–129

Singh, M., Nagrath, A.R. & Maini, P.S., (1970), Changes in Trabecular Pattern of the Upper end of the Femur as an index of osteoporosis, *The Journal of Bone and Joint Surgery*, Vol. 52, pp. 457-467

Sordia, L.H., Vazquez, J., Iglesias, J.L., Piñeyro, M.O., Vidal, O., Saldivar, D., Morales A., Merino, M. Pons, G. & Rosales, E., (2004), Low height and low weight correlates better with osteoporosis than low body mass index in postmenopausal woman, *International Congress Series*, Vol. 1271, pp. 407-410

Takata, S. & Yasui, N., (2001), Disuse osteoporosis, *The Journal of Medical Investigation*, Vol. 48, pp. 147–156

Tsubota, K., Adachi, T. & Tomita, Y., (2002), Functional adaptation of cancellous bone in human proximal femur predicted by trabecular surface remodeling simulation toward uniform stress state, *Journal of Biomechanics*, Vol. 35, pp. 1541-1551

Tsubota, K. & Adachi, T., (2006), Simulation study on local and integral mechanical quantities at single trabecular level as candidate of remodeling stimuli, *Journal of Biomechanical Science and Engineering*, Vol. 1, pp. 124-135

Tsubota, K., Suzuki, Y., Yamada, T., Hojo, M., Makinouchi, M. & Adachi, T., (2009), Computer simulation of trabecular remodeling in human proximal femur using large-scale voxel FE models: Approach to understanding Wolff's law, *Journal of Biomechanics*, Vol. 42, pp. 1088-1094

Vailas, A.C., Zernicke, R.F., Grindeland, R.E., Kaplansky, A., Durnova, G.N., Li, K.C. & Martinez, D.A., (1990), Effects of spaceflight on rat humerus geometry, biomechanics, and biochemistry. *The FASEB Journal*, Vol. 4, pp. 47–54

Vico, L., Collet, A., Guignandon, P., Lafage-Proust, M.H., Thomas T., Rehailia M. & Alexandre C., (2000), Effects of long term microgravity exposure on cancellous and cortical weight-bearing bones of cosmonauts, *The Lancet*, Vol. 355, pp. 1607-1611

Weinreb, M., Rodan, G.A. & Thompson, D.D.,(1989), Osteopenia in the immobilized rat hind limb is associated with increased bone resorption and decreased bone formation, *Bone*, Vol. 10, pp.187–194

Wolff, J., (1986), The Law of Bone Remodeling, *Springer (Trans. Maquet, P., Furlong, R.)*

Apoptosis and Pathology

Apoptosis as a Therapeutic Target in Cancer and Cancer Stem Cells: Novel Strategies and Futures Perspectives

María A. García, Esther Carrasco, Alberto Ramírez, Gema Jiménez,
Elena López-Ruiz, Macarena Perán, Manuel Picón,
Joaquín Campos, Houria Boulaiz and Juan Antonio Marchal

Additional information is available at the end of the chapter

1. Introduction

Apoptosis is an essential part of the normal development. The homeostatic balance between cell proliferation and cell death rate is critical for maintaining normal physiological processes. Aberrant regulation of apoptotic cell death mechanisms is one of the hallmarks of cancer development and progression, and many cancer cells exhibit significant resistance to apoptosis signalling [1]. Triggering of apoptosis can be achieved via the activation of two distinct molecular pathways, the extrinsic or death receptor pathway or via the intrinsic or mitochondrial apoptotic cascades. Both pathways lead to the hierarchical activation of a family of cysteine proteases called caspases [2], that cleave a series of cellular substrates which induce changes including chromatin condensation, internucleosomal DNA fragmentation, membrane blebbing and cell shrinkage [3]. Extrinsic pathway is activated from outside the cell by proapoptotic ligands that interact with specialized cell surface death receptors, including CD95 and TNF-related apoptosis-inducing ligand (TRAIL) receptors [4]. After binding to receptors apoptosis is triggered by the intracellular formation of a death-inducing signalling complex (DISC) that consists of FAS-associated death domain (FADD) and procaspase-8 and 10 [5,6]. As a result, this protein complex activates procaspase-8 and 10 inside itself, hence triggering procaspase-3 to execute the apoptosis process [7]. The mitochondria (intrinsic) pathway is activated from inside the cell by severe cell stress, such as DNA or cytoskeletal damage, inducing mitochondrial outer membrane permeabilization and transcription or post-translational activation of BH3-only proapoptotic B-cell leukemia/lymphoma 2 (Bcl-2) family proteins [4]. This permeabilization allows the release of apoptogenic proteins, including cytochrome c and second mitochondria-derived activator of caspase (Smac; also known as DIABLO), from the mitochondrial intermembrane space into the cytosol [8]. Cytochrome c assembles with apoptotic protease-activating factor-1 (Apaf-1) to activate caspase 9. This caspase, in turn, activates the effector caspases 3, 6, and 7, which carry out apoptosis [4]. Smac promotes caspase activation and apoptosis by neutralization of several IAP proteins, including XIAP, c-IAP1 and c-IAP2 [9,10]

Chemotherapeutic agents act by inhibiting tumour cell proliferation and survival and most of them can kill tumour cells by activating common apoptotic pathways [11]. Therefore, apoptosis plays an important role in the treatment of cancer as it is a popular target of many treatment strategies. 5-fluorouracil (5-FU), an antimetabolite analogue of uracil employed primarily in the treatment of a variety of solid malignant tumours, leads to a wide range of biological effects which can act as triggers for apoptotic cell death [12,13]. However, resistance to the drug remains a major clinical problem. Given that many of the apoptotic regulators altered in multidrug resistant tumours have been identified, one new approach to therapy is to restore apoptotic potential through genetic or pharmacological methods [14]. Moreover, since defects in the mediators of apoptosis may account for chemo-resistance, the identification of new targets involved in drug-induced apoptosis is of main clinical interest. Recently, we have identified the ds-RNA-dependent protein kinase (PKR) as a key molecular target of 5-FU involved in apoptosis induction, in a p53 -independent manner. These results suggest the clinical importance that the PKR status could play in response to chemotherapy based on 5-FU. Moreover, the effectiveness of 5-FU cytotoxic activity induced by IFNα, especially in cancer cells expressing a mutated form or lacking p53, but with a functional PKR, might have relevant clinical application in patients [15] .

The increased knowledge of some of the molecular components of the apoptosis signalling pathways has paved the way for the generation of more specific agents that target one crucial signalling component. This has allowed a change in anticancer therapy trends, from classic cytotoxic strategies to the development of new non-harmful therapies which target the apoptosis response selectively only in tumour cells. Moreover, these strategies overcome the adverse effects associated with cytotoxic drugs and increase their anti-cancer activity. Novel antitumour drugs have been synthesised such as 5-FU O,N-acetals and benzo-fused seven-membered O,N-acetal in which the 5-FU moiety was changed for the naturally-occurring pyrimidine base uracil, which induced cell cycle-mediated apoptosis in breast and colon cancer cells [16,17,18,19]. The mechanism of action of these drugs was mainly centred on positive apoptosis regulatory pathway genes, and the repression of genes involved in carcinogenesis, proliferation and tumour invasion. In addition, these drugs were more selective against tumour cells with lower toxic effects in non-tumour cells [20,21].

As over-expression of IAP proteins frequently occurs in various human cancers and has been linked to tumour progression, chemo-resistance and poor prognosis, it is not surprising that IAP proteins are considered to be attractive targets for improve outcomes for patients with solid tumours and hematologic malignancies [22,23]. IAPs are also attractive as therapeutic targets because their inhibition does not appear to be toxic to normal adult cells [24]. Several therapeutic strategies have been designed to target IAP, including a small-molecule approach that is based on mimicking the IAP-binding motif of the endogenous IAP antagonist Smac [25]. Other therapies involve antisense strategies and short-interfering RNA (siRNA) molecules [26,27]. There is increasing interest in therapeutic drug development targeting the IAP family.

MicroRNAs (miRNAs) are small RNA gene products that regulate the activity of messenger RNAs by antisense base pairing. They are involved in stem-cell self-renewal, cellular development, differentiation, proliferation, and apoptosis [28]. In cancer, miRNAs function as regulatory molecules, acting as oncogenes or tumour suppressors [29] and they are strongly related to the apoptosis. New insights indicate that many miRNAs are anti-apoptotic and mediate this effect by targeting pro-apoptotic mRNAs or positive regulators of pro-apoptotic mRNAs [30]. Conversely, many pro-apoptotic miRNAs target anti-apoptotic mRNAs or their positive regulators [31]. Therefore, their inhibition leads to the induction of programmed cell death, suggesting a promising miRNA treatment for cancer. Several drugs may have the ability to modulate the expression of miRNAs by targeting signalling pathways that ultimately converge on the activation of transcription factors that regulate miRNA encoding genes.

Evasion of apoptosis is one of the main mechanisms involved in tumourigenesis and drug resistance. Most cancers have a small population of tumour cells, as few as 1%, with stem cell characteristics and the capacity for self-renewal, termed cancer stem cells (CSCs). A malignant tumour can be viewed as an abnormal organ in which small populations of tumourigenic CSCs have escaped the normal limits of self-renewal, giving rise to abnormally differentiated cancer cells that contribute to tumour progression and growth [32]. These CSC express high levels of ATP-binding cassette (ABC) drug transporters, providing for a level of resistance [33]; are relatively quiescent; have higher levels of DNA repair and a lowered ability to enter apoptosis . Several cancer therapy approaches targeting ABC transporters and increasing apoptosis could be employed to selectively and more efficiently kill CSC.

The current review will focus on recent development of several therapeutic strategies, which interfere with apoptosis and are currently used or tested for treatment of cancer. They induce cancer cell death or enhance the responsiveness of cancer cells and CSCs to certain cytotoxic drugs. Some of them such as caspases activators, indirectly modulators of apoptosis or agents targeting apoptosis-related proteins are still in their preclinical or clinical trials. We also include future approaches directed to target apoptotic pathways with promising application in patients with cancer.

2. Novel apoptotic markers/targets in cancer

The main goal in cancer therapy is the abrogation of tumour cell growth and proliferation, and ultimately the complete elimination of tumour cells. It is commonly accepted that tumour cells treated with anticancer agents undergo apoptosis, and that cells resistant to apoptosis often do not respond to anticancer therapy [34]. Moreover, it is widely demonstrated that some oncogenic mutations suppressing apoptosis may lead to tumour initiation, progression or metastasis [11]. Apoptosis plays a major control role in cell death when DNA damage is irreparable and multiple stress-inducible molecules have been implied in transmitting the apoptotic signal [35]. Because of the potential detrimental effects on cell survival in case of inappropriate activation of apoptosis programs, apoptosis pathways have to be tightly controlled. However, the concept that apoptosis represents the major mechanism by which cancer cells are eliminated may not universally apply and caspase-independent apoptosis or other modes of cell death have also to be considered as cellular response to anticancer therapy [36].

Different determinants of drug resistance exist, including loss of cell surface receptors or transporters, altered metabolism, or mutation of specific apoptotic target [37]. The apoptotic signalling pathways are regulated by numerous hub proteins such as p53, Bcl-2, NFkB and MAPKs which function in common. In the following sections, we will summarize some of the cellular proteins considered as potential apoptotic biomarkers in cancer.

1- The tumour suppressor p53 is an important pro-apoptotic factor and tumour inhibitor, and numerous anti-tumour drugs would exert their functions through targeting p53-related signalling pathways. Some clinical investigations indicated that under abnormal situations such as chemotherapy and UV or DNA damage may occur and activate the expression of p53. Activated p53 protein binds to the regulatory sequences of a number of target genes to initiate a program of cell cycle arrest, DNA repair, apoptosis, and angiogenesis [38]. If the damage cannot be repaired completely, over-activation of p53 leads to the tumour growth stagnation or even apoptosis [39,40]. Loss of p53 function is critical in tumourigenesis, and alterations to the *p53* gene (mutations, often resulting in protein over-expression) are frequent events in cancer. Associations of *p53* tumour alterations with patient prognosis and response to adjuvant chemotherapy have been widely studied, and findings are contradictory. The fluoropyrimidine 5-FU is widely used in the treatment of a range of cancers including colorectal cancer and breast cancer [41,42,43], but resistance to the drug remains a major clinical problem. P53 was the first target described for 5-FU- induced apoptosis, however, although several reports have demonstrated that 5-FU-induced apoptosis is dependent on the tumour suppressor p53 protein, apoptosis can also occur in mutant p53 cell lines [44,45,46,47]. Moreover, the relationship between p53 status and sensitivity to chemotherapeutic drugs, including 5- FU, is still controversial. In clinical studies in which adjuvant chemotherapy– treated and non-treated groups could be analyzed, stage III colorectal patients whose tumours demonstrated no *p53* alterations experienced significantly longer survival following 5-FU–based chemotherapy than patients whose tumours over-expressed p53 [48,49]. However, other studies in colon cancer patients failed to demonstrate correlations between *p53* alterations and benefit from adjuvant therapy [50,51]. The identification of new targets involved in 5-FU-induced apoptosis could contribute to clarify the controversy results obtaining in clinic.

2- Recently, we have identified the interferon-induced protein kinase PKR, as a molecular target of 5-FU with an interesting role in the apoptosis induced by this chemotherapeutic drug [15]. The double-stranded RNA (dsRNA)-dependent kinase PKR was initially identified as an innate immune anti-viral protein approximately 35 years ago [52,53]. Since then, PKR has been linked to normal cell growth and differentiation, inflammation, cytokine signalling, and apoptosis [54]. PKR is a serine/threonine kinase, characterized by two distinct kinase activities: autophosphorylation, which represents the activation reaction, and phosphorylation of eIF-2α [55,56], which impairs eIF-2 activity, resulting in inhibition of protein synthesis [57]. In addition to its translational regulatory function, PKR has a role in signal transduction and transcriptional control through the IkB/NF-kB pathway [58,59]. PKR is activated in response to dsRNA of cellular, viral, or synthetic origin. PKR can also be activated by polyanions such as heparin, dextran sulfate, chondroitin sulfate, and poly-L-glutamine [60]. A range of cellular stresses can also activate PKR, such as arsenite, thapsigargin, H_2O_2, ethanol and ceramide [61,62,63] presumably through the PKR-associated activator (PACT)/RAX protein [64]. Moreover, it is induced by interferon type I and mediates in part, several functions of these cytokines. Altered PKR activity has been shown to play a role in neurodegenerative diseases (Alzheimer's, Creutzfeldt–Jakob, Huntington's, and Parkinson's) and cancer [65,66,67,68,69,70,71,72,73,74].

Over-expression or continued activation of PKR leads to apoptosis [75,76,77]. PKR mediates the apoptosis induced by several viruses and cellular stresses [78] by activation of intrinsic and extrinsic apoptosis pathways through the FADD/caspase 8 and mitochondrial APAF/caspase 9 activation pathways [79,80]. Recently, PKR has been shown to play an important role in apoptotic cancer cell death induced by 5-FU, doxorubicin and etoposide [15,81,82] and the antitumour activity of tumour suppressors like p53 and PTEN [81,83]. Preclinical studies in mice have shown than in tumours which do not express sufficient levels of PKR are more resistant to doxorubicin and etoposide that tumour expressing higher PKR levels. We have demonstrated that PKR is up-regulated and activated in colon and breast cancer cell lines by inducing apoptotic cell death in response to 5-FU treatment. In addition, cancer cell lines deficient in PKR expression were more resistant to the cytotoxic effect of 5-FU with an IC_{50} being 2-3 fold higher than cells expressing an active PKR protein. Moreover, apoptosis mediated by PKR in response to 5-FU occurred independently on p53 status highlighted the importance that both p53 and PKR play in the 5-FU-induced cancer cell death, and the relevance acquired by PKR in tumour cells where p53 is mutated. Such results raise the importance of determining PKR status in tumours from patients treated with 5-FU-based chemotherapy [15].

3- One pathway being targeted for antineoplastic therapy is the Bcl-2 family of proteins (Bcl-2, Bcl-XL, Bcl-w, Mcl-1, Bfl1/A-1, and Bcl-B) that bind to and inactivate BH3-domain pro-apoptotic proteins. It is controversial whether some BH3-domain proteins (Bim or tBid) directly activate multidomain pro-apoptotic proteins (e.g., Bax and Bak) or act via inhibition of those anti-apoptotic Bcl-2 proteins (Bcl-2, Bcl-XL, Bcl-w, Mcl-1, Bfl1/A-1, and Bcl-B) that stabilize pro-apoptotic proteins [84]. Since the anti-apoptotic properties of Bcl-2 were discovered, the over-expression of Bcl-2 conferring chemoresistance was reported, and the 3-D protein structure of Bcl-XL was determined. These properties have contributed to the development of protein inhibitors. The first agent targeting Bcl-2 that entered clinical trials was a Bcl-2 antisense (oblimersen sodium), which showed chemosensitizing effects when combined with conventional chemotherapy drugs in chronic lymphocytic leukemia (CLL) patients, leading to improved survival [85,86]. More recent advances include the discovery of small molecule inhibitors of the Bcl-2 family proteins. They are designed to bind the hydrophobic groove of anti-apoptotic Bcl-2 proteins in place of BH3-only proteins (i.e., BH3-mimetics). They can oligomerize Bax or Bak, which can subsequently depolarize mitochondrial membrane potential to release cytochrome *c*. To date, one Bcl-2 antisense and three small molecule Bcl-2 protein inhibitors are being tested in clinical trials. Preclinical studies seem promising, especially in combination with additional chemotherapy agents. Ongoing and planned phase II clinical trials to define the activity of single agents and drug combinations will determine the direction of future clinical development of the Bcl-2 inhibitors [84].

4- PUMA (p53 upregulated modulator of apoptosis) is a BH3-only Bcl-2 family member and a critical mediator of p53-dependent and -independent apoptosis induced by a wide variety of stimuli, including genotoxic stress, deregulated oncogene expression, toxins, altered redox status, growth factor/cytokine withdrawal and infection. It serves as a proximal signalling molecule whose expression is regulated by transcription factors in response to these stimuli. PUMA transduces death signals primarily to the mitochondria and acts indirectly on the Bcl-2 family members Bax and/or Bak by relieving the inhibition imposed by anti-apoptotic members. It directly binds and antagonizes all known anti-apoptotic Bcl-2 family members to induce mitochondrial dysfunction and caspase activation [87]. Several lines of evidence suggest that the function of PUMA is compromised in cancer cells. PUMA expression was found to be reduced in malignant cutaneous melanoma, and PUMA expression appears to be an independent predictor of poor prognosis in patients [88]. In addition, approximately 40% of primary human Burkitt's lymphomas do not express detectable levels of PUMA, which is attributable, in part, to DNA methylation [89]. Evidence of PUMA induction by therapeutic agents in patients has just begun to emerge. Analysis of tissue biopsies from breast cancer patients showed that *PUMA* mRNA was induced within 6 h of chemotherapy [90]. Increased expression of *PUMA* and *Bim* is associated with better prognosis in patients receiving 5-FU-based therapy in stage II and III colon cancer, and is an independent prognostic marker for overall and disease-free survival [91]. PUMA ablation or inhibition leads to apoptosis deficiency underlying increased risks for cancer development and therapeutic resistance. Although elevated PUMA expression elicits profound chemo- and radio-sensitization in cancer cells, inhibition of PUMA expression may be useful for curbing excessive cell death associated with tissue injury and degenerative diseases. Therefore, PUMA is a general sensor of cell death stimuli and a promising drug target for cancer.

5- The apoptosome, a complex of cytochrome-c and APAF-1, recruits and activates the initiator pro-caspase 9, leading to the activation of the effector caspases, in particular pro-caspase 3, culminating in those biochemical and morphological changes associated with apoptotic cell death [92]. This pathway is further regulated by the inhibitor of caspase protein XIAP, which works through the direct inhibition of active caspases 9 and 3 [93] and is also implicated in the ubiquitination of caspases, targeting them for proteasomal degradation [94]. Moreover there is a direct interaction between XIAP and its antagonist Smac [95].

Because of the importance of this pathway in cancer progression and chemotherapy-induced cell death, apoptosome-associated proteins may be important markers for colorectal cancer prognosis and chemotherapy response. Several studies have examined the immunohistochemical expression of individual proteins associated with apoptosis execution in colorectal cancer [96]. Increased expression of APAF-1 has been shown to be associated with longer patient survival in rectal cancer patients [97] and loss of APAF-1 has been implicated in tumour progression and more aggressive disease [98]. Similarly, longer overall survival has been associated with increased Smac [99] and caspase 9 [100]. The anti-apoptotic XIAP has also been implicated as a potential prognostic marker for colorectal cancer, with increased expression correlating with poor patient outcome [101]. However, no study to date has provided a comprehensive analysis of these key regulatory proteins as markers for colorectal cancer prognosis or chemotherapy response. Recently, the pro-caspase 3 expression in colorectal tumours has been associated with better recurrence-free and overall survival, and serves as an independent prognostic marker in localised Stage II disease [102]. This result is in agreement with previous studies that demonstrated as caspase 3 expression is a positive prognostic marker in hepatocellular carcinoma [103] and diffuse large B-cell lymphoma [104].

6- The extrinsic apoptosis pathway is triggered by the binding of death ligands of the tumour necrosis factor (TNF) family to their appropriate death receptors (DRs) on the cell surface. One TNF family member, TRAIL or Apo2L, seems to preferentially cause apoptosis of transformed cells and can be systemically administered in the absence of severe toxicity. Therefore, there has been enthusiasm for the use of TRAIL or agonist antibodies to the TRAIL DR4 and DR5 in cancer therapy. Nonetheless, many cancer cells are very resistant to TRAIL apoptosis in vitro. Therefore, there is much interest in identifying compounds that can be combined with TRAIL to amplify its apoptotic effects [105]. The combination of TRIAL DR agonists with numerous conventional and investigational anticancer drugs has been reported. Synergy has been described for the combination of TRAIL with a variety of cytotoxic agents including irinotecan, camptothecin, 5-FU, carboplatin, paclitaxel, doxorubicin, and gemcitabine in diverse preclinical models [106,107]. Many human and mouse cancer cells lines can be sensitized by proteasome inhibitors such as bortezomib (VELCADE) to the apoptotic effects of TRAIL DR agonists. Interestingly, non-transformed cells seem to be much more resistant to the apoptotic effects of bortezomib and TRAIL than are cancer cells. This suggests that a therapeutic window may exist in vivo where this combination may have a therapeutic benefit in the absence of accompanying toxicity. However, the molecular mechanism(s) of action whereby proteasome inhibition in cancer cells results in sensitization to TRAIL apoptosis remains unclear [108].

3. Selective antitumour-drug inducers of apoptosis

Improved clinical response may be obtained by identifying therapies that are particularly effective in activating apoptosis and determining how those therapies may be modified to effect maximum apoptosis induction. The cell cycle apparatus and apoptosis have attracted the attention of researchers intent on developing new types of anticancer therapy [109,110].

We will concentrate in this part of the review on the evolution of the chemical structures and on the biological properties, whilst the chemical syntheses will be referred to through the corresponding original references.

3.1. Benzo-fused seven-membered derivatives linked to 5-fluorouracil (5-FU)

We have reported the synthesis and anticancer activities of compounds 1-5 [16] and *trans*-6 [17] (Figure 1). In all cases, the linkage between the 5-FU moiety and the seven-membered ring was carried out through its N-1 atom. The structural nature of 5 implies that this compound cannot be considered as a 5-FU prodrug and it was suspected that the remaining compounds (1-6) would not be 5-FU prodrugs.

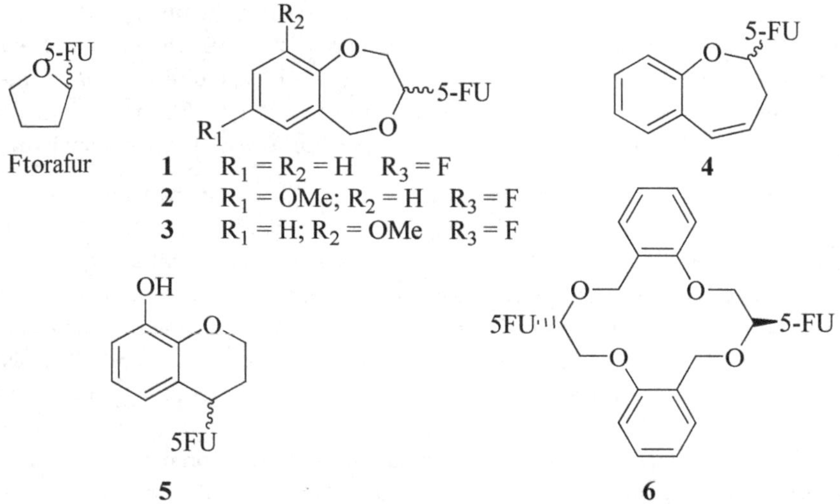

Ftorafur	1 $R_1 = R_2 = H$ $R_3 = F$
	2 $R_1 = OMe; R_2 = H$ $R_3 = F$
	3 $R_1 = H; R_2 = OMe$ $R_3 = F$

Figure 1. Several 5-FU derivatives showing interesting anti-tumour activities.

The IC_{50} values of the 5-FU cyclic O,N-acetals are shown in Table 1 (entries 4-9). The most active compounds are 1, 2 and 6 (entries 4, 5 and 9). On comparing structures 4 and 1, it is worth emphasizing that the bioisosteric change of carbon for oxygen and the saturation of the double bond in compound 1 increases the anti-proliferative activity two fold in ($IC_{50} = 7.00 \pm 0.61$ μM, entry 4). The introduction of a methoxy group into the benzene ring of 1 provokes different influences on the anti-proliferative activities. Thus, the C-7 substitution produces an increase of the anti-proliferative activity (2, $IC_{50} = 4.50 \pm 0.33$ μM, entry 5), whilst if C-9 is the substituted position it gives rise to a decrease in the anti-proliferative activity of 3 ($IC_{50} = 22.0 \pm 0.93$ μM, entry 6).

Apoptosis has been studied in terms of cancer development and treatment with attempts made to identify its role in chemotherapeutic agent-induced cytotoxicity. Cytotoxic agents often induce only a fraction of the cells to become apoptotic. To fully exploit apoptosis as a mechanism of anti-neoplastic agent response, a larger proportion of cells needs to be recruited into apoptosis. Paclitaxel (Taxol®), cyclophosphamide and cytosine arabinoside are the only commonly used cytotoxic agents shown to elicit apoptosis in breast cancer cells [111,112]. Quantitation of apoptotic cells was done by monitoring the binding of fluorescein isothiocyanate (FITC)-labelled annexin V (a phosphatidylserine-binding protein) to cells in response to our title compounds as described [113]. The apoptosis study shows that **3**, **4** and **5**, at their IC_{50} concentrations, provoke early apoptosis in the cells treated for 24 and 48 h. It is worth pointing out that **3** (entry 6) induces greater apoptosis at 48 h (46.73%) than at 24 h (40.08%). The compounds that show the most important apoptotic indexes at 24 h are **4** (57.33%, entry 7) and **5** (54.33%, entry 8), whereas at 48 h is **4** (51.37%, entry 7). These compounds are more potent as apoptosis inductors against the MCF-7 human breast cancer cells than paclitaxel (Taxol®), which induced programmed cell death of up to 43% of the cell population [114]. Accordingly, the early apoptotic inductions and the low IC_{50} values give rise to a significant anti-tumour activity.

Since the synthesized compounds induce very important apoptosis, we have carried out studies of the expression of some of the genes that intervene in this phenomenon, among which p53 and the family bcl-2 are outstanding. The tumour suppressor gene p53 protects the integrity of the genome so that if the DNA of the cell is damaged by an agent, an over-expression of it is produced inducing the stopping in G_1 for the repair of the damage, or if this is not possible, enter apoptosis [115]. On the other hand, the members of the family of proteins Bcl-2 work as regulators of apoptosis, Bcl-2 and Bcl-XL protecting against apoptosis. Bax, Bak and Bad induce such a phenomenon [116]. The treatment of the MCF-7 cells (wild-type p53) with these compounds provoked in general an increase in the protein expression of p53, mainly for 5-FU and **4**, and a marked decrease of the levels of bcl-2 for all of them. These data show that p53 activity is restored with the compounds, allowing the entrance of the tumour cells in apoptosis, which permits their elimination by this mechanism. In the same way bcl-2 inhibition facilitates the entrance of cells into the programmed cell death.

Entry	Compound	IC_{50} (µM)[a]	Cell Cycle (48 h)[b]			Apoptosis[c]	
			G_0/G_1	G_2/M	S	24 h	48 h
1	Control		68.39	12.04	19.57	1.24	1.24
2	**5-FU**	2.75	58.07	2.10	39.38	56.75	52.81
3	**Ftorafur**	3.00 ± 0.11	45.62	0.00	54.38	52.20	58.06
4	1	7.00 ± 0.61	74.41	15.77	9.82	8.45	12.17
5	2	4.50 ± 0.33	73.41	13.15	13.44	1.50	3.50
6	3	22.0 ± 0.93	71.76	10.08	18.16	40.08	46.73
7	4	14.0 ± 1.02	86.14	1.60	12.26	57.33	51.37
8	5	69.0 ± 2.31	68.61	9.60	21.79	54.33	35.49
9	6	5.50 ± 0.58	82.48	5.13	12.40	14.37	19.05

[a]See [117]. [b]Determined by flow cytometry: see [16]. [c]Apoptosis was determined using an annexin V-based assay [113]. The data indicate the percentage of cells undergoing apoptosis in each sample. All experiments were conducted in duplicate and gave similar results. The data are means ± SEM of three independent determinations.

Table 1. Anti-proliferative activitiy, cell cycle dysregulation, and apoptosis induction in the MCF-7 human breast cancer cell line after treatment for 24 and 48 h for the compounds.

3.2. 1-[2-(2-hydroxymethylphenoxy)-1-methoxyethyl]-5-fluorouracils 7a-f: antiproliferative activity, cell cycle dysregulation and apoptotic induction against breast cancer cells

Acyclic 5-FU O,N-acetals **7a-f** were previously reported by us (Figure 2) [17]. The aminalic bond is established through N-1 of the 5-FU moiety.

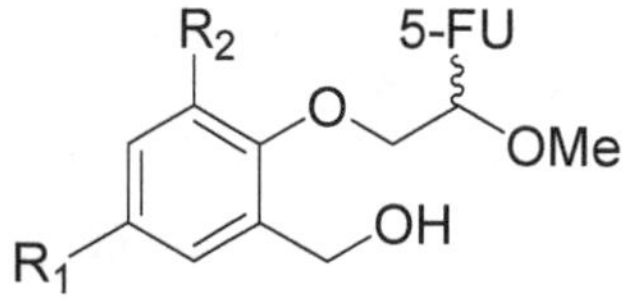

7a $R_1 = R_2 = H$
7b $R_1 = H$; $R_2 = OMe$
7c $R_1 = Cl$; $R_2 = H$
7d $R_1 = Br$; $R_2 = H$
7e $R_1 = NO_2$; $R_2 = H$
7f $R_1 = NH_2$; $R_2 = H$

Figure 2. Several acyclic 5-FU O,N-acetals previously reported by us [17].

The IC_{50} values of compounds are shown in Table 2. The most active compound is **7e** (5.42 ± 0.26 μM), with an anti-proliferative activitiy in the same order as that of Ftorafur (3.00 ± 0.11 μM). Cell cycle regulation has attracted a great deal of attention as a promising target for cancer research and treatment. The use of cell-cycle-specific treatments in cancer therapy has greatly benefited from the major advances that have been recently made in the identification of the molecular actors regulating the cell cycle and from the better understanding of the connections between cell cycle and apoptosis. As more and more 'cell cycle drugs' are being discovered, their use as anticancer drugs is being extensively investigated [118]. To study the mechanisms of the anti-tumour and anti-proliferative activities of the compounds, the effects on the cell cycle distribution were analyzed by flow cytometry. DMSO-treated cell cultures contained 68.39% G_0/G_1-phase cells, 12.04% G_2/M-phase cells and 19.57% S-phase cells. In contrast, MCF-7 cells treated during 48 h with the IC_{50} concentrations of **7a–f** showed important differences in cell cycle progression compared with DMSO-treated control cells. The treatment with Ftorafur showed a decrease of the G_0/G_1-phase cells and a corresponding accumulation of S-phase cells (45.62% G_0/G_1-phase cells and 54.38% S-phase cells). Moreover, there was an almost total disappearance in the G_2/M population of the cells treated with this drug. In general the cell cycle regulatory activities for the newly synthesized compounds can be divided into the following three groups: (a) **7c** and **7d** accumulated the cancerous cells in the G_2/M-phase, in the former compound at the expense of the S-phase cells, and (b) **7e** induced a S-phase cell cycle arrest (50.24%) in a similar percentage to that caused by Ftorafur (54.28%, Table 2). Therefore, it can be affirmed that the nitro derivative (**7e**) may act as a 5-FU prodrug. Nevertheless, this hypothesis needs to be corroborated by further assays. In response to **40a**, the percentage of apoptotic cells increased, from 1.24% in control cells to a maximum of 59.9% apoptotic cells (24 h) at a concentration equal to its IC_{50} against the MCF-7 cell line. This is a remarkable property because the demonstration of apoptosis in MCF-7 breast cancer cells by known apoptosis-inducing agents has proved to be difficult and only few cytotoxic agents act preferentially through an apoptotic mechanism in human breast cancer cells [113,114]. Finally, a fact worth emphasizing is that **7e** (the only compound tested) induced neither toxicity nor death in mice after one-month treatment when administered intravenously twice a week, with a 50 mg/kg dose each time.

Compound	IC$_{50}$ (µM)[a]	Cell cycle (48 h)[b]			Apoptosis (h)[c]	
		G$_0$/G$_1$	G$_2$/M	S	24	48
Control		68.39	12.04	19.57	1.24	1.24
5-FU						
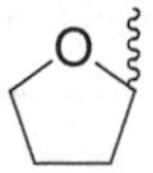 ![cyclic structure] Ftorafur	3.00 ± 0.11	45.62	0	54.38	52.20	58.06
7a	18.5 ± 0.95	67.18	4.67	28.16	59.90	40.23
7b	29.0 ± 1.63	62.72	1.59	35.69	33.35	37.87
7c	18.0 ± 0.85	71.01	28.99	0.00	44.36	50.64
7d	16.0 ± 1.18	51.45	20.66	27.88	42.24	36.37
7e	5.42 ± 0.26	46.92	2.84	50.24	40.73	48.22
7f	21.0 ± 1.02	67.32	9.40	23.28	41.15	37.81

[a]See [117]. [b]Determined by flow cytometry: see [16]. [c]Apoptosis was determined using an annexin V-based assay [113]. The data indicate the percentage of cells undergoing apoptosis in each sample. All experiments were conducted in duplicate and gave similar results. The data are means ± SEM of three independent determinations.

Table 2. Anti-proliferative activitiy, cell cycle dysregulation, and apoptosis induction in the MCF-7 human breast cancer cell line after treatment for 24 and 48 h for the compounds.

3.3. Benzo-Fused Seven-Membered Derivatives Linked To Purines

Compounds **8-10** (Figure 3) were synthesized as reported.

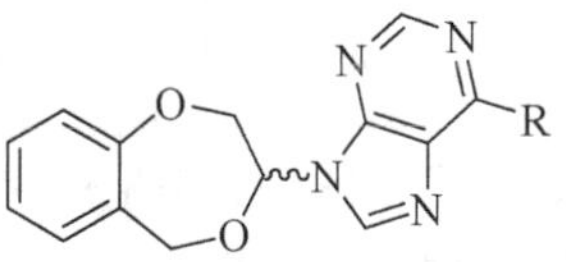

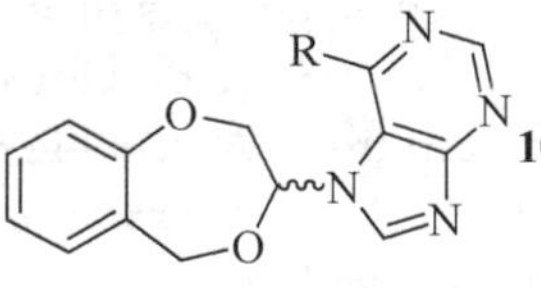

Figure 3. Several cyclic O,N-acetals reported by us [119].

The anti-tumour potential of the target molecules is stated against the MCF-7 human breast cancer cell line (Table 3). The most active compound (**8**), that presents an allyloxy group as substituent at position 6 of the purine ring, shows an IC$_{50}$ = 5.04 ± 1.68 µM nearly equipotent as 5-FU. Compounds, **9** and **10**, present bulky substituents as the phenylthio and 2,4-diclorophenylthio ones, respectively.

To study the mechanisms of the anti-tumour and anti-proliferative activities **8-10**, the effects on the cell cycle distribution were analyzed by flow cytometry (Table 3). DMSO-treated cell cultures contained a 58.62 ± 0.74 of the G$_0$/G$_1$-phase cells, a 33.82 ± 0.72 of the S-phase cells and a 7.55 ± 1.34 of the G$_2$/M-phase cells. In contrast, MCF-7 cells treated during 48 h with the IC_{50} concentrations of **8, 9** and **10** showed important differences in cell cycle progression compared with DMSO-treated control cells. The cell cycle regulatory activities can be divided into the following two groups: (a) the breast cancer cells showed an accumulation in the S-phase, up to 37.00 ± 2.00 of the cells, mainly at the expense of the G_0/G_1-phase population that decreased to a percentage of 55.63 ± 1.57 of the cells; (b) compounds **9** and **10** accumulated the cancerous cells in the G_2/M-phase (11.08 ± 1.01 and 19.16 ± 0.56, respectively) at the expense of the S-phase cells (26.82 ± 1.26 and 22.73 ± 0.37, respectively). In response to **9** (and **10**), the percentage of apoptotic cells increased, from 0.22 ± 0.31 in control cells to a maximum of 73.37 ± 0.12 (and 65.28 ± 1.92) apoptotic cells at a concentration equal to their IC_{50} against the MCF-7 cell line. This is a remarkable property because the demonstration of apoptosis in MCF-7 breast cancer cells by known apoptosis-inducing agents has proved to be difficult.

Compound	Cell cycle[a]				Apoptosis[b]
	IC_{50} (μM)	G_0/G_1	S	G_2/M	
Control		58.62 ± 0.74	33.82 ± 0.72	7.55 ± 1.34	0.22 ± 0.16
8	5.04 ± 1.68	55.63 ± 1.57	37.00 ± 2.00	7.37 ± 0.43	44.47 ± 2.98
9	7.12 ± 0.46	59.10 ± 1.28	26.82 ± 1.26	11.08 ± 0.01	73.37 ± 0.12
10	8.40 ± 0.91	58.10 ± 0.19	22.73 ± 0.37	19.16 ± 0.56	65.28 ± 1.92

[a] Determined by flow cytometry.[3] [b] Apoptosis was determined using an Annexin V-based assay [120]. The data indicate the percentage of cells undergoing apoptosis in each sample. All experiments were conducted in duplicate and gave similar results. The data are means ± SEM of three independent determinations.

Table 3. Anti-proliferative activity, cell cycle distribution and apoptosis induction in the MCF-7 human breast cancer cell line after treatment for 48 h for the three most active compounds.

3.4. Anti-cancer activity of 9-(2,3-Dihydro-1,4-benzoxathiin-3-ylMethyl)-9H-purines

Compounds **1-6** may be considered as drugs with their own entity and anti-tumour activity independent of that of 5-FU. If the previously described compounds are not prodrugs, it is not necessary to maintain the *O,N*-acetalic characteristic with the corresponding weakness of the *O,N*-acetalic bond. Therefore, molecules are being designed in which both structural entities (such as the benzo-heterocyclic ring and the purine base) are linked by a heteroatom-C-C-base-N-atom bond. The design, synthesis and biological evaluation of a series of 2- and 6-disubstituted 9-(2,3-dihydro-1,4-benzoxathiin-3-ylmethyl)-9*H*-purine derivatives **11-13** were described (Figure 4, Table 4) [121].

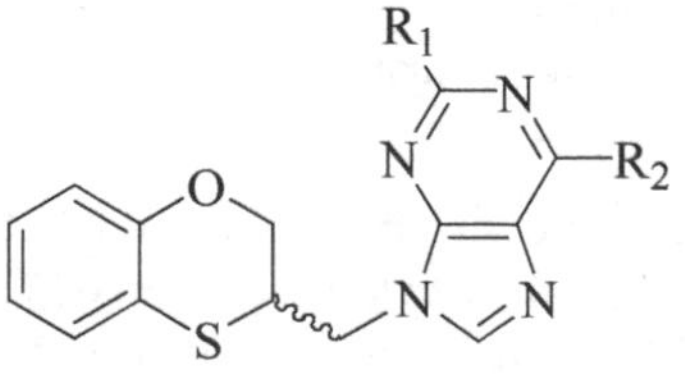

11 $R_1 = H, R_2 = Cl$
12 $R_1 = H; R_2 = Br$
13 $R_1 = Cl; R_2 = Cl$

Figure 4. Several non-acetalic purine derivatives reported by us [121].

Compounds **11-13** were subjected to cell cycle and apoptosis studies on the MCF-7 human breast cancer cell line (Table 4). The following two consequences can be stated: (a) in contrast to 5-FU, the six-membered compounds **11-13** provoked a G_0/G_1-phase cell cycle arrest when the MCF-7 cells were treated during 48 h with the IC_{50} of the compounds, mainly at the expense of the S-phase populations. The fact that at similar doses the novel derivatives exhibit different sequences of cell cycle perturbations in comparison with 5-FU indicates that these compounds act by different pathways [12]. In the case of **12** it is worth pointing out that, moreover, there is an increase in the G_2/M-phase of the cancerous cells; and (b) the apoptotic indices of the target compounds are very important, especially for **13** (58.29% for **11**, 63.05% for **12**, and 76.22% for **13**). Up to now and according to our knowledge, compound **13** is the most important apoptotic inducer against the MCF-7 human breast cancer cell line so far reported.

Compound		Cell cycle[a]			Apoptosis[b]
	IC_{50} (µM)	G_0/G_1	S	G_2/M	
Control		58.62 ± 0.74	33.82 ± 0.72	7.55 ± 1.34	0.22 ± 0.16
5-FU[c]	4.32 ± 0.02	58.07 ± 0.11	39.38 ± 0.98	2.10 ± 0.12	52.81 ± 1.05
11	10.6 ± 0.66	69.71 ± 1.50	23.73 ± 1.65	6.56 ± 0.17	58.29 ± 0.75
12	6.18 ± 1.70	62.85 ± 0.87	26.71 ± 1.25	10.43 ± 0.38	63.05 ± 0.26
13	8.97 ± 0.83	70.30 ± 0.32	23.67 ± 2.40	6.06 ± 2.72	76.22 ± 2.02

[a]Determined by flow cytometry [12]. [b]Apoptosis was determined using an Annexin V-based assay [12]. The data indicate the percentage of cells undergoing apoptosis in each sample. [c]Data were taken from [117]. All experiments were conducted in duplicate and gave similar results. The data are means ± SEM of three independent determinations.

Table 4. Anti-proliferative activity, cell cycle distribution and apoptosis induction in the MCF-7 human breast cancer cell line after treatment for 48 h for the three most active compounds as anti-proliferative agents.

3.5. Anti-cancer activity of cyclic and acyclic O,N-acetals derived from purines and 5-FU

We have recently published two *O,N*-acetals with outstanding anti-proliferative activities. The most potent antiproliferative agent against the MCF-7 adenocarcinoma cell line belongs to the benzoxazepine *O,N*-acetalic family is 9-[1-(9*H*-fluorenyl-9-methoxycarbonyl)-1,2,3,5-tetrahydro-4,1-benzoxazepine-3-yl]-6-chloro-9*H*-purine (**14**, IC_{50} = 0.67 ± 0.18 ⊚M), whilst 7-{2-(*N*-hydroxymethylphenyl)-2-nitrobenzenesulfonamido]-1-methoxyethyl} -6-chloro-7*H*-purine (**15**) shows the lowest IC_{50} value among the family of acyclic *O,N*-acetals (IC_{50} = 3.25 ± 0.23 ⊚M) (Figure 5). The global apoptotic cells caused by **14** and **15** against MCF-7 were 80.08% and 54.85% of cell population after 48 h, respectively. cDNA microarray technology reveals potential drug targets, which are mainly centred on positive apoptosis regulatory pathway genes, and the repression of genes involved in carcinogenesis, proliferation and tumour invasion [21]. We demonstrated later on that, when the anthranilic alcohol-derived acyclic 5-fluorouracil *O,N*-acetal **16** was administered to human breast cancer cells MCF-7, had no activity against classic pro-apoptotic genes such as *p53*, and even induced the down-regulation of anti-apoptotic genes such as *Bcl-2*. In contrast, several pro-apoptotic genes related with the endoplasmic reticulum (ER)-stress-induced apoptosis, such as *BBC3* and *Noxa*, appeared up-regulated. These results seem to show that the mechanism of action and selectivity of **16** was via the activation of the ER-stress-induced apoptosis. The linkage between the 5-FU moiety and the chain of **16** is through the *N*-1 atom of the pyrimidine [20] (Figure 5).

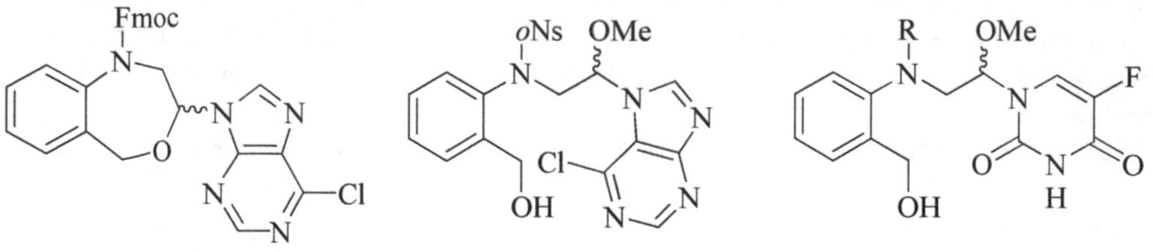

14　　　　　　　　　　　**15**　　　　**16**, R = SO_2-C_6H_4-oNH$_2$

Figure 5. Several *O,N*-acetalic purine derivatives reported by us [20,21]. Fmoc is the fluorenylmethyloxycarbonyl group; *o*Ns is the -SO_2-C_6H_4-oNO$_2$ group.

4. Inhibitor of apoptosis protein (IAP) family as therapeutic target

In the apoptotic pathway, the balance between pro and anti-apoptotic proteins is tightly regulated, so an imbalance directed to the anti-apoptotic regulation involves the apparition of survival advantages onto the initiating cancerous cells. In this way, the role of the IAPs family proteins is important. This family was first identified in baculovirus and the main characteristic is the presence of a baculovirus IAP repeat (BIR) domain, that mediates interactions with a number of proteins, including caspases and also other structural domains such as a RING or caspase activation recruitment domain (CARD) [122] (Figure 6).

In the IAP family we can find proteins involved in the inhibition of apoptosis through inhibition of caspases or modulation of the nuclear factor kB (NF-kB) signalling pathways. This family includes eight members in mammals (BIRC1– BIRC8), also known as neuronal apoptosis inhibitory protein (NAIP), cellular IAP1 (cIAP1) and IAP2 (cIAP2), Xchromosome linked IAP (XIAP), survivin, Apollon (BRUCE), livin (ML-IAP) and ILP2 [123] (Figure 6).

4.1. Role of the IAPs in apoptosis

As long as apoptosis pathway must be tightly regulated, the IAPs are the only known regulatory proteins that control the activity of both initiator and effectors caspases in the process. In this way, there are studies where XIAP was found to prevent caspase-3 processing in response to caspase-8 activation. So it is suggested that XIAP is able to inhibit this extrinsic apoptotic signaling by blocking the downstream effectors caspases, avoiding the direct interference of the caspase-8 activation [124].

On the other hand, other IAPs proteins are not potent inhibitors of caspases, such as c-IAP proteins. They are able to bind with Smac having high affinities to prevent it from blocking XIAP-mediated inhibition of caspases. These IAPs also show a E3 ubiquitin ligase activity, allowing the regulation of several cell death effectors and modulators, playing a key role in cellular survival. c-IAP1 and c-IAP2 are able to interact with TNFα-receptor-associated factors -1 and -2 (TRAFs) trough their association in a complex with TNFα receptor 2. The TNFα receptor can mediate survival and death cell signals. In this case, c-IAP1 and c-IAP2 have been proposed to reduce the level of caspase-8 activation and protect cells from apoptosis in a TNFα relationship [10].

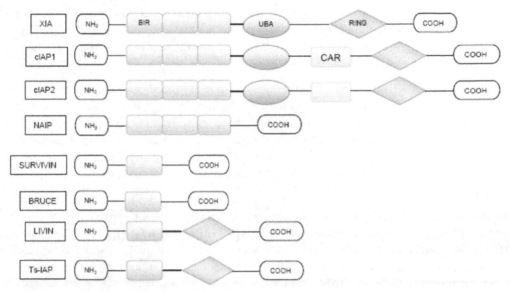

Figure 6. Identifying some of the structural motifs of IAPs. The baculoviral IAP repeat (BIR) domains; 70–80-amino-acid cysteine- and histidine-rich domains that chelate zinc ions. The presence of at least one BIR domain is the defining characteristic of the IAP family. The number of BIR domains in a given IAP varies from one to three. Another motif is the really interesting new gene (RING) zinc finger, a caspase recruitment domain and ubiquitin-associated (UBA) domains are found in individual IAPs. Finally, the CARD motif (Caspase Recruitment Domain) is a protein–protein interaction domain that mediates oligomerization with other CARD-containing proteins and is found in a number of proteins involved in the regulation of cell death.

There is another family member of the IAPs, called Livin, which are expressed in high levels in melanoma and colon cancer, in embryonic tissues and transformed cells. The over-expression of Livin isoforms α and β blocks apoptosis induced via the extrinsic death receptor pathway [125].

Survivin is involved in the control of cell proliferation and cell death and gene expression is regulated in a cell cycle dependent manner in mitosis. The spliced transcription of the survivin gene gives rise to wild-type survivin, survivin-2α, survivin-2B, survivin-ΔEx-3 and survivin-3B. This protein is stabilized by phosphorylation thanks to a p34cdc2–cyclin B1 complex during mitosis. The anti-apoptotic function of survivin has been linked with its interaction with Smac/DIABLO, with the stabilization of XIAP protein through its binding to

XIAP and with the inhibition of mitochondrial and apoptosis-inducing factor-dependent apoptotic pathways [126].

Other IAPs is Bruce/Apollon which is a membrane bounded protein involved in protein ubiquination-mediated degradation by its ability to target proteins thanks to the presence of a C-terminus E2 motif. It is also shown that Bruce is able to bind to caspase -3, -7 and -9 [127]. In Bruce regulation are involved the E2 UbcH5 and the E3 Nrdp1, which ubiquitinate the epidermal growth factor receptor family member, ErbB3. It has been shown that a decrease in Bruce content by Nrdp1 over-expression induces apoptosis in different cell lines. These studies suggest that this IAP protein play a critical role in apoptosis inhibition in certain cell types avoiding pro-caspase-9 cleavage when it binds to this protein [128].

Naip has been studied because of the clinical relevance in tumors such as prostate or breast cancer. Naip has two different functions; in the first one it is involved in the inflammatory process by caspase-1, -4 and -5 activation, and the second one is via apoptosis regulation by caspase-3 and -7 inhibition [129]. Davoody *et al* showed the cleavage inhibition of procaspase-3 by apoptosome activated caspase-9 and the inhibition of the autocatalytic processing of procaspase-9 in the apoptosome complex. This fact indicates that unlike other IAPs, Naip is an inhibitor of procaspase-9 [130].

Finally, Ts-IAP, also known as ILP-2, is the product of a human testis-specific mRNA and is related to the C-ter region of another member of the IAPs protein family, XIAP. It is showed that this protein is a weak caspase-9 inhibitor, and also a highly unstable molecule. However, a stabilized form of this protein containing nine additional N-ter residues may form a complex with Smac/DIABLO [131].

4.2. Use of IAPs into clinic: prognostic and therapeutic values

The expression and/or function of IAPs are deregulated in many human cancers because of genetic aberrations, an increase in their mRNA or protein expression or the loss of endogenous inhibitors such as Smac. The expression levels of IAPs and their antagonists have been correlated with clinical parameters and cancer prognosis in several retrospective trials. However, these results should be interpreted with caution due to the low numbers of samples studied in some reports and the limitations of currently available reagents for analyzing the expression of IAP proteins in tissue specimens. Therefore, additional studies are required to evaluate the prognostic value of IAPs in human malignancies [10].

The issue of primary or acquired resistance to current chemotherapeutic-based treatments is a major impediment to effective cancer treatment. Although there are many genetic and biochemical alterations that occur in cancer cells, in vitro experiments demonstrate that the up-regulation of IAPs expression increases resistance to chemotherapy and radiation. The fundamental role of the IAPs in apoptosis regulation and their elevated expression in many tumour types suggest that there is value in exploiting the inhibition of IAP expression and function as a direct therapeutic strategy [125].

Novel drugs have been developed and some of them are in current clinical trials. Several strategies have been chosen for anticancer drug development targeting IAP molecules: (1) small-molecule IAP antagonists, (2) antisense oligonucleotides (ASOs), (3) Smac mimetic molecules and others [132]. Small-molecule IAP antagonists and antisense oligonucleotides have garnered the most attention. Also, IAP antagonists have been extensively studied in combination with other cytotoxic agents including anticancer drugs, small-molecule signal transduction inhibitors, proteasome inhibitors and death receptor ligands as well as with radiation therapy.

4.2.1. Small-molecule IAP antagonists

Owing to the differences in the mechanism of caspase inhibition by the BIR 2 and BIR 3, domain molecules have been developed to specifically target the BIR 2 or BIR 3 region of XIAP. The structural data surrounding the interaction between the BIR 3 domain of XIAP and caspase-9, suggests that small molecules that bind the BIR 3 pocket of XIAP could mimic the action of Smac and inhibit the interaction between XIAP and caspase-9. These structural studies have facilitated a variety of chemical biology approaches including fluorescent polarization, nuclear magnetic resonance, 'in silico' virtual screening and computer modelling to identify BIR 3 inhibitors [133]. The first small-molecule XIAP inhibitors were reported by [134]. These inhibitors were identified using a high throughput enzymatic derepression assay in which recombinant XIAP was combined with active caspase-3 to inhibit caspase-mediated cleavage of the fluorogenic peptide substrate. With this assay, they screened 160.000 compounds in 1536-well format and identified potent XIAP inhibitors including the compounds TWX006 and TWX024, aryl sulphonamides with flexible acyclic diamines in the first and third fragments. These compounds derepressed XIAP-mediated inhibition of caspase-3 more potently than Smac. In addition, these molecules bound the BIR 2-linker region of XIAP, and, in enzymatic assays, relieved the repression of caspase-3 more potently than Smac peptides.

Recently, small-molecule phenylurea-based chemical inhibitors of XIAP were identified by large-scale combinatorial library screening. Subsequent studies have confirmed that the active XIAP inhibitors, but not their inactive structural analogues, could induce apoptosis in a variety of human cancer cell lines and xenograft. Furthermore, it was determined that these XIAP inhibitors act by binding to its BIR-2 domain, resulting in elevated activity of the downstream caspase-3 and caspase-7. Thus, the action of these exogenous XIAP inhibitors was found to be mechanistically distinct from that of the endogenous inhibitor second modulator of apoptotic proteases, which predominantly binds to the BIR-3 domain [135].

4.2.2. Antisense oligonucleotides

The single-stranded antisense oligodeoxynucleotides (AS ODNs) are short stretches of synthetic DNA, approximately 12–30 nucleotides long, and are complementary to a specific mRNA strand. Hybridization of the AS ODNs to the mRNA by Watson–Crick base pairing prevents the target gene from being translated into protein, thereby blocking the action of the gene, and resulting in the degradation of the target mRNA. The specificity in the AS ODN approach is based on the fact that any sequence of approximately 13 bases in RNA and 17 bases in DNA is estimated to be represented only once in the human genome. AS ODNs targeting survivin expression in human lung adenocarcinoma cell lines decreased survivin protein levels in a dose-dependent manner, induced apoptosis, stimulated higher levels of caspase-3 activation, and increased the sensitivity of cells to chemotherapeutics. XIAP AS ODNs effectively down-regulated both specific mRNA and protein levels in human non-small cell lung cancer growth both in vitro and in vivo. XIAP AS ODNs effectively induced apoptosis on their own and sensitized the tumor cells to the cytotoxic effects of several chemotherapeutics, including Taxol, etoposide and doxorubicin [26]. Furthermore, the administration of XIAP AS ODNs in a xenograft model of human non-small cell lung cancer results in a significant down-regulation of XIAP protein [124].

AEG35156 is a 19-mer oligonucleotide targeting XIAP. Its sequence was designed to achieve maximal stability and potency and to minimize immunostimulation through avoiding CpG motifs. Phase I studies in patients with refractory malignancies established safety [136]. A phase I/II study of AEG35156 in combination with idarubicin and high-dose cytarabine in patients with relapsed or refractory AML demonstrated a dose-dependent knock-down of XIAP mRNA and protein and a promising response rate [137]. The molecule may rapidly enter randomized studies in AML while being also tested in lymphomas. The dose-limiting toxicity of an antisense oligonucleotide designed to inhibit surviving mRNA expression (LY2181308) was headache and the compound demonstrated some biological efficacy in decreasing survivin expression [138]. A phase II study has opened in solid tumours. The locked nucleic acid strategy was used to design other survivin-targeting antisense oligonucleotides, including SPC3042 [139] and EZN-3042 [140], which are currently being evaluated in Phase I clinical trials as single agents and in combination with cytotoxic drugs.

4.2.3. Smac-mimicking IAP antagonist

Smac mimetics may be a useful therapeutic target as over-expression of Smac may potentiate apoptosis by neutralizing the caspase-inhibitory function of IAPs. Following the discovery that an IAP-binding motif consisting of four NH2-terminal amino acid residues was sufficient to bind to the BIR3 domain of XIAP, Smac-peptide mimetics were constructed which were capable of competing with caspase-9, displacing it from the BIR3 domain of XIAP. The three members of the IAP family, XIAP, cIAP1, and cIAP2, are structurally homologous (XIAP amino acid sequence identity to cIAP1 and cIAP2 of 36% and 39%, respectively, the amino acid sequence identity between cIAP1 and cIAP2 is 70%). In particular, the BIR3 IBM region is well conserved among the three IAPs. The XIAP BIR3 residues involved in van der Waals contacts (Val298, Lys299, and Trp310) and hydrogen bonds (Gly306, Leu307, and Trp323) with the inhibitory compounds are conserved. Minor exceptions are Leu292, replaced by Val in the cIAPs, Glu314 substituted by Asp in both cIAPs, and Gln319, which is Glu325 in cIAP1, and Gln311 in cIAP2. Finally, residues Thr308 and Asp309 that were found relevant for Smac-mimetics interaction with XIAP BIR3, are replaced by Arg314/Arg300 and Cys315/ Cys307 in cIAP1/cIAP2, respectively [127].

Smac interacts with IAPs via its N-terminal tetrapeptide, Ala1- Val2-Pro3-Ile4 (AVPI). Using this structural information, several groups designed small molecules mimicking AVPI to derive proteolytically stable compounds. Monovalent Smac mimetics were designed to mimic the Smac AVPI-binding motif and so target the XIAP-BIR3 domain. They exhibit high affinities, not only to XIAP-BIR3 but also to cIAP1, cIAP2 and MLIAP proteins. Bivalent Smac mimetics, containing two AVPI-binding motifs, bind to XIAP-BIR2–BIR3 with an extremely high affinity, exceeding that of Smac protein. Preclinical profiling studies have shown that Smac mimetics effectively sensitize cancer cells to other therapeutic agents, but they are also capable as single agents of inducing apoptosis in some but not all human cancer cell lines. To date, two Smac mimetics have reached phase I clinical development, and approximately ten are in an advanced preclinical development stage and are expected to enter human clinical testing soon [141].

Smac mimetic binds to XIAP and induces cIAP degradation. There are two cellular events that result from cIAP degradation: (1) activation of the non-canonical NF-κB pathway and subsequently increased production of autocrine TNF-α and (2) release of RIP1 from the TNF-α receptor complex, leading to caspase-8 activation, which requires TNF-α receptor to be activated in the first place. Accumulating evidence suggests that whether or not Smac mimetic induces autocrine TNF-α production is the key factor in deciding the cell's fate upon Smac mimetic treatment. Depending on their response to Smac mimetics cells can be grouped into three general classes: in class I cells, Smac mimetic induces autocrine TNF-α production and massive cell loss; in class II cells, Smac mimetic still induces autocrine TNF-α production but this does not have a major effect on the cell population; and in class III cells, Smac mimetic has no effect on autocrine TNF-α production or on cell death. The ability of cells to produce TNF-α is necessary but not sufficient for Smac mimetic to induce cell death as a single agent [142].

5. miRNA-based therapy

The miRNAs, is a class of endogenous, small, non-coding RNAs of 18–25 nucleotides in length, that negatively regulates gene expression by degradation of mRNA or suppression of mRNA translation. Mature miRNA products are formed in from a longer primary miRNA (pri-miRNA) transcript through sequential processing by the ribonucleases Drosha and Dicer1 [143,144]. miRNAs are known to repress thousands of target genes because only partial complementarity to the target mRNA is required. Thus, one miRNA may be simultaneously targeting a complexity of mRNAs as well as the expression of a single mRNA may be regulated by many miRNAs [143].

The miRNAs are involved in normal processes, including cellular development, differentiation, proliferation, apoptosis, and stem cell self-renewal [28].The aberrant expression or alteration of miRNAs contributes to a range of human pathologies, including cancer. Furthermore, the deregulation of miRNA causes evasion of apoptosis which involved tumourigenesis and drug resistance [122]. During tumour initiation and progression, the functionality of aberrant miRNAs may act as oncogenes (OncomiRs) or tumour suppressors (TSmiRs), a numbers of them are strongly related to the apoptosis phenomenon. Therefore, manipulation of miRNA expression levels which target genes and pathways that are involved in apoptosis could be a potential therapeutic strategy for developing efficient therapies against cancer. In addition, given that cancer cells often exhibit a distinctive pattern of miRNA expression, unique profiles of altered miRNAs expression could be useful as molecular biomarkers for tumour diagnosis, prognosis of disease-specific outcomes, and evaluation of tumour aggressiveness. Based on this, several anticancer therapies focusing on restoring miRNA activities and repairing gene regulatory networks or drug sensitivity are being developed.

There are two strategies of molecular therapy targeted at miRNA, one by the inhibition of oncogenic miRNAs and the other the over-expression of tumour suppressor miRNAs.

5.1. Targeting oncogenic miRNAs

The oncomiRs could be blocked by different approaches such as (i) antisense oligonucleotides, (ii) antagomirs, (iii) locked nucleic acid (LNA) constructs or (iv) sponges [145]. Antisense oligonucleotides work as competitive inhibitors of miRNA, leading to the up-regulation of tumour suppressor proteins, inducing apoptosis and blocking tumour formation in vitro and in vivo. Some of the potential ways to enhance miRNA stability include chemical modifications. In order to improve their effectiveness and stability, antisense oligonucleotides, have been modified on their 50 end, by adding 20-O-methyl and 20-O-methoxyethyl groups [146]. Oligonucleotides with 2'-O-methyl groups have proved to be effective inhibitors of miRNA expression in several cancer cell lines [30,147,148,149]. The utility of anti-miRNA oligonucleotide in vivo through intravenous injection of modified anti-miRNA oligonucleotide O-methyl-modified cholesterol-conjugated single stranded RNA analogues has been studied, with phosphorothioate linkages, an 'antagomirs', to target the liver-specific miR-122. Specific miR-122 silencing for up to 23 days was conferred with only a single injection of 240 mg•kg-1 body weight [150].

On the other hand, the locked-nucleic-acid antisense oligonucleotides exhibit relatively low toxicity and have been optimized by reducing their molecular size which has increased their therapeutic potential [151]. LNA anti-miR' constructs have been used successfully in several in vitro studies to knock down specific miRNA expression [151,152]. Also it has showed that miR-221 and miR-222 knockdown through antisense LNA oligonucleotides increases p27Kip1 in human prostate cancer (PC3) cells and strongly reduces their clonogenicity in vitro [153]. In vivo, the use of LNAs has achieved unexpected success for the treatment of hepatitis C in non-human primates [154]. These finding demonstrate the impressive potential of this strategy to overcome a major hurdle for clinical miRNA therapy.

Moreover, other techniques have emerged as an effective way to repress expression levels of miRNA families. A new form of miRNA inhibitors that can be transiently expressed in cultured mammalian cells, "miRNA sponges", was developed. Sponges are ectopically expressed mRNAs that contain multiple miRNA target sites of miRNA that share the same seed sequence [155]. In contrast to miRNA sponges, Xiao et al. designed alternative strategy called "miRNA masking" which covers up the miRNA-binding site to depress its target mRNA [156].

Although several in vitro and in vivo technologies have been developed to inhibit the oncomiRs, it is still a long and arduous way to go for substantial applications of miRNAs in cancer treatments. To date, many OncomiRs seem to have a role in apoptosis (Table 5), therefore some have been regarded as hallmarks in tumour progressions and hot targets in cancer therapy. Thus, miR-21 is functionally considered oncogenic because it is overexpressed in various tumours [157,158]. In fact, antisense inhibition of miR-21 leads to the induction of programmed cell death in neuroepithelial cells, through activation of caspases [159]. This apoptosis induction was also confirmed in breast cancer, colon cancer, pancreas cancer, lung cancer, liver cancer, prostate cancer, stomach cancer and oral squamous cell carcinoma (OSCC) [31,160,161,162,163]. So far, multiple targets of miR-21 have been identified and mapped to anti-apoptotic signalling pathways which suggest a promising miRNA treatment for cancer [31,164]. Other examples have been reported; in breast cancer cells, the overexpresion of the anti-apoptotic Bcl-2 is restored by the silencing of miR-15a and miR-16 through the use of specific inhibitors [165]. As the anti-apoptotic Bcl-2 is frequently overexpressed in a number of human cancers, such as Hodgkin's lymphoma, cell lymphoma and breast, miR-15 and miR-16, it could be used for therapy of cancer-associated phenotypes.

miRNA	Function	Gene target	Cancer Type	Reference(s)
miR-221, miR-222	OncomiRs	PTEN p27kip1 Bim PUMA	Hepatocarcinoma, melanoma, glioblastoma, lung, prostate cancer, leukemia, gastric carcinoma	[153,166,167,168, 169,170]
miR-21	OncomiRs	PTEN PDCD4 Bcl-2 Fas L	Non-small cell lung cancer, breast, prostate, gastric, hepatocellular cancer, colorectal cancer, glioblastoma and leukemia	[31,157,164,165,1 71,172,173,174,17 5]
miR-17-92 cluster	OncomiRs	PTEN, BIM, p21	Lymphoma, lung, breast, stomach, colon and pancreatic cancer	[176,177]
miR-29a, miR-29b and miR-29c	Tumour suppressor	CDC42 and p85a (up-regulating p53), Bcl-2	Lymphocytic leukaemia, cholangiocarcinomahepatoc arcinoma, colon, breast, and lung cancer	[148,149,178,179]
miR-34 family	Tumour suppressor	Bcl-2, SIRT1	Prostate cancer non-small-cell lung cancer and neuroblastoma	[180,181,182,183]
miR-15a, miR-16-1,	Tumour suppressor	Bcl-2, Mcl1, PDCD6IP	Leukemia, gastric cancer cells and prostate cancer	[30,184,185]
Let 7	Tumour suppressor	Bcl-Xl	Lung, colon, stomach, ovarian and breast cancer.	[186,187,188,189, 190]

Table 5. Key microRNAs involved in apoptosis

5.2. Restoration of tumour suppressor miRNAs

The restoration of tumour suppressor miRNAs, as a therapeutic strategy, includes viral delivery or synthetic miRNA mimics. Elevation of the expression levels of miRNAs can restore tumour inhibitory functions in cancer cells. Adeno-associated virus delivery of miRNAs or miRNA antagonists has the advantage of being efficient and because the virus does not integrate into the genome, non-mutagenic. In Myc-induced liver tumours, intravenous injection of adeno-associated virus 8 (AAV8)-expressing miR-26 resulted in the suppression of tumourigenicity by inducing tumour-apoptosis, without signs of toxicity [191]. These findings indicate a possibility strategy for the treatment of liver cancer, however, before this approach achieve widespread clinical use, the delivery and safety of different treatments needs to be improved.

Another strategy to increase the expression of a tumour-suppressor miRNA in cancer could be overcome by miRNA mimics, which are small, chemically modified double-stranded RNA molecules designed to mimic endogenous mature miRNAs [192]. Introduction of synthetic miRNA mimics with tumour-suppressor function in cancer cells have been implicated to induce cell death and block proliferation in several studies [30,147,178,184,186]. In prostate and AMl cell lines mimics of miR-15a and miR-29 respectively, induced apoptosis by repression of anti-apoptotic genes Mcl-1 and Bcl-2 [184].

Multiple miRNAs have been found to inhibit the apoptotic pathway following their over expression during cancer development. Reduced expression of miR-15, miR-16, and let-7 has been observed in different types of cancers and as one consequence anti-apoptotic genes and apoptotic signalling pathways have been activated in these cancer cells [30,187,193]. As well, transfection of anti-miR-24 oligonucleotides has been proved to induce apoptosis in several cell lines [194]. It has been reported that miR-195, miR-24-2 and miR-365-2 act as negative regulators of the anti-apoptotic proto-oncogene Bcl2. The overexpression of these miRNAs caused an increase in apoptosis and also augmented the apoptotic effect of etoposide in breast cancer MCF7 cells [195]

5.3. miRNAs and drug resistance

Several miRNA, some of them related to apoptosis, have been associated with drug resistance. Deregulation of miR-214 is a recurrent event in human ovarian cancer and it has been shown that miR-214 induces cell survival and cisplatin resistance primarily through targeting the PTEN/Akt pathway [196]. Also, is known that the let-7 family of miRs plays a role in a host of cellular functions such as modulation of drug sensitivity. The miRNA let-7a which directly targeting caspase-3 is over-expressed in some human cancers and has been shown to induce resistance to a variety of drugs caspase-3-dependent apoptosis, including doxorubicin, paclitaxel and interferon-gamma. Let-7e was up-regulated in some ovarian cancer cell lines with increased resistance to doxorubicin. On other hand, it has been reported that let-7i is down-regulated in chemotherapy-resistant ovarian cancer, and reintroduction of let-7i can sensitize resistant ovarian cancer cell lines to platinum-based chemotherapy [188]. In a non small cell lung cancer cell line, the down-regulation of miR-186* which increased the expression of its direct target, Caspase-10, has been indicated the cause of the apoptosis induced by the chemopreventive agent curcumin [189]. Thus, the effect of anti-cancer drugs that modulate cell proliferation apoptosis on miRNA expression profiles was explored and could help for predicting apoptosis resistance. As a result, the knowledge of potential miRNAs implicated in apoptosis resistance could avoid unnecessary morbidity and may represent a novel class of biomarkers for facilitating personalized treatment.

6. Cancer stem cells and apoptosis

Some cancers are originated in cells with intrinsic self-renewal activity or in differentiated cells in which self-renewal is activated by oncogenic mechanisms; hence, the study of normal self-renewal is important to improve our understanding of these mechanisms. Cancers express a spectrum of aberrantly differentiated cells, ranging from those that appear well differentiated to those that appear undifferentiated, and these phenotypes are commonly evident in the same tumour. This suggests that the transformation process can induce defects throughout the multistep differentiation process. Recent data suggest that cancers arise from rare self-renewing stem cells that are biologically distinct from their more numerous differentiated progeny. A small number of cells identified as cancer stem cells (CSCs) from solid tumours usually express organ-specific markers, contribute to chemotherapy resistance and are able to generate a new tumour in immunodeficient mice. Moreover, there is growing evidence that pathways regulating normal stem cell self-renewal and differentiation are also present in cancer cells and CSCs [197].

Currently, there are two theories on the origin of cancer: the classic clonal evolution theory or stochastic model, by which malignant transformation results from random mutations and subsequent clonal selection of cancer cells with similar potential to regenerate the tumour growth [198,199]; and the CSC hypothesis, which considers the tumour to be formed by a small population of cells with stem cell-like properties. The features in common with stem cells are: indefinite self-replication, asymmetric cell division, and resistance to toxic agents, owing, in part, to elevated expression of ABC transporters [199,200,201]. In addition, they are also characterized by genetic instability (chromosomal and microsatellite), changes of chromatin, transcription and epigenetics, mobilization of cellular resources, and modified microenvironment interactions (tumour cells, stromal cells, extracellular, endothelium) [202]. Both paradigms of tumour propagation are likely to exist in human cancer but only the CSC model is hierarchical. It is important to note that the two models are not mutually exclusive, as CSCs undergo clonal evolution, as shown for leukaemia stem cells [203].

The theory of cancer stem cells is not new, having started wonderings in the 19th century when comparing cancerous and embryonic tissue in the microscope and certain similarities were observed, annotating the idea that tumours arising from embryonic-like cells. This theory continued to evolve, and the isolation of four different tumour subpopulations from a single breast cancer in a mouse was reported in the decade of the 80s[204]. Tumour heterogeneity is reflected in different phenotypic aspects such as cell morphology, gene expression, metabolism, motility and proliferation, immunogenic, angiogenic and metastatic potential [202].

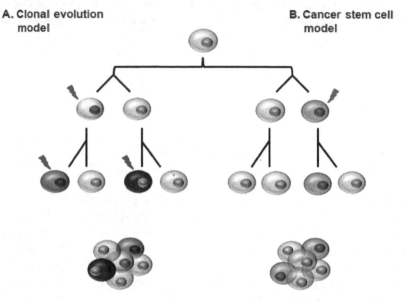

Figure 7. A. The clonal evolution model of cancer is based on the fact that accumulation of a series of mutations (inhibition of apoptosis, angiogenesis,…) in any somatic cell can cause a tumour. **B.** The cancer stem cell model is based on the principle that a progenitor cell capable of self-renewal and proliferation (stem cell characteristics) is the cause of formation of a tumour.

TUMOUR TYPE	MARKERS	Reference
Hematological malignancies	CD34$^+$ / CD38$^-$	[207]
CNS	CD133$^+$	[208]
Colon	CD133$^+$/ EpCAM^{hi}/ CD44$^+$/ CD166$^+$	[209,210]
Breast	CD24$^{-/low}$/CD44$^+$	[211]
Lung	CD133$^+$	[212]
Pancreas	CD44$^+$ / CD24$^+$ / EpCAM$^+$/ CD133$^+$	[213]
Liver	CD90$^+$	[214]
Prostate	CD44$^+$ / CD133 / $\alpha2\beta1^{hi}$	[215]
Bladder	CD44$^+$/ CK5$^+$/ CD20$^-$	[216]
Ovaries	CD44$^+$/ CD117$^+$	[217]
Head and neck	CD44$^+$	[218]
Melanoma	ABCB5$^+$	[219]

Table 6. Phenotypic markers of CSCs in various tumours.

Molecular characterization of CSCs is necessary to develop a targeted therapy (Table 6). They have been isolated from several tumour types including haematological malignancies (the first evidence) [205], breast, brain, colon, lung, head and neck, prostate, pancreas and liver cancers and melanoma. The subpopulation of CSCs self renews, differentiates, and regenerates a phenocopy of the original tumour when injected into immunodeficient mice [206].

Tumour stem cells may display multidrug resistance that is conferred by ABC transporters. These ABC transporters have been reported as CSCs markers in melanoma and osteosarcoma, among others. These transporters are up-regulated in CSCs, and the low-staining fraction of cells with the ability to efflux Hoechst 33342 dye is commonly known as the side population (SP) [33]. Targeted inactivation of ABC transporters could reinstate drug sensitivity in CSCs, resulting in their death. Moreover, pathways that regulate self-renewal of normal stem cells, such as Wnt, Notch and Hedgehog, tumor suppressor genes, such as PTEN and p53, have been implicated in the control of CSCs self-renewal. These pathways are believed to be deregulated in CSCs, leading to uncontrolled self-renewal and generation of tumours that are resistant to conventional therapies [220]. The above mentioned characteristics and the ability of CSCs to evade cell death signals contribute to the failure of existing therapies to eradicate malignant tumours, causing resistance to treatment and an increased morbidity and mortality [201,221]. Apoptosis is one of the most critical and well-studied mechanisms, governing tissue development and homeostasis through a complex network of molecules that mediate death and survival signals. Escape from death program is a prerequisite for any tumour-initiating cell and may support the survival of CSCs in response to chemo- or radio-therapy. Thus, manipulating the apoptotic machinery to eradicate tumour-initiating cells holds enormous therapeutic potential [206].

Several studies have focused on apoptosis induction in CSCs by intervening in the extrinsic or intrinsic pathways to treat cancer.

6.1. TRAIL

TRAIL has been demonstrated to induce apoptosis in a wide range of cancer types both in vitro and also in various mouse models of human cancers [106]. In certain types of tumours a correlation has been established between the expression of DR and CSCs. Chemotherapy-resistant colon cancer SP cells express high levels of DR4 [222]; in some bladder cancer cell lines there is an increased expression of DR5 [223], and, in glioblastoma and lung CSCs express DRs [206]. More importantly, several clinical trials have explored the response to the use of DRs as a treatment. Studies concluded that SP cells of colon cancer displayed higher sensitivity to TRAIL compared to the non-SP cells [222]. In bladder cancer treatment sensitivity was greater in those showing increased expression of DR5 [223]. Moreover, isolated neurospheres from glioblastomas with characteristics of stem cells showed differences in apoptosis after treatment with TRAIL and it was effective in those who keep the route intact for caspase 8. Genomic heterogeneity in glioblastomas suggests the presence of multiple mechanisms in TRAIL resistance in both CSCs and non-stem cells. Future clinical trials of TRAIL apoptotic pathway targeted therapies may consider genomic analysis of tumour tissue to identify the genomic status of TRAIL apoptotic genes such as *caspase 8* and use it as a genomic marker to predict tumour resistance to TRAIL apoptotic pathway-targeted therapies [224]. In addition, recent studies are focused on combining mesenchymal stem cells (MSCs) expressing TRAIL and chemotherapy. These MSCs migrated to tumours and reduced the growth of primary cancers and metastases by induction of apoptosis, death and reduced colony formation of the SP and were synergistic when combined with traditional chemotherapy in apoptosis induction [225].

6.2. cFLIP

cFLIP is overexpressed in many types of cancers like melanoma, colon lymphoma and thyroid cancer [226]. The CD133+ populations within the T-cell acute leukemia cell line Jurkat and the breast cancer cell line MCF7 were reported to express higher levels of cFLIP, which was associated with TRAIL resistance. The down-regulation of cFLIP using siRNA restored TRAIL signalling in both cell lines resulting in a dramatic reduction in experimental metastases and the loss of CSC self-renewal [227]. This suggests that a combined TRAIL/FLIPi therapy could prevent metastatic disease progression in cancers.

6.3. IAP

In CD133 + cells isolated from glioblastoma an increased mRNA expression of livin, survivin and the multidrug resistance-associated protein 1 (MRP1) was detected. Therefore, the effects of etoposide, a pro-apoptosis agent, on these associated protein genes in glioblastoma stem-like cells have been studied. Results showed that after etoposide treatment, glioblastoma CSCs displayed a stronger resistance to apoptosis and death. The anti-apoptotic gene livinβ was more related with the high survival rate and MRP1 was more related with transporting chemotherapeutic agent out of glioblastoma stem-like cells [228]. In pancreatic cancer it has been demonstrated that targeting XIAP by RNAi inside the cancer cells, the combination of TRAIL with MSCs suppressed metastatic growth in these tumours [229]. Recently, it has been demonstrated that survivin is regulated by the interleukin-4 (IL-4) pathway in colon CSCs. Blockage of IL-4-mediated signaling pathway with leflunomide, Stat6 inhibitor, increased the nuclear survivin pool suggesting that the IL-4/STAT-6 pathway could escape cell death. IL-4 neutralization, mediated by STAT-6, could down-regulate survivin expression and localization, increasing the nuclear pool and in this way inducing chemo-sensitivity of CSCs [230].

6.4. Bcl-2

The proteins Bcl-2 family members are anti-apoptotic molecules known to be over-expressed in most cancers, and are associated directly with the CSCs [231]. Therefore, recent studies are aimed to target these proteins. A representative example is found in pancreatic cancer, against which the most potent and clinically acceptable Bcl-2 inhibitor AT-101 is currently in 20 different clinical trials around the world [232]. In glioblastoma, high expression levels of the anti-apoptoticc Bcl-2 protein, Mcl-1, were associated with resistance to treatment with Bcl-2 inhibitor ABT-737 in glioma stem cells [233]. In hematopoietic malignances, it has been showed that despite the over-expression of Bcl-2 this is not the critical point for the generation, selection and maintenance of leukemia stem cells [234].

6.5. NF-kB

The transcription factor NF-κB has been connected to multiple aspects of oncogenesis, including inhibition of apoptosis by increasing the expression of survival factor. In fact, aberrant regulation of NF-κB has been observed in many cancers, including both solid and hematopoietic tumours [235]. A fairly representative example is found in pancreatic cancer, where there is a clear correlation between the basal activity of NF-kB and the ability to generate angiogenesis and metastasis of pancreatic tumour cells [236] and, more recently it has been found that not only in this way, but also in the non-canonical NF-kB is also activated and functional [237]. A very interesting study by [238] showed that in CD44+ breast CSCs the expression of CD24 potentiated DNA-induced apoptosis by suppressing anti-apoptotic NF-κB signaling. Several therapies are being developed to inhibit this factor because there are many tumours which relate the decrease in the activity of NF-kB with a decrease in the size and tumor growth [239,240,241,242,243].

6.6. DNA repair capacity

A classic mechanism involved in the induction of apoptosis is in response to DNA damage by p53 action. This gene is mutated in most human cancers and inactivated in about 50% of cancers [244]. p53 was found to repress the CSC marker gene CD44 in an experimental breast tumour model and the over-expression of CD44 blocked p53-dependent apoptosis, leading to expansion of tumour-initiating cells [245]. Moreover, glioma CSCs resist radiation through preferential activation of the retainer DNA damage response and an increase in DNA repair capacity. In addition, the radioresistance of CD133+ glioma stem cells could be reversed with a specific inhibitor of Chk1 and Chk2 checkpoint kinases [208]. Another report reinforces the tumour-promoting effect of DNA damage response activation in leukemia stem cells by demonstrating that cell-cycle inhibitor p21 was indispensable for maintaining self-renewal of these CSCs [246].

6.7. miRNAs, apoptosis and CSCs

Recently, miRNAs have emerged as key regulators of "stemness", collaborating in the maintenance of pluripotency, control of self-renewal, and differentiation of stem cells. Moreover, certain miRNAs involved in apoptosis appear to influence the CSC fate by controlling self-renewal. It has been shown that restoration of miR-34 modulates self-renewal in pancreatic CSCs by directly regulating down-stream target gene Notch and Bcl-2 [247]. Also the restoration of miR-34, inhibit p53-mutant gastric cancer tumourspheres growth in vitro and tumour formation in vivo, which is reported to be correlated to the self-renewal of CSCs [248]. As miR-34 is a significant tumour suppressor of CSCs by regulation of both apoptosis and self-renewal properties, restoration of miR-34 may hold significant promise for a novel molecular therapy. Data also suggest that let-7 regulates apoptosis and CSC differentiation, which is considered as a key "keeper" of the differentiated state. In this context decreased expression of these TSmiRs implicated in self-renewal could lead to further cancer progression.

7. Conclusion

In this chapter we have summarized the research in the discovery of molecules, biomarkers for predicting therapeutic response and regulatory pathways implicated in apoptosis induction. These strategies are contributing to design cancer-targeted therapies that diminish or circumvent toxicity and improve life quality and overall survival of patients. The selective eradication of cancer cells can be achieve with small molecules and monoclonal antibodies, used as single agents or in combination with conventional chemotherapy, that interfere with the deregulated cellular signals that promote proliferation and survival to block tumor growth or sensitize cancer cells to apoptosis while leaving normal cells unaffected. Moreover, targeted therapies reactivating death program in CSCs may synergize with established therapies and increase efficacy in the clinic.

Author details

María A. García[1], Esther Carrasco[2], Alberto Ramírez[3], Gema Jiménez[2],
Elena López-Ruiz[3], Macarena Perán[3], Manuel Picón[2], Joaquín Campos[4],
Houria Boulaiz[2] and Juan Antonio Marchal[2,*]

[1]Research Unit, Hospital Universitario Virgen de las Nieves, Granada, Spain
[2]Biopathology and Regenerative Medicine Institute (IBIMER), Department of Human Anatomy and Embryology, Universidad de Granada, Granada,Spain
[3]Department of Health Sciences, Universidad de Jaén, Jaén,Spain
[4]Department of Pharmaceutical and Organic Chemistry, Universidad de Granada,Granada, Spain

Acknowledgement

This work was supported in part by grants from the Instituto de Salud Carlos III (Fondo de Investigación Sanitaria, FEDER, grant number PI10/02295 and PI10/00592) and the Consejería de Economía, Innovación y Ciencia (Junta de Andalucía, excellence project number CTS-6568)

8. References

[1] Hanahan D, Weinberg RA (2000) The hallmarks of cancer. Cell 100: 57-70.
[2] Taylor RC, Cullen SP, Martin SJ (2008) Apoptosis: controlled demolition at the cellular level. Nat Rev Mol Cell Biol 9: 231-241.

* Corresponding Author

[3] Kerr JF, Wyllie AH, Currie AR (1972) Apoptosis: a basic biological phenomenon with wide-ranging implications in tissue kinetics. Br J Cancer 26: 239-257.

[4] Ashkenazi A (2008) Targeting the extrinsic apoptosis pathway in cancer. Cytokine Growth Factor Rev 19: 325-331.

[5] Kischkel FC, Lawrence DA, Chuntharapai A, Schow P, Kim KJ, et al. (2000) Apo2L/TRAIL-dependent recruitment of endogenous FADD and caspase-8 to death receptors 4 and 5. Immunity 12: 611-620.

[6] Kischkel FC, Lawrence DA, Tinel A, LeBlanc H, Virmani A, et al. (2001) Death receptor recruitment of endogenous caspase-10 and apoptosis initiation in the absence of caspase-8. J Biol Chem 276: 46639-46646.

[7] Lavrik I, Golks A, Krammer PH (2005) Death receptor signaling. J Cell Sci 118: 265-267.

[8] Fulda S, Galluzzi L, Kroemer G (2010) Targeting mitochondria for cancer therapy. Nat Rev Drug Discov 9: 447-464.

[9] Wu G, Chai J, Suber TL, Wu JW, Du C, et al. (2000) Structural basis of IAP recognition by Smac/DIABLO. Nature 408: 1008-1012.

[10] 10. Fulda S, Vucic D (2012) Targeting IAP proteins for therapeutic intervention in cancer. Nat Rev Drug Discov 11: 109-124.

[11] Lowe SW, Lin AW (2000) Apoptosis in cancer. Carcinogenesis 21: 485-495.

[12] Marchal JA, Boulaiz H, Suarez I, Saniger E, Campos J, et al. (2004) Growth inhibition, G(1)-arrest, and apoptosis in MCF-7 human breast cancer cells by novel highly lipophilic 5-fluorouracil derivatives. Invest New Drugs 22: 379-389.

[13] Grivicich I, Regner A, da Rocha AB, Grass LB, Alves PA, et al. (2005) Irinotecan/5-fluorouracil combination induces alterations in mitochondrial membrane potential and caspases on colon cancer cell lines. Oncol Res 15: 385-392.

[14] Johnstone RW, Ruefli AA, Lowe SW (2002) Apoptosis: a link between cancer genetics and chemotherapy. Cell 108: 153-164.

[15] Garcia MA, Carrasco E, Aguilera M, Alvarez P, Rivas C, et al. (2011) The chemotherapeutic drug 5-fluorouracil promotes PKR-mediated apoptosis in a p53-independent manner in colon and breast cancer cells. PLoS One 6: e23887.

[16] Saniger E, Campos JM, Entrena A, Marchal JA, Suarez I, et al. (2003) Medium benzene-fused oxacycles with the 5-fluorouracil moiety: synthesis, antiproliferative activities and apoptosis induction in breast cancer cells. Tetrahedron 59: 5457-5467.

[17] Saniger E, Campos JM, Entrena A, Marchal JA, Boulaiz H, et al. (2003) Neighbouring-group participation as the key step in the reactivity of acyclic and cyclic salicyl-derived O,O-acetals with 5-fluorouracil. Antiproliferative activity, cell cycle dysregulation and apoptotic induction of new O,N-acetals against breast cancer cells. Tetrahedron 59: 8017-8026.

[18] Conejo-Garcia A, Nunez MC, Marchal JA, Rodriguez-Serrano F, Aranega A, et al. (2008) Regiospecific microwave-assisted synthesis and cytotoxic activity against human breast cancer cells of (RS)-6-substituted-7-or 9-(2,3-dihydro-5H-1,4-benzodioxepin-3-yl)-7H-or-9H-purines. European Journal of Medicinal Chemistry 43: 1742-1748.

[19] Diaz-Gavilan M, Conejo-Garcia A, Cruz-Lopez O, Nunez MC, Choquesillo-Lazarte D, et al. (2008) Synthesis and anticancer activity of (R,S)-9-(2,3-dihydro-1,4-benzoxathiin-3-ylmethyl)-9H-purines. ChemMedChem 3: 127-135.

[20] Caba O, Rodriguez-Serrano F, Diaz-Gavilan M, Conejo-Garcia A, Ortiz R, et al. (2012) The selective cytotoxic activity in breast cancer cells by an anthranilic alcohol-derived acyclic 5-fluorouracil O,N-acetal is mediated by endoplasmic reticulum stress-induced apoptosis. Eur J Med Chem 50: 376-382.

[21] Caba O, Diaz-Gavilan M, Rodriguez-Serrano F, Boulaiz H, Aranega A, et al. (2011) Anticancer activity and cDNA microarray studies of a (RS)-1,2,3,5-tetrahydro-4,1-benzoxazepine-3-yl]-6-chloro-9H-purine, and an acyclic (RS)-O,N-acetalic 6-chloro-7H-purine. Eur J Med Chem 46: 3802-3809.

[22] Schimmer AD, Dalili S (2005) Targeting the IAP family of caspase inhibitors as an emerging therapeutic strategy. Hematology Am Soc Hematol Educ Program: 215-219.

[23] Ndubaku C, Varfolomeev E, Wang L, Zobel K, Lau K, et al. (2009) Antagonism of c-IAP and XIAP proteins is required for efficient induction of cell death by small-molecule IAP antagonists. ACS Chem Biol 4: 557-566.

[24] Harlin H, Reffey SB, Duckett CS, Lindsten T, Thompson CB (2001) Characterization of XIAP-deficient mice. Mol Cell Biol 21: 3604-3608.

[25] Flygare JA, Fairbrother WJ (2010) Small-molecule pan-IAP antagonists: a patent review. Expert Opinion on Therapeutic Patents 20: 251-267.

[26] Hu Y, Cherton-Horvat G, Dragowska V, Baird S, Korneluk RG, et al. (2003) Antisense oligonucleotides targeting XIAP induce apoptosis and enhance chemotherapeutic activity against human lung cancer cells in vitro and in vivo. Clin Cancer Res 9: 2826-2836.

[27] Ohnishi K, Scuric Z, Schiestl RH, Okamoto N, Takahashi A, et al. (2006) siRNA targeting NBS1 or XIAP increases radiation sensitivity of human cancer cells independent of TP53 status. Radiat Res 166: 454-462.

[28] Ambros V (2003) MicroRNA pathways in flies and worms: growth, death, fat, stress, and timing. Cell 113: 673-676.

[29] Wu W, Sun M, Zou GM, Chen J (2007) MicroRNA and cancer: Current status and prospective. Int J Cancer 120: 953-960.

[30] Cimmino A, Calin GA, Fabbri M, Iorio MV, Ferracin M, et al. (2005) miR-15 and miR-16 induce apoptosis by targeting BCL2. Proc Natl Acad Sci U S A 102: 13944-13949.

[31] Selcuklu SD, Donoghue MT, Spillane C (2009) miR-21 as a key regulator of oncogenic processes. Biochem Soc Trans 37: 918-925.

[32] Marchal JA BH, Prados J et al, editor (2009) Therapeutic potential of differentiation in cancer and normal stem cells. New York: Nova Science Publishers. 116 p.

[33] Chikazawa N, Tanaka H, Tasaka T, Nakamura M, Tanaka M, et al. (2010) Inhibition of Wnt signaling pathway decreases chemotherapy-resistant side-population colon cancer cells. Anticancer Res 30: 2041-2048.

[34] Brown JM, Wouters BG (1999) Apoptosis, p53, and tumor cell sensitivity to anticancer agents. Cancer Res 59: 1391-1399.

[35] Fulda S, Debatin KM (2006) Extrinsic versus intrinsic apoptosis pathways in anticancer chemotherapy. Oncogene 25: 4798-4811.

[36] Liu JJ, Lin M, Yu JY, Liu B, Bao JK (2011) Targeting apoptotic and autophagic pathways for cancer therapeutics. Cancer Lett 300: 105-114.

[37] Gottesman MM, Fojo T, Bates SE (2002) Multidrug resistance in cancer: role of ATP-dependent transporters. Nat Rev Cancer 2: 48-58.

[38] Vogelstein B, Lane D, Levine AJ (2000) Surfing the p53 network. Nature 408: 307-310.

[39] Kastan MB, Onyekwere O, Sidransky D, Vogelstein B, Craig RW (1991) Participation of p53 protein in the cellular response to DNA damage. Cancer Res 51: 6304-6311.

[40] Vermeulen K, Berneman ZN, Van Bockstaele DR (2003) Cell cycle and apoptosis. Cell Prolif 36: 165-175.

[41] Ajani J (2006) Review of capecitabine as oral treatment of gastric, gastroesophageal, and esophageal cancers. Cancer 107: 221-231.

[42] Ershler WB (2006) Capecitabine monotherapy: safe and effective treatment for metastatic breast cancer. Oncologist 11: 325-335.

[43] Longley DB, Harkin DP, Johnston PG (2003) 5-fluorouracil: mechanisms of action and clinical strategies. Nat Rev Cancer 3: 330-338.

[44] Petak I, Tillman DM, Houghton JA (2000) p53 dependence of Fas induction and acute apoptosis in response to 5-fluorouracil-leucovorin in human colon carcinoma cell lines. Clin Cancer Res 6: 4432-4441.

[45] Bunz F, Hwang PM, Torrance C, Waldman T, Zhang Y, et al. (1999) Disruption of p53 in human cancer cells alters the responses to therapeutic agents. J Clin Invest 104: 263-269.

[46] Backus HH, Wouters D, Ferreira CG, van Houten VM, Brakenhoff RH, et al. (2003) Thymidylate synthase inhibition triggers apoptosis via caspases-8 and -9 in both wild-type and mutant p53 colon cancer cell lines. Eur J Cancer 39: 1310-1317.

[47] Bressac B, Galvin KM, Liang TJ, Isselbacher KJ, Wands JR, et al. (1990) Abnormal structure and expression of p53 gene in human hepatocellular carcinoma. Proc Natl Acad Sci U S A 87: 1973-1977.

[48] Russo A, Bazan V, Iacopetta B, Kerr D, Soussi T, et al. (2005) The TP53 colorectal cancer international collaborative study on the prognostic and predictive significance of p53 mutation: influence of tumor site, type of mutation, and adjuvant treatment. J Clin Oncol 23: 7518-7528.

[49] Ahnen DJ, Feigl P, Quan G, Fenoglio-Preiser C, Lovato LC, et al. (1998) Ki-ras mutation and p53 overexpression predict the clinical behavior of colorectal cancer: a Southwest Oncology Group study. Cancer Res 58: 1149-1158.

[50] Allegra CJ, Paik S, Colangelo LH, Parr AL, Kirsch I, et al. (2003) Prognostic value of thymidylate synthase, Ki-67, and p53 in patients with Dukes' B and C colon cancer: a National Cancer Institute-National Surgical Adjuvant Breast and Bowel Project collaborative study. J Clin Oncol 21: 241-250.

[51] Elsaleh H, Powell B, Soontrapornchai P, Joseph D, Goria F, et al. (2000) p53 gene mutation, microsatellite instability and adjuvant chemotherapy: impact on survival of 388 patients with Dukes' C colon carcinoma. Oncology 58: 52-59.

[52] Roberts WK, Hovanessian A, Brown RE, Clemens MJ, Kerr IM (1976) Interferon-mediated protein kinase and low-molecular-weight inhibitor of protein synthesis. Nature 264: 477-480.

[53] Metz DH, Esteban M (1972) Interferon inhibits viral protein synthesis in L cells infected with vaccinia virus. Nature 238: 385-388.

[54] Garcia MA, Gil J, Ventoso I, Guerra S, Domingo E, et al. (2006) Impact of protein kinase PKR in cell biology: from antiviral to antiproliferative action. Microbiol Mol Biol Rev 70: 1032-1060.

[55] Galabru J, Hovanessian A (1987) Autophosphorylation of the protein kinase dependent on double-stranded RNA. J Biol Chem 262: 15538-15544.

[56] Hovanessian AG (1989) The double stranded RNA-activated protein kinase induced by interferon: dsRNA-PK. J Interferon Res 9: 641-647.

[57] Rhoads RE (1993) Regulation of eukaryotic protein synthesis by initiation factors. J Biol Chem 268: 3017-3020.

[58] Kumar A, Yang YL, Flati V, Der S, Kadereit S, et al. (1997) Deficient cytokine signaling in mouse embryo fibroblasts with a targeted deletion in the PKR gene: role of IRF-1 and NF-kappaB. EMBO J 16: 406-416.

[59] Gil J, Alcami J, Esteban M (2000) Activation of NF-kappa B by the dsRNA-dependent protein kinase, PKR involves the I kappa B kinase complex. Oncogene 19: 1369-1378.

[60] Hovanessian AG, Galabru J (1987) The double-stranded RNA-dependent protein kinase is also activated by heparin. Eur J Biochem 167: 467-473.

[61] Ito T, Yang M, May WS (1999) RAX, a cellular activator for double-stranded RNA-dependent protein kinase during stress signaling. J Biol Chem 274: 15427-15432.

[62] Ruvolo PP, Gao F, Blalock WL, Deng X, May WS (2001) Ceramide regulates protein synthesis by a novel mechanism involving the cellular PKR activator RAX. J Biol Chem 276: 11754-11758.

[63] Ke ZJ, Wang X, Fan Z, Luo J (2009) Ethanol promotes thiamine deficiency-induced neuronal death: involvement of double-stranded RNA-activated protein kinase. Alcohol Clin Exp Res 33: 1097-1103.

[64] Patel RC, Sen GC (1998) PACT, a protein activator of the interferon-induced protein kinase, PKR. EMBO J 17: 4379-4390.

[65] Peel AL, Rao RV, Cottrell BA, Hayden MR, Ellerby LM, et al. (2001) Double-stranded RNA-dependent protein kinase, PKR, binds preferentially to Huntington's disease (HD) transcripts and is activated in HD tissue. Hum Mol Genet 10: 1531-1538.

[66] Peel AL, Bredesen DE (2003) Activation of the cell stress kinase PKR in Alzheimer's disease and human amyloid precursor protein transgenic mice. Neurobiol Dis 14: 52-62.

[67] Onuki R, Bando Y, Suyama E, Katayama T, Kawasaki H, et al. (2004) An RNA-dependent protein kinase is involved in tunicamycin-induced apoptosis and Alzheimer's disease. EMBO J 23: 959-968.

[68] Peel AL (2004) PKR activation in neurodegenerative disease. J Neuropathol Exp Neurol 63: 97-105.

[69] Bando Y, Onuki R, Katayama T, Manabe T, Kudo T, et al. (2005) Double-strand RNA dependent protein kinase (PKR) is involved in the extrastriatal degeneration in Parkinson's disease and Huntington's disease. Neurochem Int 46: 11-18.

[70] Abraham N, Jaramillo ML, Duncan PI, Methot N, Icely PL, et al. (1998) The murine PKR tumor suppressor gene is rearranged in a lymphocytic leukemia. Exp Cell Res 244: 394-404.

[71] Beretta L, Gabbay M, Berger R, Hanaoh SM, Sonenberg N (1996) Expression of the protein kinase PKR in modulated by IRF-1 and is reduced in 5q- associated leukemias. Oncogene 12: 1593-1596.

[72] Li S, Koromilas AE (2001) Dominant negative function by an alternatively spliced form of the interferon-inducible protein kinase PKR. J Biol Chem 276: 13881-13890.

[73] Murad JM, Tone LG, de Souza LR, De Lucca FL (2005) A point mutation in the RNA-binding domain I results in decrease of PKR activation in acute lymphoblastic leukemia. Blood Cells Mol Dis 34: 1-5.

[74] Shimada A, Shiota G, Miyata H, Kamahora T, Kawasaki H, et al. (1998) Aberrant expression of double-stranded RNA-dependent protein kinase in hepatocytes of chronic hepatitis and differentiated hepatocellular carcinoma. Cancer Res 58: 4434-4438.

[75] Balachandran S, Kim CN, Yeh WC, Mak TW, Bhalla K, et al. (1998) Activation of the dsRNA-dependent protein kinase, PKR, induces apoptosis through FADD-mediated death signaling. EMBO J 17: 6888-6902.

[76] Gil J, Esteban M (2000) Induction of apoptosis by the dsRNA-dependent protein kinase (PKR): mechanism of action. Apoptosis 5: 107-114.

[77] Gil J, Alcami J, Esteban M (1999) Induction of apoptosis by double-stranded-RNA-dependent protein kinase (PKR) involves the alpha subunit of eukaryotic translation initiation factor 2 and NF-kappaB. Mol Cell Biol 19: 4653-4663.

[78] Kibler KV, Shors T, Perkins KB, Zeman CC, Banaszak MP, et al. (1997) Double-stranded RNA is a trigger for apoptosis in vaccinia virus-infected cells. J Virol 71: 1992-2003.

[79] Gil J, Esteban M (2000) The interferon-induced protein kinase (PKR), triggers apoptosis through FADD-mediated activation of caspase 8 in a manner independent of Fas and TNF-alpha receptors. Oncogene 19: 3665-3674.

[80] Gil J, Garcia MA, Esteban M (2002) Caspase 9 activation by the dsRNA-dependent protein kinase, PKR: molecular mechanism and relevance. FEBS Lett 529: 249-255.

[81] Yoon CH, Lee ES, Lim DS, Bae YS (2009) PKR, a p53 target gene, plays a crucial role in the tumor-suppressor function of p53. Proc Natl Acad Sci U S A 106: 7852-7857.

[82] Peidis P, Papadakis AI, Muaddi H, Richard S, Koromilas AE (2011) Doxorubicin bypasses the cytoprotective effects of eIF2alpha phosphorylation and promotes PKR-mediated cell death. Cell Death Differ 18: 145-154.

[83] Mounir Z, Krishnamoorthy JL, Robertson GP, Scheuner D, Kaufman RJ, et al. (2009) Tumor suppression by PTEN requires the activation of the PKR-eIF2alpha phosphorylation pathway. Sci Signal 2: ra85.

[84] Kang MH, Reynolds CP (2009) Bcl-2 inhibitors: targeting mitochondrial apoptotic pathways in cancer therapy. Clin Cancer Res 15: 1126-1132.

[85] Muchmore SW, Sattler M, Liang H, Meadows RP, Harlan JE, et al. (1996) X-ray and NMR structure of human Bcl-xL, an inhibitor of programmed cell death. Nature 381: 335-341.

[86] Rai KR (2008) The natural history of CLL and new prognostic markers. Clin Adv Hematol Oncol 6: 4-5; quiz 10-12.

[87] Yu J, Zhang L (2008) PUMA, a potent killer with or without p53. Oncogene 27 Suppl 1: S71-83.

[88] Karst AM, Dai DL, Martinka M, Li G (2005) PUMA expression is significantly reduced in human cutaneous melanomas. Oncogene 24: 1111-1116.

[89] Garrison SP, Jeffers JR, Yang C, Nilsson JA, Hall MA, et al. (2008) Selection against PUMA gene expression in Myc-driven B-cell lymphomagenesis. Mol Cell Biol 28: 5391-5402.

[90] Middelburg R, de Haas RR, Dekker H, Kerkhoven RM, Pohlmann PR, et al. (2005) Induction of p53 up-regulated modulator of apoptosis messenger RNA by chemotherapeutic treatment of locally advanced breast cancer. Clin Cancer Res 11: 1863-1869.

[91] Sinicrope FA, Rego RL, Okumura K, Foster NR, O'Connell MJ, et al. (2008) Prognostic impact of bim, puma, and noxa expression in human colon carcinomas. Clin Cancer Res 14: 5810-5818.

[92] Li P, Nijhawan D, Budihardjo I, Srinivasula SM, Ahmad M, et al. (1997) Cytochrome c and dATP-dependent formation of Apaf-1/caspase-9 complex initiates an apoptotic protease cascade. Cell 91: 479-489.

[93] Deveraux QL, Reed JC (1999) IAP family proteins--suppressors of apoptosis. Genes Dev 13: 239-252.

[94] Suzuki Y, Nakabayashi Y, Nakata K, Reed JC, Takahashi R (2001) X-linked inhibitor of apoptosis protein (XIAP) inhibits caspase-3 and -7 in distinct modes. J Biol Chem 276: 27058-27063.

[95] Verhagen AM, Ekert PG, Pakusch M, Silke J, Connolly LM, et al. (2000) Identification of DIABLO, a mammalian protein that promotes apoptosis by binding to and antagonizing IAP proteins. Cell 102: 43-53.

[96] Hector S, Prehn JH (2009) Apoptosis signaling proteins as prognostic biomarkers in colorectal cancer: a review. Biochim Biophys Acta 1795: 117-129.

[97] Zlobec I, Steele R, Terracciano L, Jass JR, Lugli A (2007) Selecting immunohistochemical cut-off scores for novel biomarkers of progression and survival in colorectal cancer. J Clin Pathol 60: 1112-1116.

[98] Paik SS, Jang KS, Song YS, Jang SH, Min KW, et al. (2007) Reduced expression of Apaf-1 in colorectal adenocarcinoma correlates with tumor progression and aggressive phenotype. Ann Surg Oncol 14: 3453-3459.

[99] Endo K, Kohnoe S, Watanabe A, Tashiro H, Sakata H, et al. (2009) Clinical significance of Smac/DIABLO expression in colorectal cancer. Oncol Rep 21: 351-355.

[100] Strater J, Herter I, Merkel G, Hinz U, Weitz J, et al. (2010) Expression and prognostic significance of APAF-1, caspase-8 and caspase-9 in stage II/III colon carcinoma: caspase-8 and caspase-9 is associated with poor prognosis. Int J Cancer 127: 873-880.

[101] Xiang G, Wen X, Wang H, Chen K, Liu H (2009) Expression of X-linked inhibitor of apoptosis protein in human colorectal cancer and its correlation with prognosis. J Surg Oncol 100: 708-712.

[102] Hector S, Conlon S, Schmid J, Dicker P, Cummins RJ, et al. (2012) Apoptosome-dependent caspase activation proteins as prognostic markers in Stage II and III colorectal cancer. Br J Cancer.

[103] Huang H, Zhang XF, Zhou HJ, Xue YH, Dong QZ, et al. (2010) Expression and prognostic significance of osteopontin and caspase-3 in hepatocellular carcinoma patients after curative resection. Cancer Sci 101: 1314-1319.

[104] Provencio M, Martin P, Garcia V, Candia A, Sanchez AC, et al. (2010) Caspase 3a: new prognostic marker for diffuse large B-cell lymphoma in the rituximab era. Leuk Lymphoma 51: 2021-2030.

[105] Sayers TJ (2011) Targeting the extrinsic apoptosis signaling pathway for cancer therapy. Cancer Immunol Immunother 60: 1173-1180.

[106] Ashkenazi A, Herbst RS (2008) To kill a tumor cell: the potential of proapoptotic receptor agonists. J Clin Invest 118: 1979-1990.

[107] Mahalingam D, Szegezdi E, Keane M, de Jong S, Samali A (2009) TRAIL receptor signalling and modulation: Are we on the right TRAIL? Cancer Treat Rev 35: 280-288.

[108] Shanker A, Brooks AD, Tristan CA, Wine JW, Elliott PJ, et al. (2008) Treating metastatic solid tumors with bortezomib and a tumor necrosis factor-related apoptosis-inducing ligand receptor agonist antibody. J Natl Cancer Inst 100: 649-662.

[109] Lundberg AS, Weinberg RA (1999) Control of the cell cycle and apoptosis. Eur J Cancer 35: 1886-1894.

[110] Qin LF, Ng IO (2002) Induction of apoptosis by cisplatin and its effect on cell cycle-related proteins and cell cycle changes in hepatoma cells. Cancer Lett 175: 27-38.

[111] Meyn RE, Stephens LC, Hunter NR, Milas L (1995) Kinetics of cisplatin-induced apoptosis in murine mammary and ovarian adenocarcinomas. Int J Cancer 60: 725-729.

[112] Milas L, Hunter NR, Kurdoglu B, Mason KA, Meyn RE, et al. (1995) Kinetics of mitotic arrest and apoptosis in murine mammary and ovarian tumors treated with taxol. Cancer Chemother Pharmacol 35: 297-303.

[113] Chadderton A, Villeneuve DJ, Gluck S, Kirwan-Rhude AF, Gannon BR, et al. (2000) Role of specific apoptotic pathways in the restoration of paclitaxel-induced apoptosis by valspodar in doxorubicin-resistant MCF-7 breast cancer cells. Breast Cancer Res Treat 59: 231-244.

[114] Saunders DE, Lawrence WD, Christensen C, Wappler NL, Ruan H, et al. (1997) Paclitaxel-induced apoptosis in MCF-7 breast-cancer cells. Int J Cancer 70: 214-220.

[115] Agarwal ML, Agarwal A, Taylor WR, Stark GR (1995) p53 controls both the G2/M and the G1 cell cycle checkpoints and mediates reversible growth arrest in human fibroblasts. Proc Natl Acad Sci U S A 92: 8493-8497.

[116] Reed JC (1997) Double identity for proteins of the Bcl-2 family. Nature 387: 773-776.

[117] Campos J, Saniger E, Marchal JA, Aiello S, Suarez I, et al. (2005) New medium oxacyclic O,N-acetals and related open analogues: biological activities. Curr Med Chem 12: 1423-1438.

[118] Gali-Muhtasib H, Bakkar N (2002) Modulating cell cycle: current applications and prospects for future drug development. Curr Cancer Drug Targets 2: 309-336.

[119] Conejo-Garcia A, Nunez MC, Marchal JA, Rodriguez-Serrano F, Aranega A, et al. (2008) Regiospecific microwave-assisted synthesis and cytotoxic activity against human breast cancer cells of (RS)-6-substituted-7- or 9-(2,3-dihydro-5H-1,4-benzodioxepin-3-yl)-7H- or -9H-purines. Eur J Med Chem 43: 1742-1748.

[120] Boulaiz H, Prados J, Melguizo C, Garcia AM, Marchal JA, et al. (2003) Inhibition of growth and induction of apoptosis in human breast cancer by transfection of gef gene. Br J Cancer 89: 192-198.

[121] Diaz-Gavilan M, Conejo-Garcia A, Cruz-Lopez O, Nunez MC, Choquesillo-Lazarte D, et al. (2008) Synthesis and anticancer activity of (R,S)-9-(2,3-dihydro-1,4-benzoxathiin-3-ylmethyl)-9H-purines. ChemMedChem 3: 127-135.

[122] Indran IR, Tufo G, Pervaiz S, Brenner C (2011) Recent advances in apoptosis, mitochondria and drug resistance in cancer cells. Biochim Biophys Acta 1807: 735-745.

[123] Droin N, Guery L, Benikhlef N, Solary E (2011) Targeting apoptosis proteins in hematological malignancies. Cancer Lett.

[124] LaCasse EC, Mahoney DJ, Cheung HH, Plenchette S, Baird S, et al. (2008) IAP-targeted therapies for cancer. Oncogene 27: 6252-6275.

[125] Hunter AM, LaCasse EC, Korneluk RG (2007) The inhibitors of apoptosis (IAPs) as cancer targets. Apoptosis 12: 1543-1568.

[126] Fulda S (2009) Inhibitor of apoptosis proteins in hematological malignancies. Leukemia 23: 467-476.

[127] Dean EJ, Ranson M, Blackhall F, Holt SV, Dive C (2007) Novel therapeutic targets in lung cancer: Inhibitor of apoptosis proteins from laboratory to clinic. Cancer Treat Rev 33: 203-212.

[128] Qiu XB, Goldberg AL (2005) The membrane-associated inhibitor of apoptosis protein, BRUCE/Apollon, antagonizes both the precursor and mature forms of Smac and caspase-9. J Biol Chem 280: 174-182.

[129] Mazrouei S, Ziaei A, Tanhaee AP, Keyhanian K, Esmaeili M, et al. (2012) Apoptosis inhibition or inflammation: the role of NAIP protein expression in Hodgkin and non-Hodgkin lymphomas compared to non-neoplastic lymph node. J Inflamm (Lond) 9: 4.

[130] Davoodi J, Ghahremani MH, Es-Haghi A, Mohammad-Gholi A, Mackenzie A (2010) Neuronal apoptosis inhibitory protein, NAIP, is an inhibitor of procaspase-9. Int J Biochem Cell Biol 42: 958-964.

[131] Duckett CS (2005) IAP proteins: sticking it to Smac. Biochem J 385: e1-2.

[132] Miura K, Fujibuchi W, Ishida K, Naitoh T, Ogawa H, et al. (2011) Inhibitor of apoptosis protein family as diagnostic markers and therapeutic targets of colorectal cancer. Surg Today 41: 175-182.

[133] Schimmer AD, Dalili S, Batey RA, Riedl SJ (2006) Targeting XIAP for the treatment of malignancy. Cell Death Differ 13: 179-188.

[134] Wu TY, Wagner KW, Bursulaya B, Schultz PG, Deveraux QL (2003) Development and characterization of nonpeptidic small molecule inhibitors of the XIAP/caspase-3 interaction. Chem Biol 10: 759-767.

[135] Karikari CA, Roy I, Tryggestad E, Feldmann G, Pinilla C, et al. (2007) Targeting the apoptotic machinery in pancreatic cancers using small-molecule antagonists of the X-linked inhibitor of apoptosis protein. Mol Cancer Ther 6: 957-966.

[136] Dean E, Jodrell D, Connolly K, Danson S, Jolivet J, et al. (2009) Phase I trial of AEG35156 administered as a 7-day and 3-day continuous intravenous infusion in patients with advanced refractory cancer. J Clin Oncol 27: 1660-1666.

[137] Schimmer AD, Estey EH, Borthakur G, Carter BZ, Schiller GJ, et al. (2009) Phase I/II trial of AEG35156 X-linked inhibitor of apoptosis protein antisense oligonucleotide combined with idarubicin and cytarabine in patients with relapsed or primary refractory acute myeloid leukemia. J Clin Oncol 27: 4741-4746.

[138] Talbot DC, Ranson M, Davies J, Lahn M, Callies S, et al. (2010) Tumor survivin is downregulated by the antisense oligonucleotide LY2181308: a proof-of-concept, first-in-human dose study. Clin Cancer Res 16: 6150-6158.

[139] Hansen JB, Fisker N, Westergaard M, Kjaerulff LS, Hansen HF, et al. (2008) SPC3042: a proapoptotic survivin inhibitor. Mol Cancer Ther 7: 2736-2745.

[140] Sapra P, Wang M, Bandaru R, Zhao H, Greenberger LM, et al. (2010) Down-modulation of survivin expression and inhibition of tumor growth in vivo by EZN-3042, a locked nucleic acid antisense oligonucleotide. Nucleosides Nucleotides Nucleic Acids 29: 97-112.

[141] Mannhold R, Fulda S, Carosati E (2010) IAP antagonists: promising candidates for cancer therapy. Drug Discov Today 15: 210-219.

[142] Probst BL, Liu L, Ramesh V, Li L, Sun H, et al. (2010) Smac mimetics increase cancer cell response to chemotherapeutics in a TNF-alpha-dependent manner. Cell Death Differ 17: 1645-1654.

[143] Bartel DP (2009) MicroRNAs: target recognition and regulatory functions. Cell 136: 215-233.

[144] Lee Y, Ahn C, Han J, Choi H, Kim J, et al. (2003) The nuclear RNase III Drosha initiates microRNA processing. Nature 425: 415-419.

[145] Garzon R, Marcucci G, Croce CM (2010) Targeting microRNAs in cancer: rationale, strategies and challenges. Nat Rev Drug Discov 9: 775-789.

[146] Hutvagner G, Simard MJ, Mello CC, Zamore PD (2004) Sequence-specific inhibition of small RNA function. PLoS Biol 2: E98.

[147] Wang H, Garzon R, Sun H, Ladner KJ, Singh R, et al. (2008) NF-kappaB-YY1-miR-29 regulatory circuitry in skeletal myogenesis and rhabdomyosarcoma. Cancer Cell 14: 369-381.

[148] Mott JL, Kobayashi S, Bronk SF, Gores GJ (2007) mir-29 regulates Mcl-1 protein expression and apoptosis. Oncogene 26: 6133-6140.

[149] Garzon R, Heaphy CE, Havelange V, Fabbri M, Volinia S, et al. (2009) MicroRNA 29b functions in acute myeloid leukemia. Blood 114: 5331-5341.

[150] Krutzfeldt J, Rajewsky N, Braich R, Rajeev KG, Tuschl T, et al. (2005) Silencing of microRNAs in vivo with 'antagomirs'. Nature 438: 685-689.

[151] Obad S, dos Santos CO, Petri A, Heidenblad M, Broom O, et al. (2011) Silencing of microRNA families by seed-targeting tiny LNAs. Nat Genet 43: 371-378.

[152] Corsten MF, Miranda R, Kasmieh R, Krichevsky AM, Weissleder R, et al. (2007) MicroRNA-21 knockdown disrupts glioma growth in vivo and displays synergistic cytotoxicity with neural precursor cell delivered S-TRAIL in human gliomas. Cancer Res 67: 8994-9000.

[153] Galardi S, Mercatelli N, Giorda E, Massalini S, Frajese GV, et al. (2007) miR-221 and miR-222 expression affects the proliferation potential of human prostate carcinoma cell lines by targeting p27Kip1. J Biol Chem 282: 23716-23724.

[154] Lanford RE, Hildebrandt-Eriksen ES, Petri A, Persson R, Lindow M, et al. (2010) Therapeutic silencing of microRNA-122 in primates with chronic hepatitis C virus infection. Science 327: 198-201.

[155] Ebert MS, Neilson JR, Sharp PA (2007) MicroRNA sponges: competitive inhibitors of small RNAs in mammalian cells. Nat Methods 4: 721-726.

[156] Xiao J, Yang B, Lin H, Lu Y, Luo X, et al. (2007) Novel approaches for gene-specific interference via manipulating actions of microRNAs: examination on the pacemaker channel genes HCN2 and HCN4. J Cell Physiol 212: 285-292.

[157] Chan JA, Krichevsky AM, Kosik KS (2005) MicroRNA-21 is an antiapoptotic factor in human glioblastoma cells. Cancer Res 65: 6029-6033.

[158] Zhu S, Wu H, Wu F, Nie D, Sheng S, et al. (2008) MicroRNA-21 targets tumor suppressor genes in invasion and metastasis. Cell Res 18: 350-359.

[159] Sathyan P, Golden HB, Miranda RC (2007) Competing interactions between micro-RNAs determine neural progenitor survival and proliferation after ethanol exposure: evidence from an ex vivo model of the fetal cerebral cortical neuroepithelium. J Neurosci 27: 8546-8557.

[160] Volinia S, Calin GA, Liu CG, Ambs S, Cimmino A, et al. (2006) A microRNA expression signature of human solid tumors defines cancer gene targets. Proc Natl Acad Sci U S A 103: 2257-2261.

[161] Li J, Huang H, Sun L, Yang M, Pan C, et al. (2009) MiR-21 indicates poor prognosis in tongue squamous cell carcinomas as an apoptosis inhibitor. Clin Cancer Res 15: 3998-4008.

[162] Barker EV, Cervigne NK, Reis PP, Goswami RS, Xu W, et al. (2009) microRNA evaluation of unknown primary lesions in the head and neck. Mol Cancer 8: 127.

[163] Cervigne NK, Reis PP, Machado J, Sadikovic B, Bradley G, et al. (2009) Identification of a microRNA signature associated with progression of leukoplakia to oral carcinoma. Hum Mol Genet 18: 4818-4829.

[164] Ribas J, Lupold SE (2010) The transcriptional regulation of miR-21, its multiple transcripts, and their implication in prostate cancer. Cell Cycle 9: 923-929.

[165] Krajewski S, Krajewska M, Turner BC, Pratt C, Howard B, et al. (1999) Prognostic significance of apoptosis regulators in breast cancer. Endocr Relat Cancer 6: 29-40.

[166] Garofalo M, Quintavalle C, Di Leva G, Zanca C, Romano G, et al. (2008) MicroRNA signatures of TRAIL resistance in human non-small cell lung cancer. Oncogene 27: 3845-3855.

[167] Felicetti F, Errico MC, Bottero L, Segnalini P, Stoppacciaro A, et al. (2008) The promyelocytic leukemia zinc finger-microRNA-221/-222 pathway controls melanoma progression through multiple oncogenic mechanisms. Cancer Res 68: 2745-2754.

[168] Chun-Zhi Z, Lei H, An-Ling Z, Yan-Chao F, Xiao Y, et al. (2010) MicroRNA-221 and microRNA-222 regulate gastric carcinoma cell proliferation and radioresistance by targeting PTEN. BMC Cancer 10: 367.

[169] Zhang C, Kang C, You Y, Pu P, Yang W, et al. (2009) Co-suppression of miR-221/222 cluster suppresses human glioma cell growth by targeting p27kip1 in vitro and in vivo. Int J Oncol 34: 1653-1660.

[170] Garofalo M, Romano G, Di Leva G, Nuovo G, Jeon YJ, et al. (2012) EGFR and MET receptor tyrosine kinase-altered microRNA expression induces tumorigenesis and gefitinib resistance in lung cancers. Nat Med 18: 74-82.

[171] Meng F, Henson R, Wehbe-Janek H, Ghoshal K, Jacob ST, et al. (2007) MicroRNA-21 regulates expression of the PTEN tumor suppressor gene in human hepatocellular cancer. Gastroenterology 133: 647-658.

[172] Wang ZX, Lu BB, Wang H, Cheng ZX, Yin YM (2011) MicroRNA-21 modulates chemosensitivity of breast cancer cells to doxorubicin by targeting PTEN. Arch Med Res 42: 281-290.

[173] Zhang BG, Li JF, Yu BQ, Zhu ZG, Liu BY, et al. (2012) microRNA-21 promotes tumor proliferation and invasion in gastric cancer by targeting PTEN. Oncol Rep 27: 1019-1026.

[174] Yamanaka Y, Tagawa H, Takahashi N, Watanabe A, Guo YM, et al. (2009) Aberrant overexpression of microRNAs activate AKT signaling via down-regulation of tumor suppressors in natural killer-cell lymphoma/leukemia. Blood 114: 3265-3275.

[175] Zhang JG, Wang JJ, Zhao F, Liu Q, Jiang K, et al. (2010) MicroRNA-21 (miR-21) represses tumor suppressor PTEN and promotes growth and invasion in non-small cell lung cancer (NSCLC). Clin Chim Acta 411: 846-852.

[176] Xiao C, Srinivasan L, Calado DP, Patterson HC, Zhang B, et al. (2008) Lymphoproliferative disease and autoimmunity in mice with increased miR-17-92 expression in lymphocytes. Nat Immunol 9: 405-414.

[177] Hayashita Y, Osada H, Tatematsu Y, Yamada H, Yanagisawa K, et al. (2005) A polycistronic microRNA cluster, miR-17-92, is overexpressed in human lung cancers and enhances cell proliferation. Cancer Res 65: 9628-9632.

[178] Xiong Y, Fang JH, Yun JP, Yang J, Zhang Y, et al. (2010) Effects of microRNA-29 on apoptosis, tumorigenicity, and prognosis of hepatocellular carcinoma. Hepatology 51: 836-845.

[179] Park SY, Lee JH, Ha M, Nam JW, Kim VN (2009) miR-29 miRNAs activate p53 by targeting p85 alpha and CDC42. Nat Struct Mol Biol 16: 23-29.

[180] Bommer GT, Gerin I, Feng Y, Kaczorowski AJ, Kuick R, et al. (2007) p53-mediated activation of miRNA34 candidate tumor-suppressor genes. Curr Biol 17: 1298-1307.

[181] Hagman Z, Larne O, Edsjo A, Bjartell A, Ehrnstrom RA, et al. (2010) miR-34c is downregulated in prostate cancer and exerts tumor suppressive functions. Int J Cancer 127: 2768-2776.

[182] Gallardo E, Navarro A, Vinolas N, Marrades RM, Diaz T, et al. (2009) miR-34a as a prognostic marker of relapse in surgically resected non-small-cell lung cancer. Carcinogenesis 30: 1903-1909.

[183] Cole KA, Attiyeh EF, Mosse YP, Laquaglia MJ, Diskin SJ, et al. (2008) A functional screen identifies miR-34a as a candidate neuroblastoma tumor suppressor gene. Mol Cancer Res 6: 735-742.

[184] Bonci D, Coppola V, Musumeci M, Addario A, Giuffrida R, et al. (2008) The miR-15a-miR-16-1 cluster controls prostate cancer by targeting multiple oncogenic activities. Nat Med 14: 1271-1277.

[185] Aqeilan RI, Calin GA, Croce CM (2010) miR-15a and miR-16-1 in cancer: discovery, function and future perspectives. Cell Death Differ 17: 215-220.

[186] Akao Y, Nakagawa Y, Naoe T (2006) let-7 microRNA functions as a potential growth suppressor in human colon cancer cells. Biol Pharm Bull 29: 903-906.

[187] Johnson CD, Esquela-Kerscher A, Stefani G, Byrom M, Kelnar K, et al. (2007) The let-7 microRNA represses cell proliferation pathways in human cells. Cancer Res 67: 7713-7722.

[188] Boyerinas B, Park SM, Hau A, Murmann AE, Peter ME (2010) The role of let-7 in cell differentiation and cancer. Endocr Relat Cancer 17: F19-36.

[189] Zhang J, Du Y, Wu C, Ren X, Ti X, et al. (2010) Curcumin promotes apoptosis in human lung adenocarcinoma cells through miR-186* signaling pathway. Oncol Rep 24: 1217-1223.

[190] Shimizu S, Takehara T, Hikita H, Kodama T, Miyagi T, et al. (2010) The let-7 family of microRNAs inhibits Bcl-xL expression and potentiates sorafenib-induced apoptosis in human hepatocellular carcinoma. J Hepatol 52: 698-704.

[191] Kota J, Chivukula RR, O'Donnell KA, Wentzel EA, Montgomery CL, et al. (2009) Therapeutic microRNA delivery suppresses tumorigenesis in a murine liver cancer model. Cell 137: 1005-1017.

[192] De Guire V, Caron M, Scott N, Menard C, Gaumont-Leclerc MF, et al. (2010) Designing small multiple-target artificial RNAs. Nucleic Acids Res 38: e140.

[193] Bottoni A, Piccin D, Tagliati F, Luchin A, Zatelli MC, et al. (2005) miR-15a and miR-16-1 down-regulation in pituitary adenomas. J Cell Physiol 204: 280-285.

[194] Qin W, Shi Y, Zhao B, Yao C, Jin L, et al. (2010) miR-24 regulates apoptosis by targeting the open reading frame (ORF) region of FAF1 in cancer cells. PLoS One 5: e9429.

[195] Singh R, Saini N (2012) Downregulation of BCL2 by miRNAs augments drug induced apoptosis: Combined computational and experimental approach. J Cell Sci.

[196] Yang H, Kong W, He L, Zhao JJ, O'Donnell JD, et al. (2008) MicroRNA expression profiling in human ovarian cancer: miR-214 induces cell survival and cisplatin resistance by targeting PTEN. Cancer Res 68: 425-433.

[197] Marchal JA GM, Boulaiz H, Perán M, Álvarez P, Prados JC, Melguizo C, Aránega A (2012) Role of cancer stem cells of breast, colon, and melanoma tumors in the response to antitumor therapy. In: MA H, editor. Stem Cells and Cancer Stem Cells: Therapeutic Applications in Disease and Injury, Volume 3. New York: Springer. pp. 157-171.

[198] Nowell PC (1976) The clonal evolution of tumor cell populations. Science 194: 23-28.

[199] Tysnes BB (2010) Tumor-initiating and -propagating cells: cells that we would like to identify and control. Neoplasia 12: 506-515.

[200] Fulda S, Pervaiz S (2010) Apoptosis signaling in cancer stem cells. Int J Biochem Cell Biol 42: 31-38.

[201] Pannuti A, Foreman K, Rizzo P, Osipo C, Golde T, et al. (2010) Targeting Notch to target cancer stem cells. Clin Cancer Res 16: 3141-3152.

[202] Diaz-Cano SJ (2012) Tumor heterogeneity: mechanisms and bases for a reliable application of molecular marker design. Int J Mol Sci 13: 1951-2011.

[203] Barabe F, Kennedy JA, Hope KJ, Dick JE (2007) Modeling the initiation and progression of human acute leukemia in mice. Science 316: 600-604.

[204] Heppner GH (1984) Tumor heterogeneity. Cancer Res 44: 2259-2265.

[205] Lapidot T, Sirard C, Vormoor J, Murdoch B, Hoang T, et al. (1994) A cell initiating human acute myeloid leukaemia after transplantation into SCID mice. Nature 367: 645-648.

[206] Signore M, Ricci-Vitiani L, De Maria R (2011) Targeting apoptosis pathways in cancer stem cells. Cancer Lett.

[207] Bonnet D, Dick JE (1997) Human acute myeloid leukemia is organized as a hierarchy that originates from a primitive hematopoietic cell. Nat Med 3: 730-737.

[208] Bao S, Wu Q, McLendon RE, Hao Y, Shi Q, et al. (2006) Glioma stem cells promote radioresistance by preferential activation of the DNA damage response. Nature 444: 756-760.

[209] Ricci-Vitiani L, Lombardi DG, Pilozzi E, Biffoni M, Todaro M, et al. (2007) Identification and expansion of human colon-cancer-initiating cells. Nature 445: 111-115.

[210] Dalerba P, Dylla SJ, Park IK, Liu R, Wang X, et al. (2007) Phenotypic characterization of human colorectal cancer stem cells. Proc Natl Acad Sci U S A 104: 10158-10163.

[211] Al-Hajj M, Wicha MS, Benito-Hernandez A, Morrison SJ, Clarke MF (2003) Prospective identification of tumorigenic breast cancer cells. Proc Natl Acad Sci U S A 100: 3983-3988.

[212] Eramo A, Lotti F, Sette G, Pilozzi E, Biffoni M, et al. (2008) Identification and expansion of the tumorigenic lung cancer stem cell population. Cell Death Differ 15: 504-514.

[213] Hermann PC, Huber SL, Herrler T, Aicher A, Ellwart JW, et al. (2007) Distinct populations of cancer stem cells determine tumor growth and metastatic activity in human pancreatic cancer. Cell Stem Cell 1: 313-323.

[214] Yang ZF, Ho DW, Ng MN, Lau CK, Yu WC, et al. (2008) Significance of CD90+ cancer stem cells in human liver cancer. Cancer Cell 13: 153-166.

[215] Maitland NJ, Collins AT (2008) Prostate cancer stem cells: a new target for therapy. J Clin Oncol 26: 2862-2870.

[216] Chan KS, Espinosa I, Chao M, Wong D, Ailles L, et al. (2009) Identification, molecular characterization, clinical prognosis, and therapeutic targeting of human bladder tumor-initiating cells. Proc Natl Acad Sci U S A 106: 14016-14021.

[217] Zhang S, Balch C, Chan MW, Lai HC, Matei D, et al. (2008) Identification and characterization of ovarian cancer-initiating cells from primary human tumors. Cancer Res 68: 4311-4320.

[218] Prince ME, Sivanandan R, Kaczorowski A, Wolf GT, Kaplan MJ, et al. (2007) Identification of a subpopulation of cells with cancer stem cell properties in head and neck squamous cell carcinoma. Proc Natl Acad Sci U S A 104: 973-978.

[219] Schatton T, Murphy GF, Frank NY, Yamaura K, Waaga-Gasser AM, et al. (2008) Identification of cells initiating human melanomas. Nature 451: 345-349.

[220] Medina V, Calvo MB, Diaz-Prado S, Espada J (2009) Hedgehog signalling as a target in cancer stem cells. Clin Transl Oncol 11: 199-207.

[221] Frank NY, Schatton T, Frank MH (2010) The therapeutic promise of the cancer stem cell concept. J Clin Invest 120: 41-50.

[222] Sussman RT, Ricci MS, Hart LS, Sun SY, El-Deiry WS (2007) Chemotherapy-resistant side-population of colon cancer cells has a higher sensitivity to TRAIL than the non-SP, a higher expression of c-Myc and TRAIL-receptor DR4. Cancer Biol Ther 6: 1490-1495.

[223] Szliszka E, Mazur B, Zydowicz G, Czuba ZP, Krol W (2009) TRAIL-induced apoptosis and expression of death receptor TRAIL-R1 and TRAIL-R2 in bladder cancer cells. Folia Histochem Cytobiol 47: 579-585.

[224] Qi L, Bellail AC, Rossi MR, Zhang Z, Pang H, et al. (2011) Heterogeneity of primary glioblastoma cells in the expression of caspase-8 and the response to TRAIL-induced apoptosis. Apoptosis 16: 1150-1164.

[225] Loebinger MR, Sage EK, Davies D, Janes SM (2010) TRAIL-expressing mesenchymal stem cells kill the putative cancer stem cell population. Br J Cancer 103: 1692-1697.

[226] Dutton A, Young LS, Murray PG (2006) The role of cellular FLICE inhibitory protein (c-FLIP) in the pathogenesis and treatment of cancer. Expert Opin Ther Targets 10: 27-35.

[227] Piggott L, Omidvar N, Perez SM, Eberl M, Clarkson RW (2011) Suppression of apoptosis inhibitor c-FLIP selectively eliminates breast cancer stem cell activity in response to the anti-cancer agent, TRAIL. Breast Cancer Res 13: R88.

[228] Jin F, Zhao L, Guo YJ, Zhao WJ, Zhang H, et al. (2010) Influence of Etoposide on anti-apoptotic and multidrug resistance-associated protein genes in CD133 positive U251 glioblastoma stem-like cells. Brain Res 1336: 103-111.

[229] Mohr A, Albarenque SM, Deedigan L, Yu R, Reidy M, et al. (2010) Targeting of XIAP combined with systemic mesenchymal stem cell-mediated delivery of sTRAIL ligand inhibits metastatic growth of pancreatic carcinoma cells. Stem Cells 28: 2109-2120.

[230] Di Stefano AB, Iovino F, Lombardo Y, Eterno V, Hoger T, et al. (2010) Survivin is regulated by interleukin-4 in colon cancer stem cells. J Cell Physiol 225: 555-561.

[231] Madjd Z, Mehrjerdi AZ, Sharifi AM, Molanaei S, Shahzadi SZ, et al. (2009) CD44+ cancer cells express higher levels of the anti-apoptotic protein Bcl-2 in breast tumours. Cancer Immun 9: 4.

[232] Azmi AS, Wang Z, Philip PA, Mohammad RM, Sarkar FH (2011) Emerging Bcl-2 inhibitors for the treatment of cancer. Expert Opin Emerg Drugs 16: 59-70.

[233] Tagscherer KE, Fassl A, Campos B, Farhadi M, Kraemer A, et al. (2008) Apoptosis-based treatment of glioblastomas with ABT-737, a novel small molecule inhibitor of Bcl-2 family proteins. Oncogene 27: 6646-6656.

[234] Gonzalez-Herrero I, Vicente-Duenas C, Orfao A, Flores T, Jimenez R, et al. (2010) Bcl2 is not required for the development and maintenance of leukemia stem cells in mice. Carcinogenesis 31: 1292-1297.

[235] Xiao G, Fu J (2011) NF-kappaB and cancer: a paradigm of Yin-Yang. Am J Cancer Res 1: 192-221.

[236] Xiong HQ, Abbruzzese JL, Lin E, Wang L, Zheng L, et al. (2004) NF-kappaB activity blockade impairs the angiogenic potential of human pancreatic cancer cells. Int J Cancer 108: 181-188.

[237] Wharry CE, Haines KM, Carroll RG, May MJ (2009) Constitutive non-canonical NFkappaB signaling in pancreatic cancer cells. Cancer Biol Ther 8: 1567-1576.

[238] Ju JH, Jang K, Lee KM, Kim M, Kim J, et al. (2011) CD24 enhances DNA damage-induced apoptosis by modulating NF-kappaB signaling in CD44-expressing breast cancer cells. Carcinogenesis 32: 1474-1483.

[239] Katsman A, Umezawa K, Bonavida B (2009) Chemosensitization and immunosensitization of resistant cancer cells to apoptosis and inhibition of metastasis by the specific NF-kappaB inhibitor DHMEQ. Curr Pharm Des 15: 792-808.

[240] Rajasekhar VK, Studer L, Gerald W, Socci ND, Scher HI (2011) Tumour-initiating stem-like cells in human prostate cancer exhibit increased NF-kappaB signalling. Nat Commun 2: 162.

[241] Chen W, Li Z, Bai L, Lin Y (2011) NF-kappaB in lung cancer, a carcinogenesis mediator and a prevention and therapy target. Front Biosci 16: 1172-1185.

[242] Spiller SE, Logsdon NJ, Deckard LA, Sontheimer H (2011) Inhibition of nuclear factor kappa-B signaling reduces growth in medulloblastoma in vivo. BMC Cancer 11: 136.

[243] Nogueira L, Ruiz-Ontanon P, Vazquez-Barquero A, Moris F, Fernandez-Luna JL (2011) The NFKappaB pathway: a therapeutic target in glioblastoma. Oncotarget 2: 646-653.

[244] Whibley C, Pharoah PD, Hollstein M (2009) p53 polymorphisms: cancer implications. Nat Rev Cancer 9: 95-107.

[245] Godar S, Ince TA, Bell GW, Feldser D, Donaher JL, et al. (2008) Growth-inhibitory and tumor- suppressive functions of p53 depend on its repression of CD44 expression. Cell 134: 62-73.

[246] Viale A, De Franco F, Orleth A, Cambiaghi V, Giuliani V, et al. (2009) Cell-cycle restriction limits DNA damage and maintains self-renewal of leukaemia stem cells. Nature 457: 51-56.

[247] Ji Q, Hao X, Zhang M, Tang W, Yang M, et al. (2009) MicroRNA miR-34 inhibits human pancreatic cancer tumor-initiating cells. PLoS One 4: e6816.

[248] Ji Q, Hao X, Meng Y, Zhang M, Desano J, et al. (2008) Restoration of tumor suppressor miR-34 inhibits human p53-mutant gastric cancer tumorspheres. BMC Cancer 8: 266.

Chapter 4

Cell Death and Anti-DNA Antibodies

Yasuhiko Hirabayashi

Additional information is available at the end of the chapter

1. Introduction

Autoantibodies are characteristic features of autoimmune diseases (Table 1). In organ or tissue-specific autoimmune diseases, autoantibodies against cell-surface molecules are usually observed. These antibodies (Abs) stimulate or damage the target cells and cause organ- or tissue-specific diseases. In systemic autoimmune diseases, in addition to anti-cell-surface molecule Abs, Abs against intracellular molecules are frequently observed although B cell tolerance to intracellular molecules is strictly enforced in normal subjects. Some indicate high disease specificity with a high incidence rate. Therefore, such Abs may be closely associated with development of the disease as well as with disease activity. However, it is not known how or why Abs against intracellular molecules are generated.

2. Role of cell death in the generation of anti-intracellular molecule antibodies

Cell death, including apoptosis and necrosis, represents a possible source of exposure of intracellular molecules outside the cell. For example, low – intermediate doses (< 35 mJ/cm^2) of ultraviolet B (UVB) induce apoptosis of keratinocytes, resulting in translocation of native DNA, Ku, and Sm to the cytoplasmic membrane, while a high dose (80 mJ/cm^2) of UVB induces necrosis, resulting in discharge of all of the cell compartments [1]. Intracellular molecules are exposed on the surface blebs of apoptotic cells (apoptotic blebs) [2]. Apoptotic blebs contain fragmented endoplasmic reticulum (ER), ribosomes, ribonucleoprotein, nucleosomal DNA, Ro, La, small nuclear ribonucleoproteins, etc. Autoantigens receive various epigenetic modifications (acetylation, methylation, phosphorylation, dephosphorylation, ADP-ribosylation, ubiquitination, oxidation, transglutamination, citrullination, SUMOylation, etc) during apoptotic cell death [3]. In the case of cytotoxic granule-mediated cell death, granzymes plays an important role in cleavage of autoantigens [4]. These epigenetic modifications may alter preexisting epitopes, expose cryptic epitopes,

Anti-	Target molecule	Disease
1. Organ or tissue-specific autoimmune diseases		
a. Anti-cell-surface molecules		
TSH receptor*	TSH receptor	Graves' disease
NMDA	N-methyl-D-aspartate receptor	Encephalitis
GAD	glutamic acid decarboxylase	Diabetes mellitus type I
Ach*	Acetylcholine	Myasthenia gravis
myelin associated protein	myelin	Multiple sclerosis
ganglioside	ganglioside	Neuropathy

b. Anti-intracellular molecules

mitochondria	mitochondria	PBC
thyroid microsomal	thyroid microsomal	Hashimoto's thyroiditis
thyroid peroxidase	thyroid peroxidase	Hashimoto's thyroiditis
thyroglobulin	thyroglobulin	Hashimoto's thyroiditis

2. Systemic autoimmune diseases		
a. Anti-cell-surface molecules		
PDGF receptor	PDGF receptor	Systemic sclerosis
phospholipid*	phospholipid	APS
b. Anti-intracellular molecules		
dsDNA*	dsDNA	SLE
Sm	Smith	SLE
SS-B	La/SS-B	Sjögren's syndrome
centromere	centromere	Systemic sclerosis
topoisomerase-I	topoisomerase-I	Systemic sclerosis
Jo-1	histidyl-tRNA synthetase	DM
PR3-ANCA*	proteinase 3	WG
MPO-ANCA*	myeloperoxidase	MPA, AGA

TSH: thyroid stimulating hormone; PBC: Primary biliary cirrhosis; PDGF: Platelet-derived growth factor; APS: Anti-phospholipid syndrome; dsDNA: double-stranded DNA; SLE: systemic lupus erythematosus; DM: dermatomyositis; WG: Wegener's granulomatosis; MPA: microscopic polyarteritis; AGA: allergic granulomatous angiitis (Churg–Strauss syndrome).
*: The titer of the Ab is correlated with the disease activity in a proportion of patients.

Table 1. Examples of autoantibodies.

or form novel epitopes, and may contribute to bypassing tolerance to autoantigens [5]. Normally, apoptotic cells are quickly eliminated by professional phagocytes. Delay of apoptotic cell clearance not only increases the time of exposure of intracellular molecules to the immune system but also changes the degree of modification of these molecules, which alters their antigenicity. When clearance fails, apoptotic cells enter the stage of secondary necrosis. The ability to cause inflammation depends on the stage of cell death [6]. Damage-associated molecular patterns (DAMPs), such as HMGB1, SAP130, etc., are released from late apoptotic/necrotic cells into the extracellular space [7-9]. DAMPs activate Toll-like receptors and act as intrinsic adjuvants, resulting in inflammation and initiation of the host immune system. Thus, delay of apoptotic cell clearance increases the risk of an autoimmune response.

3. Abnormalities related to apoptosis in SLE

There are many lupus autoantibodies that bind to autoantigens of apoptotic cells [10]. In SLE, defective clearance of apoptotic cells has been reported. As a result, high levels of circulating early apoptotic cells are found in SLE [11]. T-lymphocytes [12], macrophages/monocytes [13,14], neutrophils [15], and endothelial cells [16] are included among the increased numbers of apoptotic cells. Monocytes and granulocytes, which take up autoantibody remnants of secondary necrotic cell complex, secrete inflammatory cytokines in SLE [17]. These phenomena threaten self-tolerance and are likely involved in the production of lupus autoantibodies [18,19]. The reason for defective apoptotic cell clearance in SLE has yet to be elucidated. It has been suggested that the efficacy of clearance is affected by the cell death trigger, but there have been no reports related to SLE from this viewpoint [20]. Anti-class A scavenger receptor autoantibodies from patients with SLE impair the clearance of apoptotic debris by macrophages [21]. However, the mechanism of autoantibody production is not yet known.

In SLE, the response to early apoptotic cells is also abnormal. Under normal conditions, macrophages secrete antiinflammatory cytokines (IL-10, TGF-β, PGE2, etc.) to send "tolerate me" signals after ingestion of apoptotic cells [22]. Monocytes from healthy controls showed prominent TGF-β secretion and minimal TNF-α production, but monocytes from SLE patients show prominent TNF-α production and diminished TGF-β secretion [23]. The authors speculated that this abnormal response may be an intrinsic property of lupus monocytes.

Recent studies have highlighted the role of neutrophils in the pathogenesis and manifestations of SLE [24]. NETosis is a process characterized by the formation of neutrophil extracellular traps (NETs) [25]. NETs become not only a source of intracellular molecules but also immunogens for lupus autoantibodies [26]. Low-density granulocytes (LDGs), an abnormal subset of neutrophils, were identified among the PBMCs derived from patients with SLE [27]. LDGs secrete type I interferons (INFs), have endothelial cytotoxicity, and have higher capacity to form NETs [28]. Degradation of NETs is impaired in patients with SLE [29]. NETs activate complement and deposited C1q inhibits NET degradation [30]. These phenomena increase NETs and may contribute to the production of autoantibodies, including anti-DNA Abs.

It has recently been demonstrated that the source of intracellular molecules is microparticles (MPs), which are small membrane-bound vesicles [31]. MPs, which emerge from the cell membrane during cell activation and apoptosis, contain a variety of cellular components, including nucleic acids [32]. MPs become antigenic targets of anti-DNA Abs, form huge immune complexes in the plasma of patients with SLE, and induce complement activation [33,34].

4. Anti-dsDNA Abs

4.1. Brief historical aspects of anti-dsDNA Abs

SLE is characterized by the production of a variety of autoantibodies. Especially, anti-dsDNA Abs are the most characteristic of SLE and contribute to the pathogenesis of lupus nephritis. In general, anti-dsDNA Abs are specific for SLE and the anti-dsDNA Abs titer is closely correlated to the activity of lupus nephritis [35]. A proportion of anti-dsDNA Abs are directly involved in immune complex-mediated glomerulonephritis [36]. Thus, the trigger of anti-DNA response may be closely related to the pathogenesis of SLE. However, mammalian native dsDNA is not immunogenic, suggesting that DNA itself does not act as a triggering or driving antigen [37]. The origin of anti-DNA Abs is a long-standing enigma.

Anti-dsDNA responses can be evoked by dsDNA with the aid of a carrier, such as the 27-amino acid nucleic acid-binding Fus1 peptide [38], polyoma BK virus large T Ag [39], or DNaseI-dsDNA complex [40] which have been shown to induce production of anti-dsDNA Abs in mice, suggesting a possible role of excess amounts of DNA – protein complex in disruption of tolerance to DNA. Nucleosomes have been suggested as possible Ags responsible for triggering of anti-dsDNA Abs [41,42]. Crude nucleosomes or crude histones [41] have been shown to induce production of anti-dsDNA Abs in mice. Mononucleosome-reactive Th clones augment the production of IgG autoantibodies to dsDNA, histones, and histone – DNA complex. However, immunization of SNF$_1$ mice with pure mononucleosomes did not elicit production of IgG anti-dsDNA Abs [43]. HMGB1 – nucleosome complexes derived from apoptotic cells, but not HMGB1-free nucleosomes, elicited IgG anti-dsDNA Abs in BALB/c mice although their titer was not high, suggesting that adjuvants such as HMGB1 are necessary to break tolerance to dsDNA in non-autoimmune mice [44].

Another possible mechanism is molecular mimicry. Some mouse or human monoclonal anti-DNA Abs have been shown to cross-react with non-nucleic acid self-Ags, such as extracellular matrix protein HP8 [45], heterogeneous nuclear ribonucleoprotein A2 [46], NR2 glutamate receptor [47], α-actinin [48,49], ribosomal protein S1 [50], and phospholipids, including cardiolipin [51]. However, it is not yet known whether these molecules can elicit anti-DNA responses.

The peptide, DWEYSVWLSN, is recognized by the R4A mouse monoclonal anti-dsDNA Ab [52]. Immunization with this peptide elicited anti-dsDNA Ab production and caused deposition of IgG in glomeruli in normal mice [53]. These observations indicate that a non-nucleic acid Ag can elicit production of anti-DNA Abs and cause renal disorder in normal animals. However, no proteins containing this peptide sequence have been reported to date.

It should be noted that immunization with recombinant EBNA-1 protein elicited anti-EBNA-1 Abs that cross-react with dsDNA, suggesting molecular mimicry between the viral antigen and dsDNA [54]. However, nephritogenicity of the anti-EBNA-1/dsDNA Abs has not been reported.

4.2. Cross-reactive antigen of the O-81 human nephritogenic anti-DNA mAb

We prepared human monoclonal anti-DNA Ab, O-81, which binds strongly to single-stranded DNA (ssDNA) and moderately to dsDNA, and demonstrated that the O-81 idiotype (Id) is distributed among IgG anti-DNA Abs of circulating immune complexes as well as lupus glomerular deposits [55-58]. The intravenous infusion of IgG isotype anti-DNA Abs expressing O-81 Id also caused glomerular IgG deposition in SCID mice [59]. The VH region of O-81 Ab contains many somatic mutations [60]. Similarly, the VH regions of O-81 Id-positive B cells in patients with SLE were shown to already contain somatic mutations [61]. These observations prompted us to explore the triggering Ags for human nephritogenic anti-DNA Abs using the O-81 Ab.

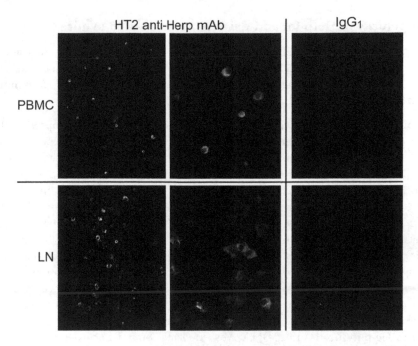

Figure 1. The expression of Herp in peripheral blood mononuclear cells (PBMCs) or the cells in a cervical lymph node (LN) from a patient who developed SLE and had yet to receive treatment.

[Methods] The cells were fixed in 50% acetone/50% methanol for 20 min at −20°C and blocked with 5% normal goat serum and 3% BSA in PBS overnight at 4°C. The cells were then incubated with HT2 mouse monoclonal IgG$_1$ anti-Herp Ab or mouse IgG$_1$ as an isotype control for 1 h at room temperature followed by incubation with FITC-conjugated goat F(ab')$_2$ anti-mouse IgG Ab (KPL, Gaithersburg, MD) for 1 h at room temperature [63].

We found that the O-81 Ab specifically cross-reacts with human homocysteine-induced endoplasmic reticulum protein (Herp) [62]. Anti-dsDNA Abs purified from the sera of SLE patients bound to Herp, and anti-Herp Abs purified from the sera of SLE bound to dsDNA [62]. The production of Herp is induced by endoplasmic reticulum (ER) stress. The PBLs from subjects in active SLE, especially at the time of onset or flare-up of the disease, tended to show Herp expression [62]. The expression of Herp was also observed in the lymph node of an untreated patient with active SLE, indicating that Herp can be exposed to the immune system in lymph nodes where Ag recognition occurs (Figure 1).

Excessive ER stress is known to induce apoptosis [64,65]. Herp can be exposed on apoptotic blebs of ER stress-induced apoptotic cells [62]. Many apoptotic cells expressing Herp were observed in the peripheral blood mononuclear cells (PBMCs) of patients with active SLE, but not normal control subjects [62]. This observation is compatible with those reported previously [11]. These results suggest that Abs against Herp on ER stress-induced apoptotic cells may become anti-Herp/dsDNA cross-reactive Abs, i.e., initial anti-dsDNA Abs.

4.3. Antigenicity of Herp for anti-dsDNA Ab production in mice

Immunization of normal BALB/c mice with Herp elicited anti-dsDNA Abs and caused glomerular IgG deposition [62]. However, urinary protein level did not increase and overt nephritis did not develop. The pathological changes in the kidneys in Herp-immunized BALB/c mice went no further than silent lupus nephritis.

Nucleosomes, which are major autoantigens in SLE, are exposed at the apoptotic cell surface [66,67]. Anti-nucleosome Abs are present in SLE at a rate of more than 50% and they have been linked to lupus nephritis [68]. Nucleosomes and histones are present in glomerular deposits [69]. Nucleosomes bind to glomerular endothelial cells and serve as targets for anti-nucleosome Abs [70]. Therefore, a portion of anti-nucleosome Abs may be involved in lupus nephritis [71]. However, even oligonucleosomes are much less effective than Herp in inducing anti-nucleosome Abs as well as anti-dsDNA Abs [62]. Therefore, to reproduce overt lupus nephritis, BALB/c mice were immunized with Herp followed by immunization with oligonucleosomes. In this procedure, both anti-dsDNA Ab and anti-nucleosome Ab-producing clones induced by Herp may be able to recognize oligonucleosomes easily. The production of anti-dsDNA Abs and glomerular IgG deposition were observed in all mice. In addition, overt nephritis with significant proteinuria occurred in one mouse (Figure 2). Although further investigations are in progress to define the mechanisms, it was speculated that (i) the Herp-induced anti-dsDNA Abs efficiently bound to nucleosomes and formed pathogenic immune complexes, and (ii) affinity maturation and epitope spreading of Herp-induced anti-dsDNA Abs occurred by nucleosomes.

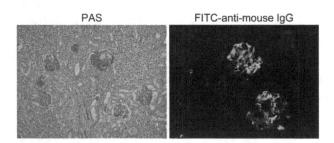

Figure 2. Overt nephritis in a BALB/c mouse immunized with Herp followed by immunization with oligonucleosomes. Left: Periodic acid Schiff (PAS) staining. Right: Immunofluorescence staining with fluorescein isothiocyanate (FITC)-conjugated anti-mouse IgG Ab.

[Methods] Five 6-week-old female BALB/c mice were immunized intraperitoneally with 100 μg of Herp on days 0 and 10 and 50 μg of Herp on day 20, followed by immunization with 10 μg of oligonucleosomes on days 30, 40, and 50. Preparation of Herp and oligonucleosomes was described previously [62]. Fresh-frozen tissue sections 4 μm thick were fixed in 100% acetone for 10 min at 4°C and blocked with 5% normal goat serum and 3% BSA in PBS overnight at 4°C. Sections were stained with FITC-conjugated goat F(ab')2 anti-mouse IgG Ab (KPL) for 1 h at room temperature.

4.4. Antigenicity of Herp for anti-dsDNA Ab production in humans

To examine whether Herp can be an antigen for anti-dsDNA Ab production in humans, ELISPOT was performed using PBMCs (representative cases are shown in Figure 3). The number of spots increased when the PBMCs were incubated with Herp, but not with dsDNA, in 4 of 6 untreated active SLE patients (Figure 3A); in 2 of these 4 positive cases, a few spots were observed even in wells without stimulation (Figure 3B). The remaining two cases showed no spots (Figure 3C). On the other hand, no spots were detected in the PBMCs from nine treated active SLE patients, eight inactive SLE patients, and five normal control subjects (data not shown). These results suggest that Herp can stimulate anti-dsDNA antibody-producing clones but this stimulation is cancelled by immunosuppressive therapy.

[Methods] Approximately 1×10^6 PBMCs in 20% fetal calf serum (FCS)-supplemented RPMI 1640 (20% FCS-RPMI 1640) were cultured for 5 days with or without 2 μg/mL Herp, or 10 μg/mL dsDNA. For preparation of dsDNA, calf thymus DNA (Invitrogen, Carlsbad, CA) was pretreated with S1 nuclease (Takara Bio, Otsu, Japan) to remove single-stranded DNA (ssDNA) according to the manufacturer's instructions. Multiscreen 96-well filtration plates (Millipore, Billerica, MA) were coated with 10 mg/mL protamine overnight at 4°C, and washed with PBS followed by coating with 10 μg/mL dsDNA in PBS for 2 h at room temperature. Following blocking with 20% FCS-RPMI 1640, the cultured PBMCs in 20% FCS-RPMI were plated at 1×10^5 cells/well and cultured for 24 h. After washing the cells with PBS, goat alkaline phosphatase-conjugated anti-human IgG antibodies (diluted 1:10000; Sigma-Aldrich, St. Louis, MO) were added and the wells were incubated for 1 h at room temperature. Following a further wash, the spots were visualized using NBT-5-bromo-4-chloro-3-indolyl phosphate substrate (Sigma-Aldrich).

5. Anti-single-stranded DNA (ssDNA) Abs

The mechanism involved in the production of anti-ssDNA Abs has yet to be elucidated. As ssDNA can have multiple conformational epitopes and all Abs that bind to ssDNA are called anti-ssDNA Abs, these Abs are highly heterogeneous and display low disease specificity. However, the susceptibility of lupus-inducing drugs to anti-ssDNA Ab production is very high. In such cases, there may be a unique mechanism of anti-ssDNA Ab production, as the chemical structures and pharmacological actions of lupus-inducing drugs are known to be highly diverse [72,73]. As higher risk drugs include procainamide and hydralazine, which inhibit DNA methylation, hypomethylation may be one of the causes of anti-ssDNA Ab production, but its precise mechanism remains unknown [74-76].

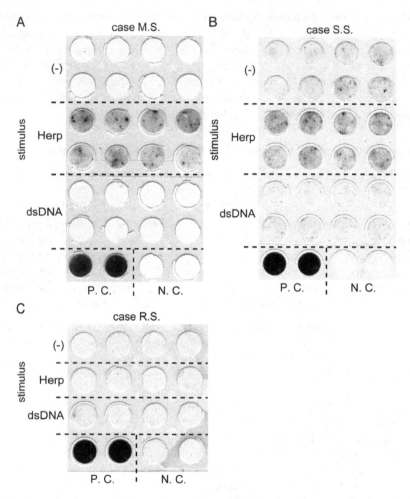

Figure 3. Herp can stimulate anti-dsDNA Ab-producing B cells in untreated patients with active SLE. The PBMCs were stimulated with Herp or dsDNA. Anti-dsDNA Ab-producing PBMCs were detected with ELISPOT. Three representative cases (A, Case M. S.; B, Case S. S.; C, Case R. S.) are shown. The lowest row is the positive control (P. C.: human serum with anti-dsDNA Abs, diluted 1:200) and negative control (N. C.; second antibody only).

Immunization with Herp elicits production of not only anti-dsDNA antibodies but also anti-ssDNA antibodies in BALB/c mice. Among several anti-Herp mAbs established in our laboratory, the HT4 anti-Herp mAb cross-reacts specifically with ssDNA [77]. The epitope of the HT4 mAb on Herp, EPAGSNR, was identified by screening a synthetic peptide library. The binding of HT4 mAb to the peptide was competitively inhibited by ssDNA. Immunization of the epitope peptide elicited anti-ssDNA Abs in BALB/c mice. Treatment with chlorpromazine, procainamide, and hydralazine induced Herp expression and apoptosis in HeLa cells. These findings suggest that (i) ER stress and apoptosis by drugs and

(ii) molecular mimicry between Herp and ssDNA are involved in anti-ssDNA antibody production in drug-induced lupus.

6. Postulated mechanism of anti-DNA Ab generation

Autoimmunity is associated with both genetic predisposition and environmental factors [78]. The monozygotic disease concordance rate ranges from 24% to 57% (and not 100%) for SLE [79]. Most patients with SLE are non-familial sporadic cases. That is, environmental etiologies of SLE may be common. It is well known that environmental factors such as viral infection, UV exposure, chemicals, etc., can trigger clinical onset or flare of SLE [80,81]. However, little is known regarding how those factors elicit anti-DNA antibody production in vivo. These factors, i.e., cell stressors, affect the expression patterns of cellular proteins, resulting in ER stress in some cases.

What is a practical model of this hypothesis? Natural infection with viruses can cause ER stress on a large scale in vivo [80]. ER stress has been shown to increase when viral proteins are produced at high levels, e.g., in virion formation during the active lytic cycle of infection. Epstein–Barr virus (EBV) infection has been suggested to have a causative role in SLE [82,83]. The titers of anti-EBV Abs in SLE patients are higher than those of healthy controls [84]. Kang et al. reported that: (i) patients with SLE had an approximately 40-fold increase in EBV viral load compared with controls; (ii) the frequency of EBV-specific CD69+ CD8+ T cells producing IFN-γ was higher in patients with SLE than in controls, but the frequency of EBV-specific CD69+ CD4+ T cells producing IFN-γ was lower in patients with SLE than in controls; and (iii) the EBV viral loads were positively correlated with the frequency of EBV-specific CD69+ CD8+ T cells but inversely correlated with the frequency of EBV-specific CD69+ CD4+ T cells [85]. Larsen et al. reported that EBV-specific CD8+ T cell responses in patients with SLE are functionally impaired [86]. The defective control of latent EBV infection in patients with SLE may result in recurrent reactivation of EBV. In fact, aberrant expression of BZLF1, which is a hallmark of EBV lytic infection, has been detected in the PBMCs of SLE patients [87]. In primary EBV infection, EBV infects tonsillar B cells in which lytic replication occurs, and differentiation of latently EBV-infected B cells to plasma cells in lymphoid tissues is associated with induction of the EBV lytic cycle [88]. Herp is expressed in BZLF1-positive EBV-infected B cells (Figure 4).

[Methods] EBV-transformed B cells were fixed in 50% acetone/50% methanol for 20 min at − 20°C and blocked with 5% normal goat serum and 3% BSA in PBS overnight at 4°C. The cells were stained with DAPI. The cells were then co-stained with HT4 mouse IgG2a anti-human Herp mAb and mouse IgG1 anti-BZLF1 mAb (Dako, Glostrup, Denmark) for 1 h at room temperature followed by co-staining with rhodamine-conjugated goat anti-mouse IgG2a Ab (Santa Cruz Biotechnology, Santa Cruz, CA) and FITC-conjugated goat anti-mouse IgG1 Ab (Santa Cruz Biotechnology) for 1 h at room temperature.

ER stress, which is induced by the production of viral proteins, causes EBV lytic replication, resulting in the release of virions and intracellular molecules [89]. The Herp produced in cells entering the lytic phase of EBV infection can be recognized by the immune system in

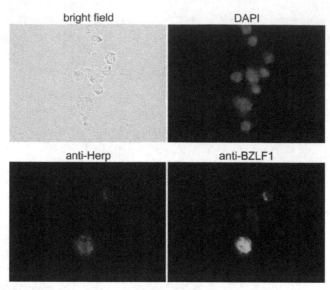

Figure 4. Herp protein is expressed in BZLF1 positive EBV infected B cells.

lymphoid tissues. In addition, EBV-encoded latent membrane protein 2A (LMP2A) induces hypersensitivity to TLR stimulation, leading to activation of autoreactive B cells through the BCR/TLR pathway [90]. Immunization with the membrane fraction of EBV-transformed B cells elicited anti-dsDNA Abs as well as anti-Herp Abs and causes glomerular IgG deposition in BALB/c mice [62]. These observations support the hypothesis that EBV infection may be a trigger of SLE.

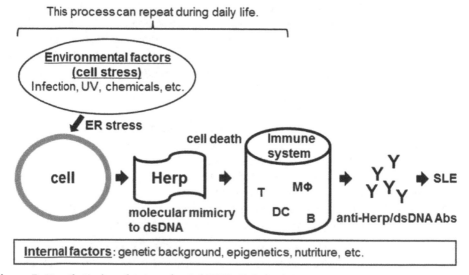

Figure 5. Hypothetical mechanism of anti-dsDNA Ab induction.

The results of the present study led to the following hypothesis in which cell stress triggers an anti-DNA response via Herp in normal individuals: ER stress by environmental factors → Herp expression → recognition by the immune system of minor epitope(s) mimicking ssDNA or dsDNA → anti-Herp/ssDNA or dsDNA cross-reactive Abs (initial anti-ssDNA or anti-dsDNA Abs) → anti-DNA Ab – DNA complex, anti-DNA Ab – nucleosome complex → ?→ tissue injury (Figure 5). After the initial production of anti-Herp/DNA Abs, the production of the Abs is stimulated whenever ER stress-induced apoptosis occurs, and the cause of ER stress is not restricted. Repeated cell stress during daily life may strengthen this pathway. Herp is a good candidate as a link between common environmental factors and the etiology of SLE.

7. Conclusions

As dead cells are not only a source of intracellular antigens but also a source of proinflammatory molecules, it is likely that they play an important role in the generation of nephritogenic anti-dsDNA Abs. Herp was identified as a molecule directly involved in cell stress/death as the cause of anti-dsDNA Ab production. Further investigations are therefore needed to clarify the relationship between cell stress/death and the etiology of SLE.

Author details

Yasuhiko Hirabayashi
Department of Rheumatology, Hikarigaoka Spellman Hospital, Japan
Department of Hematology & Rheumatology, Tohoku University Hospital, Japan

8. References

[1] Caricchio R, McPhie L, Cohen PL (2003) Ultraviolet B radiation-induced cell death: critical role of ultraviolet dose in inflammation and lupus autoantigen redistribution. J. Immunol. 171: 5778-5786.

[2] Casciola-Rosen LA, Anhalt G, Rosen A (1994) Autoantigens targeted in systemic lupus erythematosus are clustered in two populations of surface structures on apoptotic keratinocytes. J. Exp. Med. 179: 1317-1330.

[3] Utz PJ, Gensler TJ, Anderson P (2000) Death, autoantigen modifications, and tolerance. Arthritis Res. 2: 101-114.

[4] Darrah E, Rosen A (2010) Granzyme B cleavage of autoantigens in autoimmunity. Cell Death Differ. 17: 624-632.

[5] Brooks WH, Le Dantec C, Pers JO, Youinou P, Renaudineau Y (2010) Epigenetics and autoimmunity. J. Autoimmun. 34: J207-J219.

[6] Poon IK, Hulett MD, Parish CR (2010) Molecular mechanisms of late apoptotic/necrotic cell clearance. Cell Death Differ. 17: 381-397.

[7] Chen GY, Nuñez G (2010) Sterile inflammation: sensing and reacting to damage. Nat. Rev. Immunol. 10: 826-837.

[8] Pisetsky D (2011) Cell death in the pathogenesis of immune-mediated diseases: the role of HMGB1 and DAMP-PAMP complexes. Swiss Med. Wkly. 141: w13256.

[9] Harris HE, Andersson U, Pisetsky DS (2012) HMGB1: A multifunctional alarmin driving autoimmune and inflammatory disease. Nat Rev Rheumatol 8: 195-202.

[10] Huggins ML, Todd I, Cavers MA, Pavuluri SR, Tighe PJ, Powell RJ (1999) Antibodies from systemic lupus erythematosus (SLE) sera define differential release of autoantigens from cell lines undergoing apoptosis. Clin. Exp. Immunol. 118: 322-328.

[11] Perniok A, Wedekind F, Herrmann M, Specker C, Schneider M (1998) High levels of circulating early apoptic peripheral blood mononuclear cells in systemic lupus erythematosus. Lupus 7: 113-118.

[12] Gröndal G, Traustadottir KH, Kristjansdottir H, Lundberg I, Klareskog L, Erlendsson K, Steinsson K (2002) Increased T-lymphocyte apoptosis/necrosis and IL-10 producing cells in patients and their spouses in Icelandic systemic lupus erythematosus multicase families. Lupus 11: 435-442.

[13] Ren Y, Tang J, Mok MY, Chan AW, Wu A, Lau CS (2003) Increased apoptotic neutrophils and macrophages and impaired macrophage phagocytic clearance of apoptotic neutrophils in systemic lupus erythematosus. Arthritis Rheum. 48: 2888-2897.

[14] Kaplan MJ, Lewis EE, Shelden EA, Somers E, Pavlic R, McCune WJ, Richardson BC (2002) The apoptotic ligands TRAIL, TWEAK, and Fas ligand mediate monocyte death induced by autologous lupus T cells. J. Immunol. 169: 6020-6029.

[15] Courtney PA, Crockard AD, Williamson K, Irvine AE, Kennedy RJ, Bell AL (1999) Increased apoptotic peripheral blood neutrophils in systemic lupus erythematosus: relations with disease activity, antibodies to double stranded DNA, and neutropenia. Ann. Rheum. Dis. 58: 309-314.

[16] Kaplan MJ (2004) Apoptosis in systemic lupus erythematosus. Clin. Immunol. 112: 210-218.

[17] Muñoz LE, Janko C, Grossmayer GE, Frey B, Voll RE, Kern P, Kalden JR, Schett G, Fietkau R, Herrmann M and Gaipl US, (2009) Remnants of secondarily necrotic cells fuel inflammation in systemic lupus erythematosus. Arthritis Rheum. 60: 1733-1742.

[18] Muñoz LE, Lauber K, Schiller M, Manfredi AA, Herrmann M (2010) The role of defective clearance of apoptotic cells in systemic autoimmunity. Nat Rev Rheumatol 6: 280-289.

[19] Muñoz LE, Janko C, Schulze C, Schorn C, Sarter K, Schett G, Herrmann M (2010) Autoimmunity and chronic inflammation - two clearance-related steps in the etiopathogenesis of SLE. Autoimmun. Rev. 10: 38-42.

[20] Wiegand UK, Corbach S, Prescott AR, Savill J, Spruce BA (2001) The trigger to cell death determines the efficiency with which dying cells are cleared by neighbours. Cell Death Differ. 8: 734-746.

[21] Chen XW, Shen Y, Sun CY, Wu FX, Chen Y, Yang CD (2011) Anti-class a scavenger receptor autoantibodies from systemic lupus erythematosus patients impair phagocytic clearance of apoptotic cells by macrophages in vitro. Arthritis Res. Ther. 13: R9.

[22] Voll RE, Herrmann M, Roth EA, Stach C, Kalden JR, Girkontaite I (1997) Immunosuppressive effects of apoptotic cells. Nature 390: 350-351.

[23] Sule S, Rosen A, Petri M, Akhter E, Andrade F (2011) Abnormal production of pro- and anti-inflammatory cytokines by lupus monocytes in response to apoptotic cells. PLoS One 6: e17495.

[24] Kaplan MJ, (2011) Neutrophils in the pathogenesis and manifestations of SLE. Nat Rev Rheumatol 7: 691-699.

[25] Amulic B, Cazalet C, Hayes GL, Metzler KD, Zychlinsky A (2012) Neutrophil function: from mechanisms to disease. Annu. Rev. Immunol. 30: 459-489.

[26] Liu CL, Tangsombatvisit S, Rosenberg JM, Mandelbaum G, Gillespie EC, Gozani OP, Alizadeh AA, Utz PJ (2012) Specific post-translational histone modifications of neutrophil extracellular traps as immunogens and potential targets of lupus autoantibodies. Arthritis Res. Ther. 14: R25.

[27] Denny MF, Yalavarthi S, Zhao W, Thacker SG, Anderson M, Sandy AR, McCune WJ, Kaplan MJ (2010) A distinct subset of proinflammatory neutrophils isolated from patients with systemic lupus erythematosus induces vascular damage and synthesizes type I IFNs. J. Immunol. 184: 3284-3297.

[28] Villanueva E, Yalavarthi S, Berthier CC, Hodgin JB, Khandpur R, Lin AM, Rubin CJ, Zhao W, Olsen SH, Klinker M, Shealy D, Denny MF, Plumas J, Chaperot L, Kretzler M, Bruce AT, Kaplan MJ (2011) Netting neutrophils induce endothelial damage, infiltrate tissues, and expose immunostimulatory molecules in systemic lupus erythematosus. J. Immunol. 187: 538-552.

[29] Hakkim A, Fürnrohr BG, Amann K, Laube B, Abed UA, Brinkmann V, Herrmann M, Voll RE, Zychlinsky A, (2010) Impairment of neutrophil extracellular trap degradation is associated with lupus nephritis. Proc. Natl. Acad. Sci. U. S. A. 107: 9813-9818.

[30] Leffler J, Martin M, Gullstrand B, Tydén H, Lood C, Truedsson L, Bengtsson AA, Blom AM (2012) Neutrophil extracellular traps that are not degraded in systemic lupus erythematosus activate complement exacerbating the disease. J. Immunol. 188: 3522-3531.

[31] Beyer C, Pisetsky DS (2010) The role of microparticles in the pathogenesis of rheumatic diseases. Nat Rev Rheumatol 6: 21-29.

[32] Pisetsky DS, Gauley J, Ullal AJ (2011) Microparticles as a source of extracellular DNA. Immunol. Res. 49: 227-234.

[33] Ullal AJ, Reich CF, Clowse M, Criscione-Schreiber LG, Tochacek M, Monestier M, Pisetsky DS (2011) Microparticles as antigenic targets of antibodies to DNA and nucleosomes in systemic lupus erythematosus. J. Autoimmun. 36: 173-180.

[34] Pisetsky DS (2012) Microparticles as autoantigens: making immune complexes big. Arthritis Rheum. 64: 958-961.

[35] Hughes GR, Cohen SA, Christian CL (1971) Anti-DNA activity in systemic lupus erythematosus. A diagnostic and therapeutic guide. Ann. Rheum. Dis. 30: 259-264.

[36] Winfield JB, Faiferman I, Koffler D (1977) Avidity of anti-DNA antibodies in serum and IgG glomerular eluates from patients with systemic lupus erythematosus. Association of high avidity antinative DNA antibody with glomerulonephritis. J. Clin. Invest. 59: 90-96.

[37] Madaio MP, Hodder S, Schwartz RS, Stollar BD (1984) Responsiveness of autoimmune and normal mice to nucleic acid antigens. J. Immunol. 132: 872-876.

[38] Desai DD, Krishnan MR, Swindle JT, Marion TN (1993) Antigen-specific induction of antibodies against native mammalian DNA in nonautoimmune mice. J. Immunol. 151: 1614-1626.

[39] Rekvig OP, Moens U, Fredriksen K, Traavik T (1997) Human polyomavirus BK and immunogenicity of mammalian DNA: a conceptual framework. Methods 11: 44-54.

[40] Marchini B, Puccetti A, Dolcher MP, Madaio MP, Migliorini P (1995) Induction of anti-DNA antibodies in non autoimmune mice by immunization with a DNA-DNAase I complex. Clin. Exp. Rheumatol. 13: 7-10.

[41] Voynova EN, Tchorbanov AI, Todorov TA, Vassilev TL (2005) Breaking of tolerance to native DNA in nonautoimmune mice by immunization with natural protein/DNA complexes. Lupus 14: 543-550.

[42] Decker P (2006) Nucleosome autoantibodies. Clin. Chim. Acta. 366: 48-60.

[43] Mohan C, Adams S, Stanik V, Datta SK (1993) Nucleosome: a major immunogen for pathogenic autoantibody-inducing T cells of lupus. J. Exp. Med. 177: 1367-1381.

[44] Urbonaviciute V, Fürnrohr BG, Meister S, Munoz L, Heyder P, De Marchis F, Bianchi ME, Kirschning C, Wagner H, Manfredi AA, Kalden JR, Schett G, Rovere-Querini P, Herrmann M, Voll RE (2008) Induction of inflammatory and immune responses by HMGB1-nucleosome complexes: implications for the pathogenesis of SLE. J. Exp. Med. 205: 3007-3018.

[45] Zack DJ, Yamamoto K, Wong AL, Stempniak M, French C, Weisbart RH (1995) DNA mimics a self-protein that may be a target for some anti-DNA antibodies in systemic lupus erythematosus. J. Immunol. 154: 1987-1994.

[46] Sun KH, Tang SJ, Wang YS, Lin WJ, You RI (2003) Autoantibodies to dsDNA cross-react with the arginine-glycine-rich domain of heterogeneous nuclear ribonucleoprotein A2 (hnRNP A2) and promote methylation of hnRNP A2. Rheumatology (Oxford) 42: 154-161.

[47] DeGiorgio LA, Konstantinov KN, Lee SC, Hardin JA, Volpe BT, Diamond B (2001) A subset of lupus anti-DNA antibodies cross-reacts with the NR2 glutamate receptor in systemic lupus erythematosus. Nat. Med. 7: 1189-1193.

[48] Mostoslavsky G, Fischel R, Yachimovich N, Yarkoni Y, Rosenmann E, Monestier M, Baniyash M, Eilat D (2001) Lupus anti-DNA autoantibodies cross-react with a glomerular structural protein: a case for tissue injury by molecular mimicry. Eur. J. Immunol. 31: 1221-1227.

[49] Deocharan B, Qing X, Lichauco J, Putterman C (2002) Alpha-actinin is a cross-reactive renal target for pathogenic anti-DNA antibodies. J. Immunol. 168: 3072-3078.

[50] Tsuzaka K, Leu AK, Frank MB, Movafagh BF, Koscec M, Winkler TH, Kalden JR, Reichlin M (1996) Lupus autoantibodies to double-stranded DNA cross-react with ribosomal protein S1. J. Immunol. 156: 1668-1675.

[51] Lafer EM, Rauch J, Andrzejewski C, Mudd D, Furie B, Furie B, Schwartz RS, Stollar BD (1981) Polyspecific monoclonal lupus autoantibodies reactive with both polynucleotides and phospholipids. J. Exp. Med. 153: 897-909.

[52] Gaynor B, Putterman C, Valadon P, Spatz L, Scharff MD, Diamond B (1997) Peptide inhibition of glomerular deposition of an anti-DNA antibody. Proc. Natl. Acad. Sci. U. S. A. 94: 1955-1960.

[53] Putterman C, Diamond B (1998) Immunization with a peptide surrogate for double-stranded DNA (dsDNA) induces autoantibody production and renal immunoglobulin deposition. J. Exp. Med. 188: 29-38.

[54] Yadav P, Tran H, Ebegbe R, Gottlieb P, Wei H, Lewis RH, Mumbey-Wafula A, Kaplan A, Kholdarova E, Spatz L (2011) Antibodies elicited in response to EBNA-1 may cross-react with dsDNA. PLoS One 6: e14488.

[55] Tamate E, Sasaki T, Muryoi T, Takai O, Otani K, Tada K, Yoshinaga K (1986) Expression of idiotype on the surface of human B cells producing anti-DNA antibody. J. Immunol. 136: 1241-1246.

[56] Sasaki T, Hatakeyama A, Shibata S, Osaki H, Suzuki M, Horie K, Kitagawa Y, Yoshinaga K (1991) Heterogeneity of immune complex-derived anti-DNA antibodies associated with lupus nephritis. Kidney Int. 39: 746-753.

[57] Suzuki M, Hatakeyama A, Kameoka J, Tamate E, Yusa A, Kurosawa K, Saito T, Sasaki T, Yoshinaga K (1991) Anti-DNA idiotypes deposited in renal glomeruli of patients with lupus nephritis. Am. J. Kidney Dis. 18: 232-239.

[58] Shibata S, Sasaki T, Hatakeyama A, Munakata Y, Hirabayashi Y, Yoshinaga K (1992) Clonal frequency analysis of B cells producing pathogenic anti-DNA antibody-associated idiotypes in systemic lupus erythematosus. Clin. Immunol. Immunopathol. 63: 252-258.

[59] Suzuki Y, Funato T, Munakata Y, Sato K, Hirabayashi Y, Ishii T, Takasawa N, Ootaka T, Saito T, Sasaki T (2000) Chemically modified ribozyme to V gene inhibits anti-DNA production and the formation of immune deposits caused by lupus lymphocytes. J. Immunol. 165: 5900-5905.

[60] Hirabayashi Y, Munakata Y, Sasaki T, Sano H (1992) Variable regions of a human anti-DNA antibody O-81 possessing lupus nephritis-associated idiotype. Nucleic Acids Res. 20: 2601.

[61] Munakata Y, Saito S, Hoshino A, Muryoi T, Hirabayashi Y, Shibata S, Miura T, Ishii T, Funato T, Sasaki T (1998) Somatic mutation in autoantibody-associated VH genes of circulating IgM+IgD+ B cells. Eur. J. Immunol. 28: 1435-1444.

[62] Hirabayashi Y, Oka Y, Ikeda T, Fujii H, Ishii T, Sasaki T, Harigae H (2010) The endoplasmic reticulum stress-inducible protein, Herp, is a potential triggering antigen for anti-DNA response. Journal of Immunology 184: 3276-3283.

[63] Oka Y, Hirabayashi Y, Ishii T, Takahashi R, Sasaki T (2007) A monoclonal antibody against human homocysteine-induced endoplasmic reticulum protein (Herp): a useful tool for evaluating endoplasmic reticulum stress. Tohoku J. Exp. Med. 212: 431-437.

[64] Puthalakath H, O'Reilly LA, Gunn P, Lee L, Kelly PN, Huntington ND, Hughes PD, Michalak EM, McKimm-Breschkin J, Motoyama N, Gotoh T, Akira S, Bouillet P, Strasser A (2007) ER stress triggers apoptosis by activating BH3-only protein Bim. Cell 129: 1337-1349.

[65] Shore GC, Papa FR, Oakes SA (2011) Signaling cell death from the endoplasmic reticulum stress response. Curr. Opin. Cell Biol. 23: 143-149.

[66] Bruns A, Bläss S, Hausdorf G, Burmester GR, Hiepe F (2000) Nucleosomes are major T and B cell autoantigens in systemic lupus erythematosus. Arthritis Rheum. 43: 2307-2315.

[67] Radic M, Marion T, Monestier M (2004) Nucleosomes are exposed at the cell surface in apoptosis. J. Immunol. 172: 6692-6700.

[68] Gómez-Puerta JA, Burlingame RW, Cervera R (2008) Anti-chromatin (anti-nucleosome) antibodies: diagnostic and clinical value. Autoimmun. Rev. 7: 606-611.

[69] van Bruggen MC, Kramers C, Walgreen B, Elema JD, Kallenberg CG, van den Born J, Smeenk RJ, Assmann KJ, Muller S, Monestier M, Berden JH (1997) Nucleosomes and histones are present in glomerular deposits in human lupus nephritis. Nephrol. Dial. Transplant 12: 57-66.

[70] O'Flynn J, Flierman R, van der Pol P, Rops A, Satchell SC, Mathieson PW, van Kooten C, van der Vlag J, Berden JH, Daha MR (2011) Nucleosomes and C1q bound to glomerular endothelial cells serve as targets for autoantibodies and determine complement activation. Mol. Immunol. 49: 75-83.

[71] Muller S, Dieker J, Tincani A, Meroni PL (2008) Pathogenic anti-nucleosome antibodies. Lupus 17: 431-436.

[72] Rubin RL (2005) Drug-induced lupus. Toxicology 209: 135-147.

[73] Marzano AV, Vezzoli P, Crosti C (2009) Drug-induced lupus: an update on its dermatologic aspects. Lupus 18: 935-940.

[74] Lee BH, Yegnasubramanian S, Lin X, Nelson WG (2005) Procainamide is a specific inhibitor of DNA methyltransferase 1. J. Biol. Chem. 280: 40749-40756.

[75] Deng C, Lu Q, Zhang Z, Rao T, Attwood J, Yung R, Richardson B (2003) Hydralazine may induce autoimmunity by inhibiting extracellular signal-regulated kinase pathway signaling. Arthritis Rheum. 48: 746-756.

[76] Strickland FM, Richardson BC (2008) Epigenetics in human autoimmunity. Epigenetics in autoimmunity - DNA methylation in systemic lupus erythematosus and beyond. Autoimmunity 41: 278-286.

[77] Oka Y, Hirabayashi Y, Ikeda T, Fujii II, Ishill T, Harigae H (2011) A single-stranded DNA-cross-reactive immunogenic epitope of human homocysteine-inducible endoplasmic reticulum protein. Scand. J. Immunol. 74: 296-303.

[78] Christen U, von Herrath MG (2004) Initiation of autoimmunity. Curr. Opin. Immunol. 16: 759-767.

[79] Wandstrat A, Wakeland E (2001) The genetics of complex autoimmune diseases: non-MHC susceptibility genes. Nat. Immunol. 2: 802-809.

[80] He B (2006) Viruses, endoplasmic reticulum stress, and interferon responses. Cell Death Differ. 13: 393-403.

[81] Komori R, Taniguchi M, Ichikawa Y, Uemura A, Oku M, Wakabayashi S, Higuchi K, Yoshida H (2012) Ultraviolet a induces endoplasmic reticulum stress response in human dermal fibroblasts. Cell Struct. Funct. 37: 49-53.

[82] James JA, Kaufman KM, Farris AD, Taylor-Albert E, Lehman TJ, Harley JB (1997) An increased prevalence of Epstein-Barr virus infection in young patients suggests a possible etiology for systemic lupus erythematosus. J. Clin. Invest. 100: 3019-3026.

[83] Niller HH, Wolf H, Minarovits J (2008) Regulation and dysregulation of Epstein-Barr virus latency: implications for the development of autoimmune diseases. Autoimmunity 41: 298-328.

[84] Fattal I, Shental N, Mevorach D, Anaya JM, Livneh A, Langevitz P, Zandman-Goddard G, Pauzner R, Lerner M, Blank M, Hincapie ME, Gafter U, Naparstek Y, Shoenfeld Y, Domany E, Cohen IR (2010) An antibody profile of systemic lupus erythematosus detected by antigen microarray. Immunology 130: 337-343.

[85] Kang I, Quan T, Nolasco H, Park SH, Hong MS, Crouch J, Pamer EG, Howe JG, Craft J (2004) Defective control of latent Epstein-Barr virus infection in systemic lupus erythematosus. J. Immunol. 172: 1287-1294.

[86] Larsen M, Sauce D, Deback C, Arnaud L, Mathian A, Miyara M, Boutolleau D, Parizot C, Dorgham K, Papagno L, Appay V, Amoura Z, Gorochov G (2011) Exhausted cytotoxic control of Epstein-Barr virus in human lupus. PLoS Pathog 7: e1002328.

[87] Gross AJ, Hochberg D, Rand WM, Thorley-Lawson DA (2005) EBV and systemic lupus erythematosus: a new perspective. J. Immunol. 174: 6599-6607.

[88] Laichalk LL, Thorley-Lawson DA (2005) Terminal differentiation into plasma cells initiates the replicative cycle of Epstein-Barr virus in vivo. J. Virol. 79: 1296-1307.

[89] Taylor GM, Raghuwanshi SK, Rowe DT, Wadowsky RM, Rosendorff A (2011) Endoplasmic reticulum stress causes EBV lytic replication. Blood 118: 5528-5539.

[90] Wang H, Nicholas MW, Conway KL, Sen P, Diz R, Tisch RM, Clarke SH (2006) EBV latent membrane protein 2A induces autoreactive B cell activation and TLR hypersensitivity. J. Immunol. 177: 2793-2802.

Chapter 5

Cell Death and Cancer, Novel Therapeutic Strategies

Silvina Grasso, M. Piedad Menéndez-Gutiérrez,
Estefanía Carrasco-García, Leticia Mayor-López,
Elena Tristante, Lourdes Rocamora-Reverte,
Ángeles Gómez-Martínez, Pilar García-Morales,
José A. Ferragut, Miguel Saceda and Isabel Martínez-Lacaci

Additional information is available at the end of the chapter

1. Introduction

1.1. History, definition and classification

Life and death are essential parts of the natural cycle of all multicellular organisms. In metazoans, somatic cells divide normally during the process known as mitosis. Cell proliferation is tightly controlled, according to the organism needs. An increase in the number of cells takes place during growth and when one of these cells finishes its physiological function or detects DNA or cell damage, it undergoes a physiological process known as apoptosis that induces its own death. In humans about a hundred thousand cells are formed every second through mitosis, while a similar number is destroyed by apoptosis [1]. This dynamic balance between proliferation and cell death is known as homeostasis. If altered, different pathologic processes such as carcinogenesis can take place. Besides its role in embryonic development, homeostasis maintenance and aging, apoptosis is also a defence mechanism by which infected, injured or mutated cells as a result of irradiation or chemotherapeutic drugs are eliminated. This type of cell death involves the activation of an evolutionary conserved and tightly regulated intracellular machinery that requires energy consumption [2]. An important feature of apoptosis is that the cell is eliminated without triggering an immune response, avoiding thus tissue damage [3].

The term **apoptosis** to describe cell death was introduced by Kerr and colleagues in 1972 [4], from the Greek term "appo-teo-sis" which means "falling off" (as in leaves from a tree or petals from a flower). Apoptosis has been used as a synonym of programmed cell death and refers to a suicidal type of death. However, as more mechanisms of programmed cell death are being elucidated, the Nomenclature Committee on Cell Death (NCCD) recommends caution when using the term apoptosis [5]. Historically, the classical methods to define cell death rely on morphological criteria. Thus, apoptosis was termed programmed cell death type I, autophagy named as programmed cell death type II and necrosis as a type of death lacking characteristics of both type I and type II. As more biochemical methods become available over the last decades, a more definite and concise classification of types of cell death has become necessary. Therefore, cell death types can be classified according to morphological appearance, biochemical features, functional criteria or immunological aspects. In the following figure (Figure 1) the classification of different types of cell death, in accordance with to the NCCD are described:

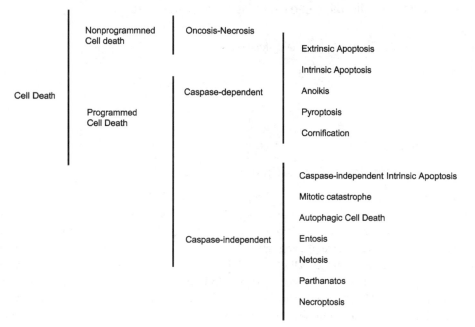

Figure 1. Classification of different types of cell death according to the NCCD (modified from Galluzzi et al. 2012)

It is important to mention that a single stimulus can trigger more than one mechanism of cell death simultaneously within a cell. However, only the most efficacious mechanism will be prevalent.

2. Apoptosis

2.1. Morphological and biochemical features

Morphological features that share both caspase-dependent and independent apoptotic pathways are: a) loss of plasma membrane symmetry and loss of cell-to-cell contact and cell round-up, b) chromatin condensation (picnosis) and fragmentation and nuclear breakdown, c) overall cell shrinkage and cytoskeleton alterations although the majority of the cell organelles remain intact, d) membrane blebbing and formation of the membrane-enclosed particles called apoptotic bodies that contain nuclear or cytoplasmic material and will be engulfed by phagocytes or neighbouring cells [6].

Because apoptotic cells are eaten so quickly, there are few dead cells left on tissue sections. This is the reason why apoptosis was neglected by pathologists for a long time, even though apoptosis is the main mechanism for discarding of harmful or unwanted cells in multicellular organisms.

Along with the morphological transformation described, there are several biochemical alterations that take place. For instance, activation of endonucleases, some of them dependent on Ca^{2+} and Mg^{2+} channels, that cleaves genomic DNA. In the apoptotic process, this event gives rise to internucleosomal DNA double-strand breaks with fragments of multiples of 180 bp, resulting in the typical pattern of *DNA ladder* that can be detected by electrophoresis on agarose gels. Other changes include loss of the inner mitochondrial transmembrane potential [7] and exposure of the phospholipid phosphatidylserine to the outer cell membrane, which allows phagocytes to detect and engulf these apoptotic cells [8].

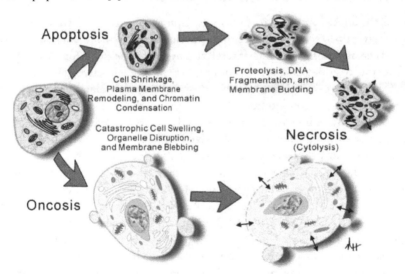

Figure 2. Features of Apoptosis, Oncosis and Necrosis (taken from Hail et al., 2006)

The most distinguishable feature of apoptosis is the formation of apoptotic bodies. On the other hand, oncosis is characterized by cytoplasmic swelling, dilation of organelles and vacuolization and plasma membrane blebbing. The cell will finally die by cytolysis, which is a typical hallmark of necrosis [9]. Apoptotic cells can also lose their plasma membrane and eventually undergo a secondary necrosis. However, this phenomenon has only been observed *in vitro*. According to Majno and Joris, necrosis is not a type of cell death, but rather, it refers to changes subsequent to any type of death. They prefer to use the term oncosis to describe a nonprogrammed or accidental type of cell death characterized by swelling [10]. Moreover, a new modality of cell death has recently emerged, termed **necroptosis** to indicate a regulated form of necrosis [11].

Apoptosis can be divided into three stages: initiation, integration/decision and execution/degradation [12]. The initiation phase mainly depends upon cell type and apoptotic stimulus. The integration/decision phase consists of the activation of proteases, nucleases and other effector molecules. The execution/degradation phase involves morphological and biochemical changes that are common to all apoptotic mechanisms, regardless of the stimulus that initiated the process. The classical apoptosis is defined as a type of programmed cell death characterized by the activation of zymogens known as caspases, which are cystein-dependent **asp**artate-directed proteases. Both caspase-dependent and caspase-independent cell death mechanisms share multiple characteristics, such as mitochondrial membrane permeabilization (MMP), DNA fragmentation, etc. Hence, (MMP) can commit cells to die even in the absence of caspase activation, by releasing factors such as apoptosis inducing-factor (AIF) or endonuclease G (endo G).

3. Caspases: Executioners of the apoptotic process

Caspases are the effector molecules of apoptosis in mammals. They were first discovered mediating programmed cell death during development in the nematode *Caenorhabditis elegans* [13]. The caspase family is characterized by their specificity for cleaving substrates after aspartic acid residues and for containing cystein in their active centre [14]. There are about fourteen caspases that used to be classified into two families: one involved in inflammation processes and the other taking part in apoptosis. However, some of the non-apoptotic caspases have some apoptotic roles, and conversely, some non-apoptotic caspases can induce pyroptosis [15]. They are synthesized as inactive precursors or zymogens (pro-caspases) that need to be proteolytically processed to become active. Once a caspase is activated, it can cleave another caspase creating thus an expansive hierarchical activating cascade that serves to amplify the apoptotic signal. Caspase activation is a complex and tightly regulated process. Within the apoptotic group, there are two types of caspases: upstream initiator or apical caspases and downstream effector or executioner caspases. The initiator group consists of caspases -2, -8, -9 and -10 and the effector group consists of -3, -6 and -7 caspases. Apical but not executioner caspases have long prodomains. Caspase-9 is considered the initiator of the mitochondrial pathway and caspase-8 is regarded as the originator of the dead receptor-mediated apoptotic pathway. Effector caspases carry out apoptotic programmes through direct processing of a variety of cellular substrates. The proteolytic cleavage of such substrates brings about a whole plethora of effects within the cell: disassembling of cytoskeleton, cytoplasm scaffolding, nuclear fragmentation as a consequence of laminin degradation, activation of endonucleases that cleave chromatin, inactivation of DNA damage repair proteins (PARP1, Rad9, etc.), phagocyte signalling, loss of cell-to-cell contact, etc. [16-18]

Substrate	Funtion
PARP, DSBs, Rad51 and Rad9, DNA-PKCs and ATM	DNA damage repair
C1, C2 and U170	mRNA processing
Actin, Gas2, FAK, PAK2, Fodrin and Gelsolin	Cellular structure
PKC and Akt	Apoptosis and cell cycle
Phosphatase 2A, Raf1 and Rb	Cell cycle
Bid, Bcl-2, Bcl-xL and XIAP	Apoptosis regulation
Pro-IL-1β	Cytokine
D4-GDP Inhibitor	Rho GTPases regulator
Laminin A/C and B and NuMA	Nuclear shape maintenance
ACINUS	Regulation of chromatin condensation
ICAD	DNA fragmentation
Huntingtin	Huntington disease
SREBP-1 and SREBP-2	Sterol regulatory element binding proteins
MDM2	Transcription factor

Table 1. Proteins substrates of caspases (Blank and Shiloh 2007; Cohen et al 1997)

Caspase activity can be regulated both in a positive and in a negative manner. Negative caspase regulators are called Inhibitors of Apoptosis Proteins (IAPs) that bind to the catalytic site of caspases neutralizing its activity [19], or targeting them to degradation by ubiquinitation [20,21]. Some examples of IAPs are: XIAP, c-IAP1, c-IAP2, NAIP and survivin. On the other hand, Smac/DIABLO and Omi/HtrA2 act as positive regulators by inactivating IAPs [22].

4. Extrinsic and intrinsic apoptotic pathways

There are at least two main well characterized apoptotic routes: the death receptor or extrinsic pathway and the mitochondrial or intrinsic pathway. The extrinsic pathway plays a major role in tissue homeostasis and responds to external cues, coming especially from the immune system, whereas the intrinsic pathway is triggered as a response to various internal insults such as DNA damage, cytosolic calcium overload, starvation, oxidative stress, radiation, cytotoxic agents, etc., and involves mitochondrial destabilization [23,24].

The death receptor pathway is initiated by extracellular stimuli that are recognized by a subgroup of the tumour necrosis factor receptor (TNF-R) family named dead receptors (Fas/CD95/APO-1, TNFR1, TRAIL R1/DR4 and TRAIL R2/DR5). Upon binding of their ligands (FAS, TNFα and TRAIL) these receptors become activated and interact via their death domain with the protein motif Fas-associated death domain (FADD) in adapter proteins, forming the Death inducing signalling complex (DISC), which binds to the prodomain of the initiator caspase-8 [25]. Thus, caspase-8 is activated by dimerization, which leads to autocatalyisis and consequently activation of executioner caspases -3, -6 and -7 [26].

The intrinsic mitochondrial pathway is characterized by the action of B-cell lymphoma (Bcl-2) proteins. This family consists of proapoptotic and antiapoptotic proteins. The proapoptotic members promote mitochondrial outer membrane permeabilization (MOMP) and the antiapoptotic members counteract this action, so that the balance between these two groups of proteins determines the final outcome [27,28]. If the balance is in favour of the proapoptotic members, the outer mitochondrial membrane is permeabilized through pore formation and cytchrome c and other proteins such as Smac/DIABLO and Omi/HtrA2 are released to the cytosol. Then, cytochrome c binds to the adaptor protein Apaf-1 and dATP, forming the apoptosome, a catalytic complex that activates caspase-9 which in turn activates the executioner caspases.

The extrinsic and intrinsic pathways are interconnected through Bid. In some cases, when DISC formation is low, caspase-8 activation can induce MOMP through Bid cleavage, which translocates to the mitochondria and induces cytchrome c release, apoptosome formation and engagement of the caspase cascade [29].

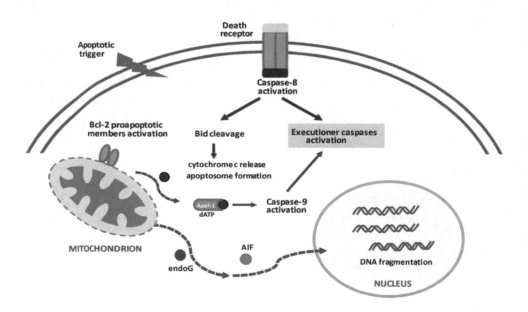

Figure 3. The extrinsic and intrinsic apoptotic pathways.

5. Death receptors

Death receptors belong to the TNF-R superfamily of receptors. They participate in proliferation, differentiation, immune response, gene expression, survival and cell death. In fact, TNF-R1 and Fas (CD95/APO-1) are involved in apoptotic processes [30]. Death receptors are stimulated by death ligands: TNF; Fas ligand (FasL) and TNF-related apoptosis-inducing ligand (TRAIL). Death receptors contain an intracellular globular interaction domain called death domain (DD). Upon ligand binding, death receptors aggregate forming trimers. As a consequence of this aggregation, they recruit adaptors such as FADD, which interacts with caspase-8 by virtue of its death effector domain (DED), forming the multi-protein complex known as DISC. However, in some instances, death receptors can oligomerize in the absence of ligand binding. For example, they can be activated by UV radiation [31]. Fas and TNF-R1 can also recruit RIP (receptor-interacting protein) -associated Ich-1/CED homologous [ICE (interleukin-1β-converting enzyme)/CED-3 (cell-death determining 3) homologue 1] protein with death domain (RAIDD) which activates caspase-2 through a caspase activation and recruitment domain (CARD). In addition, death receptors can participate in caspase-independent mechanisms. For instance, TNF-R1 can activate Extracellular signal-regulated kinase 2 (Erk2) through the mitogen-activated kinase activating death domain (MADD) protein [32]. Fas can bind Death-domain associated protein (Daxx) and activate c-Jun amino-terminal kinase (JNK) [33]. TNF-receptor associated death domain (TRADD) and RIP can activate Nuclear factor kappa-light-chain-enhancer of activated B cells (NFkB), triggering a form of regulated necrosis named **necroptosis** [34].

6. Role of mitochondria in cell death

MMP, a crucial event of the intrinsic mitochondrial apoptotic pathway, is considered a "point of no return" in the sequence of events leading to apoptosis. This phenomenon is associated with mitochondrial membrane potential loss ($\Delta\Psi_m$) that occurs as a result of assymetrical distribution of protons on both sides of the inner mitochondrial membrane. This irreversible process can take place before, during or after MOMP. The pore formation caused by Bcl-2 proteins induces MOMP, which leads to $\Delta\Psi_m$ dissipation, inhibition of ATP synthesis and $\Delta\Psi_m$-dependent transport activities. Consequently, the respiratory change ceases, causing reactive oxygen species (ROS) generation and release of proteins confined within the inner mitochondrial space [35]. The contribution of the inner mitochondrial permeabilization, however, is controversial. MOMP can also result from the phenomenon called mitochondrial permeability transition that implies the opening of a non-selective pore in the inner mitochondrial membrane known as the mitochondrial permeability transition pore complex [12].

6.1. Mechanisms of MOMP

Bax/Bak pore formation. The first proposed mechanism of MOMP is mediated by the Bcl-2 family of proteins that directly act on the outer mitochondrial membrane. This family consists of about 17 members, some of which are proapoptotic and others antiapoptotic. The antiapoptotic members (Bcl-2, Bcl-xL, Bcl-w, Mcl-1, A1, Bcl-B, etc.) contain 3-4 BH domains (Bcl-2 homology regions: BH-1, BH-2, BH-3 and BH-4) and are named "Bcl-2 like" proteins.

Among the proapoptotic members, some contain 2-3 BH domains (Bax, Bak, Bok, etc.) and the "BH3-only" proteins (Bid, Bad, Bim, Bik, Bmf, Hrk, Bnip3, Noxa, Puma, Spike, etc.) have the biggest proapoptotic potential. BH-3 only proteins exert their effects either by inhibiting antiapoptotic Bcl-2 like members or by directly activating Bak and Bax. In the first mechanism, "facilitators" such as Bad interact with antiapoptotic Bcl-2 like proteins, dissociating them and creating MMP. In the second mechanism, however, "activators" such as truncated Bid (tBid) activate proapoptotic proteins by stimulating translocation of Bax to the mitochondrial membrane or by activating Bak [36].

The interplay between antiapoptotic and proapoptotic proteins decides the final fate of the damaged cell. Antiapoptotic proteins can be found in the outer mitochondrial membrane, in the cytoplasm or in the Endoplasmic Reticulum (ER). Some propapoptotic members such as Bax or Bid reside in the cytosol, but translocate to the outer membrane upon triggering of a death stimulus, where they oligomerize to form a channel, either with themselves or with membrane anchored-Bak or tBid. The importance of the interactions between proteins and lipids is becoming more evident. In fact, it is now thought that Bak or Bax destabilize lipid bilayers instead of forming pores. BH3-only proteins activity is transcriptionally and posttranslationally regulated. Bid, for instance, is regulated by caspase-8, cathepsins, granzyme B or calpain [37-39]. The proapototic protein Bad in inhibited through phosphorylation by Akt and activated through dephosphorylation by calcineurin [40].

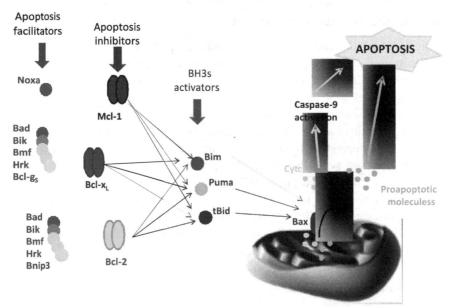

Figure 4. Proposed model of mitochondria membrane permeabilization. **Bax/Bak pore formation.** Bcl-2 family members regulation.

Permeability transition pore complex. The second proposed mechanism of MOMP is based on the formation of the permeability transition pore (PTP) complex, a high conductance channel which allows the influx of water and small solutes. PTP is a multiprotein complex located in the mitochondrial membrane that spans contact sites between the inner and the outer mitochondrial membranes. It is composed of three proteins: the voltage-dependent anion channel (VDAC), which is the most abundant protein of the outer membrane, the soluble matrix protein cyclophilin D (CypD), and the adenine nuclease translocase (ANT) located in the inner membrane. Other proposed PTP components are: hexokinase, creatine kinase and peripheral benzodiazepine receptor. The requirement of ANT is controversial. Distinct VDAC isoforms may interact with Bcl-2 members in a different manner. PTP opening leads to $\Delta\Psi_m$ dissipation, uncoupling of oxidative phosphorylation, water and ions influx, matrix swelling, outer membrane rupture and release of intermembrane space proteins, such as cytochrome c. Ca^{2+} favours PTP opening and permeability transitions are regulated by $\Delta\Psi_m$, pH of mitochondrial matrix, redox potential, adenine nucleotides and bivalent metallic ions. PTP regulation by Bcl-2 proteins is still a matter of debate. It has been proposed that proapototic members promote pore opening, whereas antiapoptotic proteins favour pore closure [41]. Bcl-2 members can indirectly regulate PTP opening through Ca^{2+} efflux from the ER. In healthy cells, Bcl-2 or Bcl-X_L can be located not only inserted in the mitochondrial outer membrane but also in the ER. Calcium ions exit the ER through inositol phosphate 3 receptors (IP3R), which can be blocked upon Bcl-2 binding. The proapoptotic members Bax and Bak induce Ca^{2+} movement from the ER to the mitochondrion [42]. However, Ca^{2+} interchange through PTP can be regulated by proapoptotic tBid [43,44].

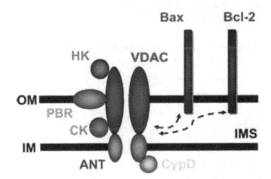

Figure 5. Proposed model of mitochondrial membrane permeabilization. **Permeability transition pore complex.** OM: outer membrane. IMS: intermembrane space. IM: inner membrane. VDAC: voltage-dependent anion channel. HK: hexokinase. PBR: peripheral benzodiazepine receptor. CK: creatine kinase. ANT: adenine nuclease translocase. CypD: cyclophilin D (modified from Kroemer et al. 2007).

The rise of mitochondrial matrix Ca^{2+} induces oxidative metabolism. However, upon apoptotic stimulus, Ca^{2+} can cause PTP opening, which can be transient and provide a fast calcium release mechanism, or persistent, giving rise to outer membrane rupture and release of apoptotic factors. Additionally, a different channel has been identified in the mitochondrion, composed of ceramide, a lipid that can form hydrogen bonds giving rise to ceramide structures that form channels which allow the efflux of proteins up to 60 kDa [42].

6.2. Cell death effectors released from mitochondria

The release of proteins from the intermembrane space and the loss of membrane-associated mitochondrial functions lead to cell death. Albeit is not clear the mechanism, mitochondrial destabilization provokes release of factors that mediate in caspase-dependent and independent pathways.

Cytochrome c is a protein localized in the intermembrane mitochondrial space where it participates in the electron transport between complex III and IV of the mitochondrial respiratory chain. As already mentioned, cytochrome c is involved in the apoptosome formation and caspase cascade activation [45]. Cytochrome c release is a crucial step in the intrinsic apoptotic pathway.

Smac/DIABLO means **S**econd **m**itochondrial **a**ctivator of **c**aspases or **D**irect **IAP B**inding protein with **LO**w p**I**. This 29 kDa protein is also localized in the intermembrane space and is released to the cytosol upon activation by certain apoptotic stimuli, where it binds IAPs, preventing their function and favouring caspase activation.

Omi/HtrA2 functions in a similar way to Smac/Diablo. This 36 kDa protein belongs to a highly conserved protein family. In healthy cells, Omi/HtraA2 is located in the intermembrane space. Upon apoptotic triggering (TRAIL, UV radiation, staurosporine, etc.) is released to the cytosol where it binds and inactivates IAPS promoting caspase activation.

In addition, to the caspase-dependent effectors already mentioned, other caspase-independent proteins can be activated:

AIF stands for Apoptosis Inducing Factor. AIF is a 57 kDa flavoprotein localized in the intermembrane space whose aminoacid sequence resembles ferrodoxin. AIF is expressed as a 67 kDa precursor that possesses two mitochondrial localization sequences in its amino-terminal end. Once in the mitochondria, AIF precursor is cleaved giving rise to the mature protein that is believed to have an oxidoreductase function based on its FAD domain, playing an important physiological role in oxidative phosphorylation. AIF translocates to the nucleus in response to apoptotic stimuli where it induces chromatin condensation and large-scale (50 Kbp) DNA fragmentation by an unknown mechanism, leading to apoptosis in a caspase-independent fashion. It has been recently suggested that Steroid receptor coactivator-interactive protein prevents AIF release from the mitochondria [46]. Conversely, calcium-activated calpain promotes AIF release from the mitochondria [47] and poly-ADP-ribose-plymerase 1 (PARP-1) activity is necessary for AIF translocation to the nucleus [48].

Endo G or Endonuclease G also participates in caspase-independent cell demise mechanisms. It belongs to a family of Mg^{2+}-dependent endonucleases. Endo G is a 30 kDa mitochondrial protein that, like AIF, is synthesized as a precursor form and its mitochondrial localization sequence is cleaved when the protein reaches the intermembrane space. Endo G is released from the mitochondria upon certain apoptotic stimuli such as UV radiation or anti-Fas antibodies, and it translocates to the nucleus, where it cleaves chromatin DNA into nucleosomal fragments. Endo G cooperates with exonucleases and

DNase I in order to facilitate DNA processing. Endo G nuclease activity has been observed even in the presence of caspase inhibitors. Therefore, similarly to AIF, Endo G participates in caspase-independent cell death mechanisms. The physiological role of Endo G, however, needs to be established.

Besides its function as IAP inhibitor, **Omi/HtrA2** has been implicated in caspase-independent cell death mechanisms by virtue of its serine protease activity. This effector molecule cleaves proteins such as ped/pea-15 [49], HAX-1 [50], or RIP-1 [51]. However, the exact mechanism as to how this occurs is still unknown.

AIF and Endo G activity together with Omi/HtrA2 serine protease activity are considered responsible for the recently called **caspase-independent intrinsic apoptosis** [11].

6.3. Afferent signals from other organelles

Mitochondria play a central role in programmed cell death pathways and integrate different signals coming from other organelles, being MMP, a point of no return, as already mentioned.

6.3.1. Nuclear DNA damage

The tumour suppressor p53 mediates DNA damage response, either by stimulating DNA damage response or by inducing apoptosis. As a transcription factor, p53 transactivates Bcl-2 proteins (Bad, Bid, Puma and Noxa) [52], which induce MMP and release of proteins from the intermembrane space [12]. After DNA damage, p53 can also induce the expression of p53-induced protein with a death domain (PIDD), which activates nuclear caspase-2. PIDD associates with RAIDD, forming a signalling platform known as the PIDDosome [53]. The PIDDosome can activate caspase-2 and the transcription factor NFkB in response to DNA damage [54]. Caspase-2 acts upstream of the mitochondria by inducing Bid cleavage, Bax translocation and cytochrome c release. The role of caspase-2 in apoptosis, however, is controversial. It is now thought that caspase-2 functions as a tumour suppressor gene regulating the cell cycle machinery [55]. p53 may also induce apoptosis by transcription-independent mechanisms, for instance, by directly interacting with Bak, Bax, Bcl-2 and Bcl-X_L at the outer mitochondrial membrane [56]. The activity of p53 can be promoted by glycogen synthase kinase 3β (GSK3β) binding both in the nucleus and in the mitochondria, which in turns promotes cytochrome c release and caspase-3 activation [57].

6.3.2. Endoplasmic reticulum: Unfolded protein response

The ER has primordial roles in normal physiological and survival processes. These include intracellular calcium homeostasis, protein secretion and lipid biosynthesis [58]. Apoptosis can be initiated as a consequence of stress in the ER. This stress condition can be caused by calcium homeostasis alteration, glucose deprivation, hypoxia, low redox potential, excessive protein synthesis or defective protein secretion. These insults can cause accumulation of unfolded proteins in the lumen of the ER, triggering an evolutionary conserved signalling pathway known as the Unfolded Protein Response (UPR) which can culminate in cell death. UPR consists of global protein synthesis reduction, synthesis induction of chaperones and other proteins related to protein folding and retro-translocation of misfolded or unfolded proteins from the ER to the cytosol, where they will be degraded by the proteasome [59]. When the ER stress is sustained and ER function cannot be restored, UPR activates a specific apoptotic pathway. Caspase-12, which is localized in the ER, is activated by calcium-dependent proteases known as calpains [60]. Once activated, caspase-12 activates caspase-9 without apoptosome intervention [61,62]. It has been postulated that ER stress can also activate caspase-8, which induces cytochrome c release through Bid processing [29,63]. In addition, ER stress can also initiate **autophagy,** a different type of cell death.

6.3.3. Lysosomes

Lysosomes are organelles that contain acidic hydrolases such as cathepsins. Rupture of lysosomes release cathepsins to the cytosol, where they can trigger apoptosis or necrosis. Cystatins, on the other hand, are cytosolic proteins that act as negative regulators of cathepsins when they are translocated from lysosomes to the cytosol. Apoptosis initiated in lysosomes follows a mitochondrion-dependent pathway associated to caspase activation. However, it has been shown that cathepsin D activates Bax and AIF release, triggering a caspase-independent apoptotic pathway [37]. Furthermore, some cathepsins induce Bid cleavage [38], interaction with Bcl-2 proteins and permeabilization of the mitochondrial membrane [64]. In addition, cathepsins can alter mitochondria functions by cleaving subunits of the oxidative phosphorylation complexes, inducing ROS generation [12]. Lysosomes are essentially involved in **autophagic cell death** mechanisms.

6.3.4. Cytosol

Several signals coming from the cytosol can induce MMP. These include: metabolites such as glucose 6-phospate and palmitate, ROS and activation of certain kinases: GSK3β, protein kinase C (PKC) δ, [65] and members of the JNK signalling pathway [38,66]. On the other hand, other molecules inactivate PTP and protect mitochondrial membrane from permeabilization, inhibiting apoptosis. These include: metabolites (ATP, glucose, NADH, UTP, etc.), antiapoptotic Bcl-2 family members, antioxidant enzymes such as glutathiones S transferase and prosurvival kinases such as Akt. In this sense, Akt can inhibit apoptosis by several mechanisms: activation of NFkB [67], inactivation of GSK3β and caspases and through hexokinase II-dependent mechanisms [12].

6.3.5. Cytoskeleton

The cytoskeleton is composed of microtubules, microfilaments and intermediate filaments that play important roles in cell motility, polarity, attachment, shape maintenance, etc. Adherent cells can undergo a specific type of caspase-dependent cell death called **anoikis** when they become detached from the extracellular matrix or neighbouring cells [68,69]. Besides its well-established role in mitosis, cytoskeleton components can modulate mitochondria. For example, microtubules sequester BH3-only proteins Bim and Bmf, which interact with dynein [70]. Gelsolin has anti-apoptotic effect by closing the VDAC channel [71]. Actin can also modulate VDAC closure [69,72].

7. Parp proteolysis as an indicator of cell death

Poly-ADP-Ribose Polymerase (PARP) is a family of 16 nuclear enzymes, among which, the best characterized is PARP-1. PARPs have several functions in cell proliferation, cell death, DNA recombination and DNA repair. PARP-1 is a 116 kDa nuclear protein involved in DNA repair mechanisms. PARP synthesis is activated when DNA is fragmented in the presence of nuclear poly-ADP ribosylated proteins. In an early apoptotic stage, caspases cleave PARP resulting in an 89 kDa and a 24 kDa fragments [73]. The smaller fragment irreversibly binds DNA fragment ends, impeding the access of DNA repair enzymes. Hence, PARP proteolysis facilitates nuclear disorganization and ensures irreversibility of the apoptotic process [74]. PARP cleaves also takes place during necrosis. However, the fragments obtained are of different size [75]. A role of PARP cleavage in autophagy induced by DNA damage has been recently suggested [76]. **Parthanatos** is a particular case of regulated necrosis in which PARP activation plays an important role. PAR polymer, the product of PARP-1 activation, translocates to the mitochondria and induces AIF release. Subsequently, AIF translocates to the nucleus and provokes chromatin condensation and DNA fragmentation [77]. Necroptosis or regulated necrosis also involves PARP activation [78]. Therefore, PARP can participate in different types of cell death and it is not exclusive of apoptosis, as previously thought.

8. Autophagy

Autophagy is a self-digestive physiological process that occurs in all eukaryotic cells, during which long-lived proteins and organelles are degraded by lysosomes to maintain cellular homeostasis [79]. To date, at least three types of autophagic pathways have been described: macroautophagy (simply called "autophagy"), microautophagy and Chaperone-Mediated Autophagy (CMA). These forms differ in the mode the cargo is delivered to the lysosome.

Macroautophagy is a dynamic process in which portions of the cellular cytoplasm and organelles are sequestered in a double-membrane bound vesicle called autophagosome, which then fuses with the lysosome [80,81]. **Microautophagy** involves the sequestering by the lysosome itself of part of the cytoplasm. During microautophagy, the membrane of the lysosome invaginates, and then pinches off to form an internal vacuole that contains material derived from the cytoplasm [82]. The notable difference between macroautophagy and microautophagy is that in the latter, part of the cytoplasm is directly taken up into the lysosome. Both macroautophagy and microautophagy are basically nonselective degradation pathways in which bulk cytoplasm is randomly sequestered. However, in some cases, autophagy can selectively eliminate some organelles, such as damaged peroxisomes, mitochondria or ER. In contrast, **CMA** does not involve vesicular traffic and is specific for the degradation of proteins. During this process, proteins are delivered to lysosomes with the help of molecular chaperones and a lysosomal receptor. Cytosolic proteins with a specific peptide sequence motif ("KFERQ" motif) are recognized by a complex of molecular chaperones (Hsc70) and then bind to a lysosomal receptor called lysosome associated membrane protein (LAMP) type 2a [83].

Macroautophagy (hereafter referred to as autophagy) is the most studied and prevalent form of autophagy in cells. This process begins with the formation of a "C" shaped double-membrane structure in the cytosol, called "omegasome", which is formed from the ER (Initiation phase). Following this, the omegasome grows to form the "isolation membrane", which elongates to engulf cytoplasmic components (Elongation phase). Then, the "isolation membrane" curves and closes to form a vacuole called the autophagosome (Maturation phase). As a result, portions of the cell cytoplasm and some organelles are sequestered in this vacuole. Finally, the outer membrane of the autophagosome fuses with the lysosomal membrane and the inner membrane (the autophagic body) carrying the cytosolic constituents enters the lysosome. The autophagic body is degraded in the lysosome by hydrolases and the resulting free amino acids and macromolecules are transported back into the cytosol for reuse [84]. In this way, autophagy contributes to the maintenance of the cellular energy homeostasis, to the clearance of damaged organelles and to adaptation to environmental stresses [85]. Accordingly, autophagy defects have been linked to a wide range of human pathologies, including cancer.

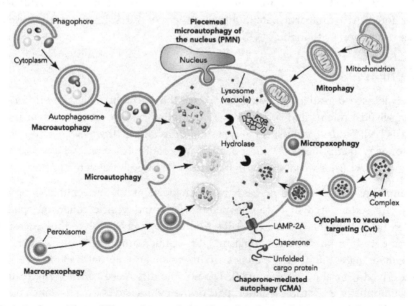

Figure 6. Autophagy. Three different forms of autophagy are depicted: Macroautophagy, Microautophagy and Chaperone-mediated autophagy (CMA) (taken from Yen and Klionsky, 2008).

9. Mitotic catastrophe

The cell death process that takes place when mitosis cannot be completed is called mitotic catastrophe. This phenomenon is triggered as a consequence of perturbations of the mitotic machinery that governs appropriate chromosome segregation. The main hallmark of mitotic catastrophe is the enlarged cell size accompanied by multinucleation [5]. Other features are chromatin condensation, DNA degradation, MMP, cyrochrome c release from the mitochondria and caspase activation [86]. Some types of mitotic catastrophe, however, take place without intervention of caspases, what has been named caspase-independent mitotic death [87]. Mitotic catastrophe results form the combination of deficient checkpoints (DNA and mitotic spindle) and DNA damage. Cells that evade the mitotic checkpoint and do not undergo apoptosis are prone to generate aneuploidy. Therefore, mitotic catastrophe is conceived as a device to avoid genomic instability. The players that take part in mitotic catastrophe are: cell cycle-dependent kinases (Cdk1, Aurora, Plk), cell cycle-check points proteins (Chk2, p53, p73), survivin, MCl-2, Blc-2 proteins, caspase-2, etc. [86]. Mitotic catastrophe is a poorly defined molecular signalling pathway that precedes apoptosis, necrosis or senescence [88].

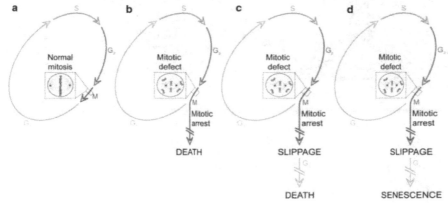

Figure 7. Mitotic catastrophe. a) In the absence of perturbations cells progress normally. A mitotic defect is detected and: **b)** cells die without exiting mitosis, **c)** cells undergo mitotic arrest, exit mitosis (mitotic slippage), reach the subsequent G₁ and die or **d)** undergo senescence (taken from Galluzzi et al. 2012).

10. Necroptosis

Necrosis is characterized by plasma membrane permeabilization, swelling and rupture. Necrosis can be accompanied in many instances by release of lysosomal hydrolases. Recently, a novel form of regulated necrosis has emerged and has been named **necroptosis** [78]. This cell death modality presents morphological features of necrosis but is regulated by signalling pathways and catabolic mechanisms. The most studied necroptotic pathway is mediated by the death receptor TNFR1 and inhibited by necrostatin-1. Upon TNF binding, TNFR1 undergoes a conformational change and recruits TRADD, TRARF2, cIAP1, cIAP2 and RIP1 to form *complex I*. RIP1 is polyubiquinated by cIPA1 and cIAP2 and activates the I Kappa B kinase (IKK) complex, triggering NFkB activation. RIP1 can be deubiquitinated by cylindromatosis D (CYLD) and together with RIP3 forms the *complex II* which also contains TRADD, FADD, and caspase-8. Subsequently, caspase-8 cleavage will induce apoptosis, whereas caspase-8 inhibition by caspase inhibitors (zVAD-fmk) for example, will favour necroptosis [89]. A novel death regulation platform named ripoptosome has recently been described. Unlike complex II, the ripoptosome forms independently of death receptors, is activated by genotoxic stress or IAP antagonists and is tightly regulated by IAPs (ciAP1, cIAP2, xIAP) or cellular FLICE-like inhibitory protein (FLIP), a catalytically inactive homologue of caspase-8 [78,90,91]. The executioner mechanisms of necroptosis are unclear. However, in some cases the necroptotic process involves ROS generation, lysosomal membrane permeabilization, AIF release from the mitochondria and PARP activation. When caspase activation is not involved, necroptosis is associated with formation of autophagic vesicles [92,93].

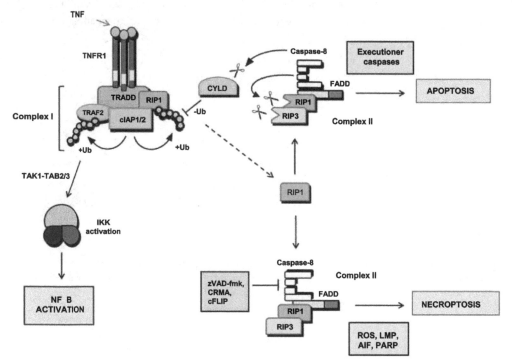

Figure 8. TNFR1-mediated apoptosis and necroptosis (taken from Long and Ryan 2012).

11. Apoptosis, chemoresistance and cancer

So far, we have reviewed generalities of different modalities of cell death that can take place after various pathologies: inflammation, stroke, ischemic injury, neurodegenerative disorders, viral infection, neoplasia, etc. The implication of apoptosis in cancer was initially observed as the type of cell death occurring in untreated tumours and in tumour regression after radiotherapy [4]. The oncogenic process requires accumulation of diverse alterations within a cell that disrupt its normal homeostasis of cell death and growth. It is well established that excessive proliferation is not only due to oncogene activation but also to failure of the pathways controlling programmed cell death mechanisms [94]. A malignant cell can be protected from cell demise though expression and/or activation of antiapoptotic factors (acting as oncogenes) or through inactivation of antiapoptotic factors (acting as tumour suppressor genes). Evasion of apoptosis is a hallmark of cancer [95]. Dysregulation of apoptotic pathways renders cells resistant to antitumour strategies since the final outcome of radio and chemotherapy is frequently apoptosis of cancer cells. Therefore, resistance to cell death- in particular apoptotic cell death- is an important aspect of carcinogenesis, as it confers resistance to anticancer agents [96]. In many tumours, chemoresistance acquisition is due to upregulation or modification of key elements of apoptosis control, such as Bcl-2, Bcl-X$_L$ and IAP family members [36]. Other mechanisms are characterized by inactivating mutations in proapoptotic proteins, such as p53.

Bcl-2. The first apoptotic gene related to neoplasia was the apoptotic inhibitor *bcl-2*. Therefore, Bcl-2 overexpression can contribute to tumour cell survival. Indeed, it was discovered at the breakpoint of the t(14;18) chromosomal translocation occurring in follicular lymphomas and leukemias [97,98]. Bcl-2 is the first oncogene that acts by inhibiting cell death instead of stimulating cell proliferation. In this respect, it was shown to cooperate with *myc* in immortalization of lymphoid cells [99] and in lymphomagenesis in transgenic mice [100]. Furthermore, it has been shown that Bcl-2 overexpression can confer multidrug resistance (MDR) phenotype and evasion of apoptosis to tumour cells exposed to serum deprivation, certain toxins or chemotherapeutic agents [101]. Mcl-1 and Bcl-X$_L$ have been shown to play important roles in tumourigenesis as well. Bcl-2 overexpression can also be the result of gene amplification or reduced expression of micro RNAs (miRNAs) in cancer cells [102].

BH3-only proteins. Loss or suppression of proaptotic proteins is frequent in cancer. Bax frameshift mutations appear in 50 % of colon carcinomas with DNA mismatch repair defects [103]. In addition, 17% of mantel cell lymphoma has homozygous Bim deletions [104], Bok and Puma suffer allelic deletions [105], and Bim and Puma are silenced by hypermethylationin Burkitt lymphoma [106,107]. Bim expression can also by silenced by miRNAs in several types of cancers [102]. Some BH-3 only proteins (Bim, Puma and Bmf) are necessary to initiate apoptosis in response to certain antineoplasic drugs [101].

XIAP. The X-linked inhibitor factor of apoptosis (XIAP) belongs to the IAP protein family. XIAP can render chemoresistance to cancer cells by inactivating caspases. XIAP overexpression is frequent in several tumour types [108]. Since XIAP is the only protein capable of inactivating both initiator and executioner caspases, it has become a putative biomarker of chemoresistance. It is the only IAP member capable of blocking active caspases [109]. However, XIAP has other roles in malignant transformation apart from preventing apoptosis. XIAP is involved in NFkB, MAPK and ubiquitin-proteasome pathways [108].

Survivin. This protein is a member of the IAP family that inhibits apoptosis by inactivating caspases [110] and stimulates DNA repair upon binding to DNA-PK in glioblastoma cell lines [111]. In addition, survivin plays a role in regulation of the mitotic checkpoint. Survivin expression is deregulated in cancer through several mechanisms: amplification of the locus on chromosome 17q25 [112], exon demethylation [113], or increased promoter activity [114]. Overall, survivin overexpression is an unfavourable prognostic marker and correlates with poor prognosis [110]. It is also involved in angiogenesis [115], tumour progression and chemoresistance. It has been shown that survivin inhibition sensitizes tumour cells to paclitaxel, cisplatin, etoposide, gamma radiation and immunotherapy [116].

p53. Inactivating mutations in the tumour suppressor p53 account for about 50% of human tumours and are associated with poor prognosis. One role of p53 is the regulation of cell cycle through the DNA damage response. Many chemotherapeutic agents cause DNA damage and activate p53. As a result, the cdk inhibitor p21 can be transcriptionally activated causing cell cycle arrest or Bax can be activated by translocation to the mitochondria inducing apoptosis [56]. When p53 is not mutated, cells showing DNA damage induced by genotoxic stress that cannot be repaired during cell cycle arrest, are induced to apoptosis through p53 activation of the mitochondrial pathway, mostly increasing transcription of the BH-3-only proteins, Bid, Noxa and Puma [52]. In addition, p53 can activate genes of the extrinsic pathway: TRAIL-R2 (DR5) and Fas (CD95/Apo-1) [117].

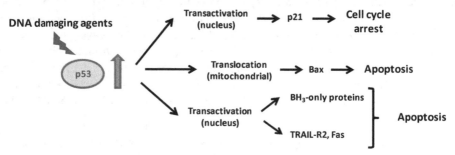

Figure 9. Transcription-dependent and independent effects of p53 activation

11.1. Defective checkpoints

p53 and the DNA damage response. DNA damaging agents can induce different types of lesions that culminate in cell death. DNA damage is detected by sensors within the cell that relay a signal causing cell cycle arrest and DNA repair or apoptosis. These networks of genome surveillance are known as replication checkpoints. When DNA damage is low the lesions can be repaired by several mechanisms. On the other hand, when the DNA damage is high or persistent, the cell undergoes apoptosis [118]. For example, as a response to genotoxic insults, p53 is phosphorylated by Checkpoint kinase 1 (Chk1) and Checkpoint kinase 2 (Chk2) and cannot be ubiquinated by mdm2 and becomes stabilized. Consequently, p21 is activated and cells are arrested either in the G_1 or the G_2 phase of the cell cycle. Conversely, if p21 is absent, apoptosis prevails. Therefore, p21 is critical in maintaining the balance between cell cycle arrest and apoptosis. The frequent loss of p53 in cancer enhances its ability to survive after DNA damage and evade apoptosis. When p53 is mutated, the response to DNA damage depends on mechanisms regulated by Ataxia telangiectasia mutated (ATM) and Ataxia telangiectasia and Rad3 related (ATR) independent of p53, which regulate Chk1 or Chk2 [119]. These kinases in turn activate NFkB, Akt and survivin [120].

Cell cycle checkpoints. Checkpoints are mechanisms that respond to internal or external stress by activating machineries that arrest the cell cycle at particular points (G_1/S, intra-S-phase, G_2/M, or mitotic spindle). Many tumour suppressor genes are components of DNA damage or cell cycle checkpoints: p53, Retinoblastoma (Rb), ATM, p16, BRCA1/2, etc. The frequency of tumour suppressor loss in cancer cells provides an advantage to their growth. The G_1 checkpoint is controlled by Rb, which in turn, is phosphorylated by cdk's. One cdk inhibitor is p16. Both Rb and p16 are frequently mutated in diverse types of cancer. The mitotic checkpoint utilizes genes (Mad, Bub, Aurora kinases) to detect spindle defects. When the cell is not able to surpass the damage, it undergoes mitotic catastrophe. In addition, loss of checkpoint controls increases genomic stability, providing cancer cells with adaptive or evolutionary advantages [121].

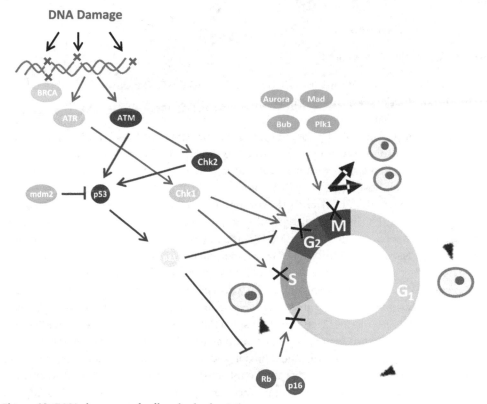

Figure 10. DNA damage and cell cycle checkpoints

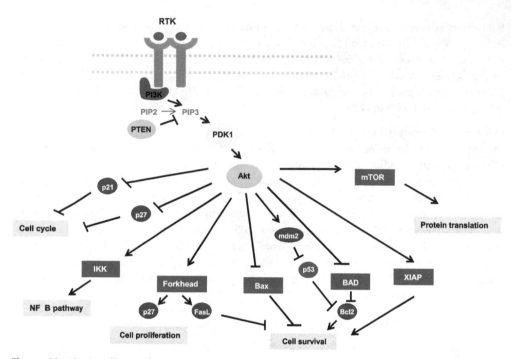

Figure 11. Akt signalling pathway

Akt. Classically, the phosphoinositide-3-kinase (PI3K)-Akt-mTOR pathway is described as a key signal transduction cascade that integrates signals from growth factors and nutrients to regulate cell growth and proliferation [122]. Following growth factor binding to cell surface receptors, PI3K is activated and phosphorylates phosphatidylinositol-4,5-biphosphate (PIP2) to generate the second messenger phosphatidylinositol-3,4,5-trisphosphate (PIP3). This process can be inhibited by the tumour suppressor protein phosphatase and tensin homolog (PTEN), which dephosphorylates PIP3 and terminates PI3K signalling. Then, the accumulated PIP3 recruits PDK1 and Akt through their Pleckstrin homology domains, and Akt is activated [123]. Akt recognizes and phosphorylates a consensus sequence that is present in many proteins. These substrates control key cellular processes such as apoptosis, cell cycle progression, transcription, and translation, all of which are critical events in cancer [124]. The Akt signal transduction pathway is probably the best survival pathway characterized. Moreover, Akt is constitutively activated in several malignant tumours, such as prostate, breast, ovary, lung and liver carcinomas [125]. Akt suppresses apoptosis through different mechanisms, including phosphorylation of forkhead transcription factors, which regulate proapoptotic proteins such as Bim and Fas ligand. The phosphorylated forkhead proteins are trapped in the cytosol and cannot enter the nucleus. Akt also phosphorylates and inactivates several proapoptotic proteins like Bad and caspase-9. Importantly, it activates IKK inducing the transcription factor NF-κB, leading to transcription of several antiapoptotic proteins such as Bcl-xL, and XIAP [96,126]. Akt also prevents the nuclear localization of p53 upon binding and phosphorylation of mdm2 [120]. Therefore, deregulation of the PI3K-Akt pathway can be considered as a cause of chemoresistance [127].

11.2. Immunogenecity of cancer cell death

It is becoming more evident that the immune response facilitates the effects of chemotherapy. Physiological death avoids autoimmunity. However cancer cell death triggered by radiotherapy or some chemotherapeutic agents such as anthracyclines can be immunogenic [128]. Immunogenic death involves changes in the composition of the cell membrane and the release of molecules called Damage Associated Molecular Patterns or alarmins. In particular, calreticulin has been shown to be crucial for immunogenic cancer cell death [129]. The immune system determines the long-term success of antitumor therapies. It seems that mitochondrial events as well as the ER response in conjunction with autophagy can establish whether cancer cells die in response to chemotherapy [130]. It has been found that calreticulin is the dominant pro-phagocytic signal on several cancers including neuroblastoma, non Hodgkin's lymphoma and bladder cancer. However, calreticulin is counterbalanced by the "don't eat me" signal CD47, which prevents cancer cell phagocytosis and is also highly expressed in these tumours [131]. Moreover, since CD47 is expressed on the surface of all human cancer cells but not in normal cells, blocking CD47 function with antibodies is emerging as a novel potential cancer strategy [132].

12. Senescence and cancer

We have already mentioned that strong p53 activation induces apoptosis. However, Leontieva and colleagues have shown that a weak and sustained p53 activation during cell cycle arrest can promote a different type of cell demise known as **senescence** [133]. Cellular senescence, induced after an irreversible cell cycle arrest, is a protective mechanism that limits proliferation of cells exposed to endogenous or exogenous insults. This cell death type can take part in processes of tumour suppression, tumour promotion, aging, and tissue repair [134]. Senescent cells become flattened and enriched with vacuoles. Some of the biochemical changes characteristic of senescence include elevation in β-galactosidase activity at an acidic pH, increase in senescence markers (p16, p15, p21, p53, ARF, etc.), heterochromatinization and, similarly to autophagy, formation of autolysosomes. It has been recently shown that chemotherapeutic agents may cause a form of "premature senescence" that can be considered as a tumour suppressor mechanism. Therefore, the induction of premature senescence as a drug-inducible arrest program has therapeutic potential for cancer treatment, but requires further investigation [135].

13. Autophagy and cancer

Regarding the role played by autophagy in cancer, mounting evidences suggest that autophagic cell death functions as a tumour suppressor mechanism. Several tumour suppressor proteins have been shown to induce autophagy. Supporting this idea, there are also several works showing that certain oncogenic proteins inhibit autophagy. Most of this oncogene products form part of the classic PI3K-Akt-mTOR pathway. In the following section, we summarize the literature on the role played by key tumour suppressors and oncogene products in the regulation of autophagy and the involvement of autophagic genes in cancer.

13.1. Tumour suppressor genes that regulate autophagy

To date, several tumour suppressor proteins that regulate autophagy have been described. These include: Beclin 1, UV irradiation-resistance-associated gene (UVRAG), PTEN, Bcl-2, and p53. The majority of them act as autophagy inducers with the exception of p53 that is able to activate or inhibit autophagy.

Role of Beclin 1 in autophagy: Beclin1 is a key regulator of autophagy. It regulates the autophagic process at different levels by combining with the enzyme class III PI3K (PI3KC3) and specific group of proteins forming unique class III PI3K-Beclin complexes. This PI3KC3-Beclin complex functions as a core complex during autophagy [136,137] by phosphorylating phosphatidylinositol to produce phosphatidylinositol 3-monophosphate, which presumably allows the recruitment of essential autophagy related (Atg) proteins to the membrane. In mammals, at least three types of class III PI3-kinase-Beclin complexes contribute to autophagy [84,138,139].

Role of Beclin 1 in tumour suppression vs. autophagy: The first link between autophagy and cancer was established in 1999, when Liang and colleagues discovered that the autophagic protein Beclin 1 was able to inhibit tumourigenesis [140]. Beclin 1 is a tumour suppressor gene [141,142] monoallelically deleted in many types of cancers, including ovary (75%), breast (50%) and prostate (40%). Several studies suggest that the activation of autophagy by Beclin 1 is tightly associated with its tumour suppression function. For example, overexpression of Beclin-1 in the human MCF-7 breast carcinoma cell line is associated with inhibition of cell proliferation and induces autophagy [140]. In contrast, studies *in vivo* demonstrate that the monoallelic disruption of Beclin 1 gene promotes cellular proliferation and reduces autophagy [142]. Furthermore, genetically engineered mouse models with heterozygous disruption of Beclin 1 have decreased autophagy and are more prone to develop tumours, including lung carcinomas, hepatocellular carcinomas, lymphomas and mammary precancerous lesions [141,143]. Taken together, these studies firmly support the notion that autophagy induction and tumour inhibition function of Beclin 1 are closely interconnected.

UVRAG is another tumour suppressor protein that is able to activate autophagy. UVRAG is a positive regulator of Beclin 1, as it promotes its binding to PI3KC3 to form the complex and enhance Beclin 1 activity. Through mediating Beclin 1-PI3KC3 complex, UVRAG can promote autophagy, thus inhibiting carcinogenesis of human colon cancer cells. It has also been proposed that UVRAG is involved in causing membrane curving [144]. Furthermore, the monoallelic deletion or mutation of UVRAG has been observed in several human cancers such as colon, gastric and breast cancer [145-147].

Bif-1 is also a tumour suppressor that promotes autophagy. Similar to UVRAG, Bif-1 is a positive regulator of Beclin 1, as it enhances its interaction with PI3KC3, thus increasing autophagy. Consistently, it has been shown that loss of Bif-1 prevented the formation of autophagosomes and induced the development of spontaneous tumours in mice [144]. Furthermore, expression of Bif-1 is significantly reduced in prostate, colon, urinary bladder and gastric tumours [146,148,149]. The precise mechanism of Bif-1 in suppressing tumourigenesis has not been fully clarified, but it is reasonable to presume that its tumour inhibitory activity is associated with its role in inducing autophagy. Some groups have also suggested that Bif-1 suppresses tumour growth through its interaction with proapoptotic Bax [150].

13.2. Oncogenic genes that regulate autophagy

Most of the oncogenic genes that regulate autophagy described to date are key proteins of the PI3K-Akt-mTOR pathway. These proteins inhibit autophagy and promote carcinogenesis.

mTOR: the best-studied downstream substrate of Akt is the serine/threonine kinase mammalian target of rapamycin (mTOR). In mammals, mTOR exists in two different complexes, known as mTORC1 and mTORC2. In the mTORC1 complex, mTOR is bound to a protein called Raptor, and in the mTORC2 complex, mTOR is bound to Rictor. mTORC1 is best known for its role in regulating protein synthesis through two of its substrates, 4E-BP1 and p70S6K [151]. Akt can directly phosphorylate and activate mTOR, or activate it indirectly by phosphorylation and inactivation of TSC2. When TSC2 is inactivated by phosphorylation, it can no longer associate with TSC1, thus abolishing the inhibitory effect of TSC1-TSC2 complex on mTORC1. As a result, mTOR is activated and signals to its downstream targets p70S6 kinase, ribosomal protein S6 and 4EBP-1/eIF-4E to control protein translation and proliferation [152]. mTORC1 is not only a key regulator of proliferation and survival, but also a master inhibitor of autophagy. Under nutrient-rich conditions, mTORC1 is activated and phosphorylates ULK1, ULK2 and mAtg13, inhibiting the initiation of autophagy. In contrast, in nutrient starvation conditions mTORC1 is inactivated and dissociates, resulting in activation of ULK1 and ULK2, which initiates the autophagy cascade [151]. The PI3K-Akt-mTOR pathway is probably the most commonly activated signalling pathway in human cancers [124]. This route is activated aberrantly in many types of tumours, and hence, inhibited autophagy and increased proliferation and protein synthesis are often observed. For example, many human cancers are characterized by activating mutations in specific components of the signalling pathway that connect Receptor tyrosine kinases to mTOR, including Ras, PI3K and Akt. Somatic mutations in PIK3CA gene encoding the catalytic subunit of class I PI3K frequently occur in cancers of colon, breast, brain and lung [153]. PTEN, another regulator of PI3KI-Akt pathway, is a tumour suppressor that is found to be mutated in many types of cancer.

Bcl-2 forms part of the Bcl-2 family proteins. For almost two decades, Bcl-2 has been regarded to function as an antiapoptotic protein that contributes to tumourigenesis. The role of Bcl-2 in autophagy was realized in 1998 when Liang identified Beclin 1 as Bcl-2 interacting protein [140]. Bcl-2 acts as a negative regulator of autophagy by inhibiting Beclin 1. It binds constitutively to Beclin 1, blocking the interaction between Beclin 1 and PI3KC3 [154-156]. As a result, PI3KC3 activity is decreased and autophagy is downregulated [155]. The binding of Bcl-2 with Beclin 1 seems to be constitutive, and its detachment from Beclin 1 is speculated to be essential in autophagy.

13.3. How does autophagy suppress tumourigenesis?

The combined data presented above strongly support the idea that autophagy functions as a tumour suppressor process. Consistent with this, oncogenes that have a role in autophagy are potent inhibitors of this process. Although the molecular mechanisms by which autophagy functions in tumour suppression are poorly defined, at least two mechanisms have been described:

a.   Autophagy maintains the integrity of the genome

The first hypothesis is that autophagy may function as a housekeeping pathway to exert the quality control of organelles, proteins and DNA. Mathews and colleague observed that in autophagy deficient tumour cells, metabolic stress promotes the accumulation of p62, damaged mitochondria and ROS generation, promoting genomic instability and leading to oncogene activation and tumour progression [157,158]. Furthermore, immortalized mouse epithelial cells with a defect in ATG genes (loss of Beclin 1 or Atg 5) display increased DNA damage, centrosome abnormalities, numerical and structural chromosomal abnormalities and gene amplification, especially after ischemic stress.

However, the mechanisms through which autophagy preserves the integrity of the genome remains elusive. One possibility could be that autophagy may contribute to cell cycle regulation, for example by degrading organelles and/or proteins involved in the cell cycle checkpoints [159]. Another possibility is that autophagy might simply function at a more general level to ensure the minimal amount of ATP and other metabolites required for DNA repair. Finally, autophagy may act by removing old and/or damaged organelles (for example, uncoupled mitochondria) which may act as a source of genotoxic chemical species such as ROS.

b.   Autophagy limits necrosis-mediated inflammation

Necrosis normally results from physical injury in which the cell lyses and releases its intracellular contents, which activate the innate immune system and a wound-healing response [160]. As a result, inflammatory cells are recruited and cytokines are released to promote cell growth to replace the damaged tissue [161,162].

In contrast, apoptosis may be the preferred means of cell demise for cells upon metabolic stress, as cells are eliminated without inflammation. However, in cancer cells with a defect in apoptosis, autophagy is induced for cell survival. Through autophagy, cells eliminate damaged organelles and may maintain their normal cellular function under adverse conditions of fluctuating oxygen and nutrient supply. However, this beneficial effect of autophagy functions during short term interruptions in nutrient availability, as in the long term (excess of autophagy) it can possibly lead to cell death.

A high proportion of tumours have been seen to present a defect both in autophagy and apoptosis. Degenhardt and colleagues have shown that the inhibition of both processes under conditions of metabolic stress generates a necrotic cell-death, suggesting that apoptosis and autophagy function to limit necrosis [163]. In these necrotic tumours a persistent inflammatory infiltration and cytokine production exists, which is thought to promote tumour growth and thus, is associated with poor prognosis.

13.4. Role of autophagy in tumour survival

The data presented above strongly supports the idea that autophagy functions as a tumour suppressor process and that inhibition of autophagy leads to carcinogenesis. However, there are some circumstances where autophagy contributes to tumour survival promoting carcinogenesis:

a.   Autophagy is induced by nutrient starvation

The survival role of autophagy during nutrient limitation is well established. When cells encounter environmental stresses such as nutrient starvation, autophagy can be activated and protects cells by preventing them from undergoing apoptosis. Through autophagy, starving cells degrade cytoplasmic material to generate both nutrients and energy [85]. Consistent with this, during nutrient starvation, inhibition of autophagy promotes apoptosis [164].

b.   Autophagy is induced by hypoxia

Hypoxia in tumours results from inadequate tumour vasculature and is associated with a more malignant phenotype, higher predisposition for metastasis, and poor prognosis. Hypoxic stress selects for cells that are resistant to apoptosis as well as poses a major barrier to chemotherapy and radiotherapy.

White and colleagues first showed that autophagy is induced specifically in the hypoxic core of tumours, where it promotes survival [163]. Further studies have unveiled the molecular connections between hypoxia and the activation of autophagy. For example, it has been reported that when oxygen concentration falls below 5% hypoxia inducible factor 1 (HIF1) is activated and this transcription factor activates key autophagy inducers (BNIP3), which in turn activate the key autophagy complex formed by PI3K III [165,166]. Further mechanistic studies have revealed that induction of BNIP3 and BNIP3L in hypoxic cells disrupts the Becn1-Bcl-2 complex, thereby releasing Becn1 to induce autophagy [167].

The role of autophagy as a key mediator of survival of hypoxic cells is emerging so that the exact mechanisms underlying this phenotype remain unclear. Because chronic hypoxia leads to major metabolic perturbations in tumour tissues, one can postulate that by recycling basic cellular components, autophagy helps stressed cells cope with the increased metabolic demand [168]. However, further studies are needed to validate this hypothesis.

c.   Autophagy is induced in metastatic cells

Epithelial cells critically depend on cell adhesion to extracellular matrix (ECM) for proper growth and survival. Upon detachment of cells from the ECM, cells undergo anoikis, a type of apoptotic cell death that serves the homeostatic function of killing cells that have lost contact with the basement membrane [68]. It has been shown that autophagy is induced in oncogene-transformed cells following matrix detachment [169]. Similarly, in three dimensional (3D) epithelial cell culture models, autophagy is significantly increased in the detached luminal cells. Furthermore, when autophagy is inhibited accelerated luminal clearance occurs [158,169]. Altogether, these results suggest that autophagy is fundamental in anoikis resistance, a process exploited by tumour cells to survive after detachment from the primary site, as well as while migrating to distant metastatic sites [170].

Debnath and colleagues have shown that detachment induces autophagy in both nontumourigenic epithelial cell lines and in primary epithelial cells. Autophagy inhibition through siRNA for ATG genes inhibits detachment-induced autophagy and increases apoptosis. Remarkably, even when apoptosis is inhibited matrix-detached cells still exhibit autophagy. Moreover, inhibition of autophagy in MCF-10 acini enhances luminal apoptosis during morphogenesis and fails to elicit long-term luminal filling [171]. Altogether, these results indicate that autophagy promotes epithelial cell survival during anoikis, including in detached cells harbouring antiapoptotic lesions.

14. Current therapeutic advances

In addition to inactivation of proliferative and prosurvival oncogenic pathways, current anticancer strategies deal with reactivation of cancer death cell signalling routes in order to induce tumour regression. The most promising cancer therapies that specifically target apoptosis, necrosis/necroptosis and autophagy are described in the following section.

14.1. Extrinsic apoptotic pathway

TRAIL: This member of the TNF family stimulates the extrinsic apoptotic pathway in a wide variety of cancers upon binding to death receptors, while having no effect on normal cells. Therefore, the use of recombinant TRAIL or agonistic TRAIL-receptor antibodies constitutes a novel therapeutic strategy [172]. TRAIL has been found to bind five receptors in human cells: TRAIL-R1, TRAIL-R2, TRAIL-R3 (DcR1), TRAIL-R4 (DcR2) and the soluble receptor osteoprogeterin. However, DcR1 and DcR2 are unable to transduce a death signal and are considered "decoy receptors". Ashkenazi and colleagues have proposed that the specificity of TRAIL for cancer cells is due to the fact that normal cells express more frequently these decoy receptors [173]. Moreover, TRAIL only weakly induces the prosurvival NFkB pathway [78]. Unfortunately, many primary tumour cell lines have shown resistance to TRAIL [174]. It has been shown that in preclinical studies, the combination of recombinant human TRAIL (rhTRAIL, AMG951, Dulanermin) or agonistic antibodies with irradiation, several chemotherapeutic agents, Akt or proteosome inhibitors induced cell death in tumours resistant to TRAIL [175-177]. Therefore, the efficacy of rhTRAIL or anti-TRAIL-R1/2 antibodies is being examined in different phase I and phase II clinical trails alone or in combination with other chemotherapeutic agents [172,178,179].

Fas activating compounds: Fas, also called CD95 or APO-1 is a member of the TNF family of death receptors involved in the extrinsic apoptotic pathway. Fas plays an important role in regulation of the immune system. This receptor belongs to a subgroup within the TNFR family which includes TNFR1, DR3, TRAIL-R4 and TRAIL-R5 [180]. Fas has been recently implicated in the activation of RIP1, resulting in necroptosis when caspase-8 is inactive [181,182]. In addition, Tschopp and colleagues have demonstrated that Fas can activate other signalling pathways involved in proliferation or differentiation [109]. Fas agonists, such as highly aggregated FasL or anti-Fas antibodies are not suitable for cancer therapy owing to their profound liver toxicity. However, weaker Fas agonists may be useful. In this respect, APO010 is undergoing clinical trials for solid tumours [109].

14.2. Intrinsic apoptotic pathway

Bcl-s inhibitors/BH3 mimetics: Many efforts in cancer therapeutics are directed towards inhibition of antiapoptotic Bcl-2 like proteins. The first class of drugs being developed was antisense nucleotides. Oblimersen has been tested in clinical trials in different types of cancers. Even though some clinical trials shows efficacy of oblimersen combined with chemotherapy in treating melanoma [183], other studies show that this Bcl-2 inhibitor does not confer significant benefit to cancer patients [178]. Small molecule Bcl-2 inhibitors are also called "BH3 mimetics", as they imitate the function of BH3-only proteins: Gossypol (AT-101) binds Bcl-2, Bcl-X$_L$ and MCl-1, inhibiting their binding to Bax and Bak. In combination with other chemotherapeutic agents it promotes apoptosis in preclinical studies and has shown efficacy in at least one clinical trial [178]. Obotoclax is a pan Bcl-2 inhibitor that induces apoptosis in leukemia cells. However, recent studies have shown that its activity is partially mediated by autophagy [184,185]. Therefore, caution should be taken when using these kinds of drugs. ABT-737 and its orally available derivative, ABT-263 (navitoclax), have shown promising results in preclinical studies. They bind Bcl-2, Bcl-X$_L$ and Bcl-w [186], but mildly inhibit Mcl-1 [187] or A1 [101]. ABT-727 synergizes with standard chemotherapeutic agents, radiation and tyrosine kinase inhibitors in cancer cells and is effective as a single agent promoting tumour regression in mice. ABT-263 is currently in phase I and II clinical trails as a single agent [78,188]. However, its use is limited because it causes trombocytophenia owing to Bcl-X$_L$ inhibition [189]. A new BH-3 mimetic agent, GDC-0199 has been developed and entered clinical trails for lymphomas [101].

IAP inhibitors and Smac mimetics: Moreover, the role of IAPs in regulating apoptosis and/or necroptosis has led to the development of small molecules that antagonize IAP, or mimic the function of Smac/DIABLO. In fact, IAPs have been found to be overexpressed in many cancers and to be associated with poor prognosis and chemoresistance [102], making them putative targets for cancer therapeutics. Smac mimetics can sensitize chemoresistant cancer cells to cisplatin, doxorubicin or TRAIL [190], either as single agents or in combination with other drugs [109]. Therefore, some of these agents (AT-406, LCL-161, HGS-1029, TL32711, GDC-0917) [73] have recently entered clinical trials [191] for the treatment of solid tumours and lymphomas [192]. However, the potential of Smac mimetics as antitumour agents has been questioned, since the inhibition of cIAP1 and cIAP2 can stabilize IKK and activate the proinflammatory NFkB pathway [102].

Other IAP-targeted therapies such as XIAP and survivin antisense oligonucleotides have been developed [102]. Antisense oligonucleotides against survivin synergize with etoposide in non small lung cancer cells [117]. LY2181308 is a second generation antisense oligonucleotide being evaluated in clinical trials. In addition, a small molecule inhibitor of survivin, YM155, has shown potency in preclinical models and has entered phase II clinical trails [110].

14.3. DNA damage and cell cycle

p53 inhibitors: Since p53 is frequently mutated in tumours, therapeutic approaches have been made to restore p53 function. Since p53 is targeted to degradation through interaction with mdm2, some drugs such as RITA, Nutlin-3 or HLI198 have been designed to disrupt p53-mdm2 interaction. These compounds bind to the p53 binding site on mdm2 or inhibit mdm2 ubiquitin ligase activity. This approach is valid in cancers with wild type p53, such as haematological malignancies [56]. A second approach is to rescue wild-type 53 function in p53-mutated tumours. PRIMA-1 restores sequence-specific DNA binding and active conformation of mutant p53 proteins. PRIMA-1 can synergize with conventional chemotherapeutic drugs and inhibit tumour growth in mice with no apparent toxicity and has recently entered clinical trials [78].

Chk1/2 inhibitors: The concept of "synthetic lethality" coined by Kaelin, by which two molecular lesions combine to have a lethal effect on the cell, although neither of them is harmful individually, has recently gained interest [193,194]. In this context, Chk inhibitors may have a therapeutic potential in p53-mutated tumours. UCN-01, the first Chk inhibitor evaluated in humans, has limited clinical value due to its toxicity. Other Chk inhibitors functioning as checkpoint abrogators that are being evaluated in clinical trials are: AZD7762, LY26303618, CBT501, PF-00477736, SCH 900776, XL844, and the wee-1 inhibitor MK-1775 [195,196].

PARP inhibitors: Other example of synthetic lethality is BRCA mutant breast cancer. BRCA1 and BRCA2 play a role in homologous recombination, an important repair pathway for

DNA double strand breaks. The current hypothesis is that since PARP is involved in repairs in DNA single strands breaks, inhibition of PARP leads to accumulation of single strand unrepaired breaks and becomes synthetically lethal in BRCA-mutated cancers [197]. Several PARP inhibitors are being evaluated in clinical trials: AG014699, AZD2281 (olaparib), ABT888 (veliparib), BSI-201 (iniparib), INO-1001, GP121016, CEP-9722 and MKI4827 [197,198]. PARP inhibitors may be useful in tumours bearing other types of genomic or functional defects in DNA damage response pathways.

14.4. Necroptosis

cFLIP: This antiapoptotic regulator is expressed as long (c-FLIPL) short (c-FLIPs) or c-FLIPR variants in human cells. In the absence of cFLIP, the ripoptosome triggers apoptotic cell death. On the contrary, expression of cFLIPL, which binds caspase-8, neutralizes its ability to engage the apoptotic machinery, but inhibits necroptosis by preventing RIP1-RIP3 association. On the other hand, c-FLIPs, although is able to bind caspase-8, does not form an active heteromer and necroptosis can develop [78,199]. Increased expression of c-FLIP has been found in several human cancers and is associated with poor prognosis [102]. Downregulation of FLIP seems to be a promising therapeutic strategy. Research efforts are being focused on the development of small interference RNA (siRNA) targeted against c-FLIP. In addition, several antitumour agents can downregulate c-FLIP at the transcriptional or posttranscriptional level [200].

14.5. Autophagy

a.   Autophagy as a protective mechanism

A large series of anticancer drugs (both clinically approved and experimental) are able to induce a significant accumulation of autophagosomes in tumour cells both *in vitro* and *in vivo* [201]. For many years, it was thought that these therapies may kill tumours through autophagy and this massive vacuolization of the cytoplasm has been considered a manifestation of **autophagic cell death**. According to this notion, autophagy would represent another mechanism of cell death and the inhibition of autophagy would protect the cell from dying. However, this perspective is nowadays rather discredited, as the majority of the literature has reported that specific inhibition of autophagy (through siRNA) contributes to cell death during cancer treatment [202-204]. Besides, it is now well accepted that acquired resistant to chemotherapeutic drugs is, in part, due to the adaptive prosurvival response conferred by autophagy. Also, several studies have shown that inhibition of autophagy by pharmacological inhibitors such as Bafilomycin A1, cloroquine, or 3-methyladenine can enhance and accelerate cytotoxic cancer therapy in several tumours [205-207]. In particular, some reports show that chloroquine as a single agent is sufficient to promote tumour regression in transplanted cancer models [208,209]. Moreover, chloroquine is a well-tolerated nontoxic drug that has entered a clinical trail as monotherapy for pancreatic cancer treatment [78]. This clearly indicates that the induction of autophagy represents an attempt of the cells to cope with the stress induced by cytotoxic drugs and suggests that the inhibition of the autophagic process might be beneficial in cancer treatment.

b.   Autophagy as a death mechanism

Although there is robust evidence indicating that autophagy has a protective role in cancer therapeutics, in certain cancer treatments, autophagy can kill cells by inducing **autophagic cell death**. For example, Abe and colleagues have demonstrated that harmol, a β-carboline alkaloid, triggered autophagic cell death in human lung carcinoma A549 cells without activation of caspase-3, caspase-8, or caspase-9 or PARP cleavage. Autophagy, but not apoptosis, was detected by electron microscopy in these cancer cells. Furthermore, pretreatment of A549 cells with the autophagy inhibitor 3-methyladenine or siRNA-mediated knockdown of LC3 suppressed harmol-induced cell death [210]. Another study shows that L929 fibrosarcoma cells die in a caspase-independent manner involving autophagy and that ATG genes are required for this cell death process [211]. In this model, caspase inhibition induces the selective autophagic degradation of catalase, a major ROS scavenger, and the resulting ROS accumulation promotes autophagic cell death [212].

14.6. What determines if autophagy is cytoprotective or cytotoxic?

Autophagy is a process that allows cells to escape death or paradoxically leads to cell death. It is not yet understood what factors determine whether autophagy is cytoprotective or cytotoxic. It has been suggested that autophagy induced under pathological conditions functions as an adaptive cell response, allowing the cell to survive bioenergetic stress. However, autophagy is a process that destroys cellular content and organelles. In this way, it has been suggested that deregulated, excessive or persistent autophagy may lead to autophagic cell death. That is, the destruction of proteins and organelles may pass a threshold, leading to cell death. However, the point at which autophagy becomes autophagic cell death remains unclear. In this perspective, the dissection of the transition from autophagy to autophagic cell death and the cross-talking between apoptosis and autophagy may help to understand this process, leading to more efficacious treatments in cancer. In contrast, a different study has reported that when cells are subjected to prolonged growth factor deprivation or shortage of glucose and oxygen they can lose the majority of their mass via autophagy. However, when these cells are placed in optimal culture conditions, they are able to fully recover [163,213]. This result suggests that cell death via autophagy may not be simply a matter of crossing a quantitative threshold of self-digestion.

14.7. Interplay among apoptosis, necrosis/necroptosis and autophagy

Cell death process *in vivo* involves a complex interaction among apoptosis, necrosis/necroptosis and autophagy [78]. In some situations, a specific stimulus triggers only one mechanism, but in other cases, the same stimulus can evoke more than one cell demise machinery. Therefore, several mechanisms can coexist within a cell, but only one will predominate over the others. The decision to undergo apoptosis, necrosis/necroptosis or autophagy depends on various factors: energy/ATP availability, extent of the stress and the damage and presence of inhibitors of particular pathways (e. g. caspases inhibitors) [214]. ATP depletion activates autophagy, but if autophagy fails to maintain energy levels, necrosis/necroptosis results [215]. Apoptosis is usually triggered with sufficient ATP levels to trigger caspases. If the damage is severe necrosis prevails over apoptosis [216,217]. Hence, apoptosis is the first choice in most circumstances and necroptosis is triggered only as a backup alternative to guarantee that cell death takes place. However, in some cases (e.g. viral infection) TNF-mediated necroptosis predominate [218]. Apoptosis can inhibit autophagy through Bcl-2-mediated sequestration [155,156] or caspase-dependent cleavage of Beclin 1 [219,220,220], and conversely, autophagy can inhibit apoptosis by caspase-8 degradation [221].

15. Concluding remarks

Programmed cell death mechanisms are intricate and usually interconnected processes. Evasion of cell death is a common feature of cancer cells leading to chemoresistance. Apoptosis, necrosis/necroptosis and autophagy are the main explored pathways that had gained interest among cancer biologists, as means to develop novel cancer therapeutics. Deeping our knowledge on the nexus between cell death and cancer will enable us to predict in a more refined manner the carcinogenic process and therefore, pave the way for a personalized approach to the disease.

Author details

Silvina Grasso, Estefanía Carrasco-García, Lourdes Rocamora-Reverte,
Ángeles Gómez-Martínez and José A. Ferragut
Instituto de Biología Molecular y Celular, Universidad Miguel Hernández, Elche (Alicante), Spain

M. Piedad Menéndez-Gutiérrez
Instituto de Biología Molecular y Celular, Universidad Miguel Hernández, Elche (Alicante), Spain

Centro Nacional de Investigaciones Cardiovasculares, Departamento de Desarrollo y Reparación Cardiovascular, Madrid, Spain

Leticia Mayor-López, Elena Tristante, Isabel Martínez-Lacaci*
Instituto de Biología Molecular y Celular, Universidad Miguel Hernández, Elche (Alicante), Spain

Unidad AECC de Investigación Traslacional en Cáncer, Hospital Universitario Virgen de la Arrixaca, Instituto Murciano de Investigación Biosanitaria, Murcia, Spain

Pilar García-Morales and Miguel Saceda
Instituto de Biología Molecular y Celular, Universidad Miguel Hernández, Elche (Alicante), Spain

Unidad de Investigación, Hospital General Universitario de Elche, Elche (Alicante), Spain

---

* Corresponding Author

16. References

[1]   Vaux DL, Korsmeyer SJ (1999) Cell death in development. Cell ;96: 245-254.

[2]   Vermeulen K, Van Bockstaele DR, Berneman ZN (2005) Apoptosis: mechanisms and relevance in cancer. Ann.Hematol.84: 627-639.

[3]   deCathelineau AM, Henson PM (2003) The final step in programmed cell death: phagocytes carry apoptotic cells to the grave. Essays Biochem.39: 105-117.

[4]   Kerr JF, Wyllie AH, Currie AR (1972) Apoptosis: a basic biological phenomenon with wide-ranging implications in tissue kinetics. Br.J.Cancer ;26: 239-257.

[5]   Kroemer G, Galluzzi L, Vandenabeele P, Abrams J, Alnemri ES, Baehrecke EH, et al. (2009) Classification of cell death: recommendations of the Nomenclature Committee on Cell Death 2009. Cell Death.Differ.16: 3-11.

[6]   Wyllie AH, Kerr JF, Currie AR (1980) Cell death: the significance of apoptosis. Int.Rev.Cytol.68: 251-306.

[7]   Zamzami N, Marchetti P, Castedo M, Zanin C, Vayssiere JL, Petit PX, et al. (1995) Reduction in mitochondrial potential constitutes an early irreversible step of programmed lymphocyte death in vivo. J.Exp.Med.181: 1661-1672.

[8]   Martin SJ, Reutelingsperger CP, McGahon AJ, Rader JA, van Schie RC, LaFace DM, et al. (1995) Early redistribution of plasma membrane phosphatidylserine is a general feature of apoptosis regardless of the initiating stimulus: inhibition by overexpression of Bcl-2 and Abl. J.Exp.Med.182: 1545-1556.

[9]   Majno G, Joris I (1995) Apoptosis, oncosis, and necrosis. An overview of cell death. Am.J.Pathol.146: 3-15.

[10] Hail N, Jr., Carter BZ, Konopleva M, Andreeff M (2006) Apoptosis effector mechanisms: a requiem performed in different keys. Apoptosis.11: 889-904.

[11] Galluzzi L, Vitale I, Abrams JM, Alnemri ES, Baehrecke EH, Blagosklonny MV, et al. (2012) Molecular definitions of cell death subroutines: recommendations of the Nomenclature Committee on Cell Death 2012. Cell Death.Differ.19: 107-120.

[12] Kroemer G, Galluzzi L, Brenner C (2007) Mitochondrial membrane permeabilization in cell death. Physiol Rev.87: 99-163.

[13] Ellis HM, Horvitz HR (1986) Genetic control of programmed cell death in the nematode C. elegans. Cell ;44: 817-829.

[14] Thornberry NA, Lazebnik Y (1998) Caspases: enemies within. Science ;281: 1312-1316.

[15] Pop C, Salvesen GS (2009) Human caspases: activation, specificity, and regulation. J.Biol.Chem.284: 21777-21781.

[16] Kumar S (2007) Caspase function in programmed cell death. Cell Death.Differ.14: 32-43.

[17] Blank M, Shiloh Y (2007) Programs for cell death: apoptosis is only one way to go. Cell Cycle ;6: 686-695.

[18] Cohen GM (1997) Caspases: the executioners of apoptosis. Biochem.J.326 (Pt 1): 1-16.

[19] Tenev T, Zachariou A, Wilson R, Ditzel M, Meier P (2005) IAPs are functionally non-equivalent and regulate effector caspases through distinct mechanisms. Nat.Cell Biol.7: 70-77.

[20] Huang H, Joazeiro CA, Bonfoco E, Kamada S, Leverson JD, Hunter T (2000) The inhibitor of apoptosis, cIAP2, functions as a ubiquitin-protein ligase and promotes in vitro monoubiquitination of caspases 3 and 7. J.Biol.Chem.275: 26661-26664.

[21] Suzuki Y, Nakabayashi Y, Takahashi R (2001) Ubiquitin-protein ligase activity of X-linked inhibitor of apoptosis protein promotes proteasomal degradation of caspase-3 and enhances its anti-apoptotic effect in Fas-induced cell death. Proc.Natl.Acad.Sci.U.S.A ;98: 8662-8667.

[22] Yang QH, Church-Hajduk R, Ren J, Newton ML, Du C (2003) Omi/HtrA2 catalytic cleavage of inhibitor of apoptosis (IAP) irreversibly inactivates IAPs and facilitates caspase activity in apoptosis. Genes Dev.17: 1487-1496.

[23] Danial NN, Korsmeyer SJ (2004) Cell death: critical control points. Cell ;116: 205-219.

[24] Meier P, Vousden KH (2007) Lucifer's labyrinth--ten years of path finding in cell death. Mol.Cell ;28: 746-754.

[25] Peter ME, Budd RC, Desbarats J, Hedrick SM, Hueber AO, Newell MK, et al. (2007) The CD95 receptor: apoptosis revisited. Cell ;129: 447-450.

[26] Lavrik I, Golks A, Krammer PH (2005) Death receptor signaling. J.Cell Sci.118: 265-267.

[27] Green DR, Kroemer G (2004) The pathophysiology of mitochondrial cell death. Science ;305: 626-629.

[28] Tait SW, Green DR (2010) Mitochondria and cell death: outer membrane permeabilization and beyond. Nat.Rev.Mol.Cell Biol.11: 621-632.

[29] Luo X, Budihardjo I, Zou H, Slaughter C, Wang X (1998) Bid, a Bcl2 interacting protein, mediates cytochrome c release from mitochondria in response to activation of cell surface death receptors. Cell ;94: 481-490.

[30] Chinnaiyan AM, Tepper CG, Seldin MF, O'Rourke K, Kischkel FC, Hellbardt S, et al. (1996) FADD/MORT1 is a common mediator of CD95 (Fas/APO-1) and tumor necrosis factor receptor-induced apoptosis. J.Biol.Chem.271: 4961-4965.

[31] Carswell EA, Old LJ, Kassel RL, Green S, Fiore N, Williamson B (1975) An endotoxin-induced serum factor that causes necrosis of tumors. Proc.Natl.Acad.Sci.U.S.A ;72: 3666-3670.

[32] Schievella AR, Chen JH, Graham JR, Lin LL (1997) MADD, a novel death domain protein that interacts with the type 1 tumor necrosis factor receptor and activates mitogen-activated protein kinase. J.Biol.Chem.272: 12069-12075.

[33] Yang J, Liu X, Bhalla K, Kim CN, Ibrado AM, Cai J, et al. (1997) Prevention of apoptosis by Bcl-2: release of cytochrome c from mitochondria blocked. Science ;275: 1129-1132.

[34] Vandenabeele P, Galluzzi L, Vanden Berghe T, Kroemer G (2010) Molecular mechanisms of necroptosis: an ordered cellular explosion. Nat.Rev.Mol.Cell Biol.11: 700-714.

[35] Galluzzi L, Zamzami N, de La Motte RT, Lemaire C, Brenner C, Kroemer G (2007) Methods for the assessment of mitochondrial membrane permeabilization in apoptosis. Apoptosis.12: 803-813.

[36] Adams JM, Cory S (2007) The Bcl-2 apoptotic switch in cancer development and therapy. Oncogene ;26: 1324-1337.

[37] Bidere N, Lorenzo HK, Carmona S, Laforge M, Harper F, Dumont C, et al. (2003) Cathepsin D triggers Bax activation, resulting in selective apoptosis-inducing factor (AIF) relocation in T lymphocytes entering the early commitment phase to apoptosis. J.Biol.Chem.278: 31401-31411.

[38] Cirman T, Oresic K, Mazovec GD, Turk V, Reed JC, Myers RM, et al. (2004) Selective disruption of lysosomes in HeLa cells triggers apoptosis mediated by cleavage of Bid by multiple papain-like lysosomal cathepsins. J.Biol.Chem.279: 3578-3587.

[39] Thomas DA, Scorrano L, Putcha GV, Korsmeyer SJ, Ley TJ (2001) Granzyme B can cause mitochondrial depolarization and cell death in the absence of BID, BAX, and BAK. Proc.Natl.Acad.Sci.U.S.A ;98: 14985-14990.

[40] Datta SR, Dudek H, Tao X, Masters S, Fu H, Gotoh Y, et al. (1997) Akt phosphorylation of BAD couples survival signals to the cell-intrinsic death machinery. Cell ;91: 231-241.

[41] Donovan M, Cotter TG (2004) Control of mitochondrial integrity by Bcl-2 family members and caspase-independent cell death. Biochim.Biophys.Acta ;1644: 133-147.

[42] Belizario JE, Alves J, Occhiucci JM, Garay-Malpartida M, Sesso A (2007) A mechanistic view of mitochondrial death decision pores. Braz.J.Med.Biol.Res.40: 1011-1024.

[43] Szabadkai G, Rizzuto R (2004) Participation of endoplasmic reticulum and mitochondrial calcium handling in apoptosis: more than just neighborhood? FEBS Lett.567: 111-115.

[44] Thomenius MJ, Wang NS, Reineks EZ, Wang Z, Distelhorst CW (2003) Bcl-2 on the endoplasmic reticulum regulates Bax activity by binding to BH3-only proteins. J.Biol.Chem.278: 6243-6250.

[45] Liu X, Kim CN, Yang J, Jemmerson R, Wang X (1996) Induction of apoptotic program in cell-free extracts: requirement for dATP and cytochrome c. Cell ;86: 147-157.

[46] Wang D, Liang J, Zhang Y, Gui B, Wang F, Yi X, et al. (2012) Steroid Receptor Coactivator-interacting Protein (SIP) Inhibits Caspase-independent Apoptosis by Preventing Apoptosis-inducing Factor (AIF) from Being Released from Mitochondria. J.Biol.Chem.287: 12612-12621.

[47] Norberg E, Gogvadze V, Ott M, Horn M, Uhlen P, Orrenius S, et al. (2008) An increase in intracellular Ca2+ is required for the activation of mitochondrial calpain to release AIF during cell death. Cell Death.Differ.15: 1857-1864.

[48] Yu SW, Wang H, Poitras MF, Coombs C, Bowers WJ, Federoff HJ, et al. (2002) Mediation of poly(ADP-ribose) polymerase-1-dependent cell death by apoptosis-inducing factor. Science ;297: 259-263.

[49] Trencia A, Fiory F, Maitan MA, Vito P, Barbagallo AP, Perfetti A, et al. (2004) Omi/HtrA2 promotes cell death by binding and degrading the anti-apoptotic protein ped/pea-15. J.Biol.Chem.279: 46566-46572.

[50] Cilenti L, Soundarapandian MM, Kyriazis GA, Stratico V, Singh S, Gupta S, et al. (2004) Regulation of HAX-1 anti-apoptotic protein by Omi/HtrA2 protease during cell death. J.Biol.Chem.279: 50295-50301.

[51] Vande WL, Wirawan E, Lamkanfi M, Festjens N, Verspurten J, Saelens X, et al. (2010) The mitochondrial serine protease HtrA2/Omi cleaves RIP1 during apoptosis of Ba/F3 cells induced by growth factor withdrawal. Cell Res.20: 421-433.

[52] Vousden KH, Lu X (2002) Live or let die: the cell's response to p53. Nat.Rev.Cancer ;2: 594-604.

[53] Berube C, Boucher LM, Ma W, Wakeham A, Salmena L, Hakem R, et al. (2005) Apoptosis caused by p53-induced protein with death domain (PIDD) depends on the death adapter protein RAIDD. Proc.Natl.Acad.Sci.U.S.A ;102: 14314-14320.

[54] Janssens S, Tinel A (2012) The PIDDosome, DNA-damage-induced apoptosis and beyond. Cell Death Differ.19: 13-20.

[55] Bouchier-Hayes L, Lartigue L, Newmeyer DD (2005) Mitochondria: pharmacological manipulation of cell death. J.Clin.Invest ;115: 2640-2647.

[56] Yu Q (2006) Restoring p53-mediated apoptosis in cancer cells: new opportunities for cancer therapy. Drug Resist.Updat.9: 19-25.

[57] Watcharasit P, Bijur GN, Song L, Zhu J, Chen X, Jope RS (2003) Glycogen synthase kinase-3beta (GSK3beta) binds to and promotes the actions of p53. J.Biol.Chem.278: 48872-48879.

[58] Kim I, Xu W, Reed JC (2008) Cell death and endoplasmic reticulum stress: disease relevance and therapeutic opportunities. Nat.Rev.Drug Discov.7: 1013-1030.

[59] Hoyer-Hansen M, Jaattela M (2007) Connecting endoplasmic reticulum stress to autophagy by unfolded protein response and calcium. Cell Death.Differ.14: 1576-1582.

[60] Nakagawa T, Yuan J (2000) Cross-talk between two cysteine protease families. Activation of caspase-12 by calpain in apoptosis. J.Cell Biol.150: 887-894.

[61] Morishima N, Nakanishi K, Takenouchi H, Shibata T, Yasuhiko Y (2002) An endoplasmic reticulum stress-specific caspase cascade in apoptosis. Cytochrome c-independent activation of caspase-9 by caspase-12. J.Biol.Chem.277: 34287-34294.

[62] Rao RV, Castro-Obregon S, Frankowski H, Schuler M, Stoka V, del RG, et al. (2002) Coupling endoplasmic reticulum stress to the cell death program. An Apaf-1-independent intrinsic pathway. J.Biol.Chem.277: 21836-21842.

[63] Jimbo A, Fujita E, Kouroku Y, Ohnishi J, Inohara N, Kuida K, et al. (2003) ER stress induces caspase-8 activation, stimulating cytochrome c release and caspase-9 activation. Exp.Cell Res.283: 156-166.

[64] Fehrenbacher N, Jaattela M (2005) Lysosomes as targets for cancer therapy. Cancer Res.65: 2993-2995.

[65] Deacon EM, Pongracz J, Griffiths G, Lord JM (1997) Isoenzymes of protein kinase C: differential involvement in apoptosis and pathogenesis. Mol.Pathol.50: 124-131.

[66] Kharbanda S, Saxena S, Yoshida K, Pandey P, Kaneki M, Wang Q, et al. (2000) Translocation of SAPK/JNK to mitochondria and interaction with Bcl-x(L) in response to DNA damage. J.Biol.Chem.275: 322-327.

[67] Ozes ON, Mayo LD, Gustin JA, Pfeffer SR, Pfeffer LM, Donner DB (1999) NF-kappaB activation by tumour necrosis factor requires the Akt serine-threonine kinase. Nature ;401: 82-85.

[68] Frisch SM, Francis H (1994) Disruption of epithelial cell-matrix interactions induces apoptosis. J.Cell Biol.124: 619-626.

[69] Gourlay CW, Ayscough KR (2005) The actin cytoskeleton: a key regulator of apoptosis and ageing? Nat.Rev.Mol.Cell Biol.6: 583-589.

[70] Day CL, Puthalakath H, Skea G, Strasser A, Barsukov I, Lian LY, et al. (2004) Localization of dynein light chains 1 and 2 and their pro-apoptotic ligands. Biochem.J.377: 597-605.

[71] Kusano H, Shimizu S, Koya RC, Fujita H, Kamada S, Kuzumaki N, et al. (2000) Human gelsolin prevents apoptosis by inhibiting apoptotic mitochondrial changes via closing VDAC. Oncogene ;19: 4807-4814.

[72] Tang HL, Le AH, Lung HL (2006) The increase in mitochondrial association with actin precedes Bax translocation in apoptosis. Biochem.J.396: 1-5.

[73] Kaufmann SH, Desnoyers S, Ottaviano Y, Davidson NE, Poirier GG (1993) Specific proteolytic cleavage of poly(ADP-ribose) polymerase: an early marker of chemotherapy-induced apoptosis. Cancer Res.53: 3976-3985.

[74] Oliver FJ, Menissier-de MJ, de MG (1999) Poly(ADP-ribose) polymerase in the cellular response to DNA damage, apoptosis, and disease. Am.J.Hum.Genet.64: 1282-1288.

[75] Guillouf C, Wang TS, Liu J, Walsh CE, Poirier GG, Moustacchi E, et al. (1999) Fanconi anemia C protein acts at a switch between apoptosis and necrosis in mitomycin C-induced cell death. Exp.Cell Res.246: 384-394.

[76] Munoz-Gamez JA, Rodriguez-Vargas JM, Quiles-Perez R, Aguilar-Quesada R, Martin-Oliva D, de MG, et al. (2009) PARP-1 is involved in autophagy induced by DNA damage. Autophagy.5: 61-74.

[77] Wang Y, Dawson VL, Dawson TM (2009) Poly(ADP-ribose) signals to mitochondrial AIF: a key event in parthanatos. Exp.Neurol.218: 193-202.

[78] Long JS, Ryan KM (2012) New frontiers in promoting tumour cell death: targeting apoptosis, necroptosis and autophagy. Oncogene .

[79] Klionsky DJ (2007) Autophagy: from phenomenology to molecular understanding in less than a decade. Nat.Rev.Mol.Cell Biol.8: 931-937.

[80] Nakatogawa H, Suzuki K, Kamada Y, Ohsumi Y (2009) Dynamics and diversity in autophagy mechanisms: lessons from yeast. Nat.Rev.Mol.Cell Biol.10: 458-467.

[81] Tanida I, Ueno T, Kominami E (2004) LC3 conjugation system in mammalian autophagy. Int.J.Biochem.Cell Biol.36: 2503-2518.

[82] Mijaljica D, Prescott M, Devenish RJ (2011) Microautophagy in mammalian cells: revisiting a 40-year-old conundrum. Autophagy.7: 673-682.

[83] Majeski AE, Dice JF (2004) Mechanisms of chaperone-mediated autophagy. Int.J.Biochem.Cell Biol.36: 2435-2444.

[84] Tanida I (2011) Autophagosome formation and molecular mechanism of autophagy. Antioxid.Redox.Signal.14: 2201-2214.

[85] Yen WL, Klionsky DJ (2008) How to live long and prosper: autophagy, mitochondria, and aging. Physiology.(Bethesda.) ;23: 248-262.

[86] Castedo M, Perfettini JL, Roumier T, Andreau K, Medema R, Kroemer G (2004) Cell death by mitotic catastrophe: a molecular definition. Oncogene ;23: 2825-2837.

[87] Niikura Y, Dixit A, Scott R, Perkins G, Kitagawa K (2007) BUB1 mediation of caspase-independent mitotic death determines cell fate. J.Cell Biol.178: 283-296.

[88] Vitale I, Galluzzi L, Castedo M, Kroemer G (2011) Mitotic catastrophe: a mechanism for avoiding genomic instability. Nat.Rev.Mol.Cell Biol.12: 385-392.

[89] Vanlangenakker N, Vanden Berghe T, Vandenabeele P (2012) Many stimuli pull the necrotic trigger, an overview. Cell Death.Differ.19: 75-86.

[90] Feoktistova M, Geserick P, Kellert B, Dimitrova DP, Langlais C, Hupe M, et al. (2011) cIAPs block Ripoptosome formation, a RIP1/caspase-8 containing intracellular cell death complex differentially regulated by cFLIP isoforms. Mol.Cell ;43: 449-463.

[91] Tenev T, Bianchi K, Darding M, Broemer M, Langlais C, Wallberg F, et al. (2011) The Ripoptosome, a signaling platform that assembles in response to genotoxic stress and loss of IAPs. Mol.Cell ;43: 432-448.

[92] Cho YS, Challa S, Moquin D, Genga R, Ray TD, Guildford M, et al. (2009) Phosphorylation-driven assembly of the RIP1-RIP3 complex regulates programmed necrosis and virus-induced inflammation. Cell ;137: 1112-1123.

[93] Dunai Z, Bauer PI, Mihalik R (2011) Necroptosis: biochemical, physiological and pathological aspects. Pathol.Oncol.Res.17: 791-800.

[94] Martinez-Lacaci I, Garcia MP, Soto JL, Saceda M (2007) Tumour cells resistance in cancer therapy. Clin.Transl.Oncol.9: 13-20.

[95] Hanahan D, Weinberg RA (2000) The hallmarks of cancer. Cell ;100: 57-70.

[96] Hersey P, Zhang XD (2003) Overcoming resistance of cancer cells to apoptosis. J.Cell Physiol ;196: 9-18.

[97] Hockenbery D, Nunez G, Milliman C, Schreiber RD, Korsmeyer SJ (1990) Bcl-2 is an inner mitochondrial membrane protein that blocks programmed cell death. Nature ;348: 334-336.

[98] Tsujimoto Y, Finger LR, Yunis J, Nowell PC, Croce CM (1984) Cloning of the chromosome breakpoint of neoplastic B cells with the t(14;18) chromosome translocation. Science ;226: 1097-1099.

[99] Vaux DL, Cory S, Adams JM (1988) Bcl-2 gene promotes haemopoietic cell survival and cooperates with c-myc to immortalize pre-B cells. Nature ;335: 440-442.

[100] Strasser A, Harris AW, Bath ML, Cory S (1990) Novel primitive lymphoid tumours induced in transgenic mice by cooperation between myc and bcl-2. Nature ;348: 331-333.

[101] Strasser A, Cory S, Adams JM (2011) Deciphering the rules of programmed cell death to improve therapy of cancer and other diseases. EMBO J.30: 3667-3683.

[102] Plati J, Bucur O, Khosravi-Far R (2011) Apoptotic cell signaling in cancer progression and therapy. Integr.Biol.(Camb.) ;3: 279-296.

[103] Rampino N, Yamamoto H, Ionov Y, Li Y, Sawai H, Reed JC, et al. (1997) Somatic frameshift mutations in the BAX gene in colon cancers of the microsatellite mutator phenotype. Science ;275: 967-969.

[104] Tagawa H, Karnan S, Suzuki R, Matsuo K, Zhang X, Ota A, et al. (2005) Genome-wide array-based CGH for mantle cell lymphoma: identification of homozygous deletions of the proapoptotic gene BIM. Oncogene ;24: 1348-1358.

[105] Beroukhim R, Mermel CH, Porter D, Wei G, Raychaudhuri S, Donovan J, et al. (2010) The landscape of somatic copy-number alteration across human cancers. Nature ;463: 899-905.

[106] Garrison SP, Jeffers JR, Yang C, Nilsson JA, Hall MA, Rehg JE, et al. (2008) Selection against PUMA gene expression in Myc-driven B-cell lymphomagenesis. Mol.Cell Biol.28: 5391-5402.

[107] Richter-Larrea JA, Robles EF, Fresquet V, Beltran E, Rullan AJ, Agirre X, et al. (2010) Reversion of epigenetically mediated BIM silencing overcomes chemoresistance in Burkitt lymphoma. Blood ;116: 2531-2542.

[108] Kashkar H (2010) X-linked inhibitor of apoptosis: a chemoresistance factor or a hollow promise. Clin.Cancer Res.16: 4496-4502.

[109] Kaufmann T, Strasser A, Jost PJ (2012) Fas death receptor signalling: roles of Bid and XIAP. Cell Death.Differ.19: 42-50.

[110] Mita AC, Mita MM, Nawrocki ST, Giles FJ (2008) Survivin: key regulator of mitosis and apoptosis and novel target for cancer therapeutics. Clin.Cancer Res.14: 5000-5005.

[111] Reichert S, Rodel C, Mirsch J, Harter PN, Tomicic MT, Mittelbronn M, et al. (2011) Survivin inhibition and DNA double-strand break repair: a molecular mechanism to overcome radioresistance in glioblastoma. Radiother.Oncol.101: 51-58.

[112] Islam A, Kageyama H, Takada N, Kawamoto T, Takayasu H, Isogai E, et al. (2000) High expression of Survivin, mapped to 17q25, is significantly associated with poor prognostic factors and promotes cell survival in human neuroblastoma. Oncogene ;19: 617-623.

[113] Hattori M, Sakamoto H, Satoh K, Yamamoto T (2001) DNA demethylase is expressed in ovarian cancers and the expression correlates with demethylation of CpG sites in the promoter region of c-erbB-2 and survivin genes. Cancer Lett.169: 155-164.

[114] Li F, Altieri DC (1999) The cancer antiapoptosis mouse survivin gene: characterization of locus and transcriptional requirements of basal and cell cycle-dependent expression. Cancer Res.59: 3143-3151.

[115] Mesri M, Morales-Ruiz M, Ackermann EJ, Bennett CF, Pober JS, Sessa WC, et al. (2001) Suppression of vascular endothelial growth factor-mediated endothelial cell protection by survivin targeting. Am.J.Pathol.158: 1757-1765.

[116] Zaffaroni N, Daidone MG (2002) Survivin expression and resistance to anticancer treatments: perspectives for new therapeutic interventions. Drug Resist.Updat.5: 65-72.

[117] Rufini A, Melino G (2011) Cell death pathology: the war against cancer. Biochem.Biophys.Res.Commun.414: 445-450.

[118] Li G, Ho VC (1998) p53-dependent DNA repair and apoptosis respond differently to high- and low-dose ultraviolet radiation. Br.J.Dermatol.139: 3-10.

[119] Gabrielli B, Brooks K, Pavey S (2012) Defective cell cycle checkpoints as targets for anti-cancer therapies. Front Pharmacol.3: 9.

[120] Roos WP, Kaina B (2012) DNA damage-induced apoptosis: From specific DNA lesions to the DNA damage response and apoptosis. Cancer Lett.

[121] Hanahan D, Weinberg RA (2011) Hallmarks of cancer: the next generation. Cell ;144: 646-674.

[122] Hers I, Vincent EE, Tavare JM (2011) Akt signalling in health and disease. Cell Signal.23: 1515-1527.

[123] Sarbassov DD, Guertin DA, Ali SM, Sabatini DM (2005) Phosphorylation and regulation of Akt/PKB by the rictor-mTOR complex. Science ;307: 1098-1101.

[124] Liu P, Cheng H, Roberts TM, Zhao JJ (2009) Targeting the phosphoinositide 3-kinase pathway in cancer. Nat.Rev.Drug Discov.8: 627-644.

[125] Vivanco I, Sawyers CL (2002) The phosphatidylinositol 3-Kinase AKT pathway in human cancer. Nat.Rev.Cancer ;2: 489-501.

[126] Haddad JJ, Abdel-Karim NE (2011) NF-kappaB cellular and molecular regulatory mechanisms and pathways: therapeutic pattern or pseudoregulation? Cell Immunol.271: 5-14.

[127] Hafsi S, Pezzino FM, Candido S, Ligresti G, Spandidos DA, Soua Z, et al. (2012) Gene alterations in the PI3K/PTEN/AKT pathway as a mechanism of drug-resistance (review). Int.J.Oncol.40: 639-644.

[128] Zitvogel L, Apetoh L, Ghiringhelli F, Andre F, Tesniere A, Kroemer G (2008) The anticancer immune response: indispensable for therapeutic success? J.Clin.Invest ;118: 1991-2001.

[129] Garg AD, Krysko DV, Verfaillie T, Kaczmarek A, Ferreira GB, Marysael T, et al. (2012) A novel pathway combining calreticulin exposure and ATP secretion in immunogenic cancer cell death. EMBO J.31: 1062-1079.

[130] Tesniere A, Panaretakis T, Kepp O, Apetoh L, Ghiringhelli F, Zitvogel L, et al. (2008) Molecular characteristics of immunogenic cancer cell death. Cell Death.Differ.15: 3-12.

[131] Chao MP, Jaiswal S, Weissman-Tsukamoto R, Alizadeh AA, Gentles AJ, Volkmer J, et al. (2010) Calreticulin is the dominant pro-phagocytic signal on multiple human cancers and is counterbalanced by CD47. Sci.Transl.Med.2: 63ra94.

[132] Chao MP, Weissman IL, Majeti R (2012) The CD47-SIRPalpha pathway in cancer immune evasion and potential therapeutic implications. Curr.Opin.Immunol.24: 225-232.

[133] Leontieva OV, Gudkov AV, Blagosklonny MV (2010) Weak p53 permits senescence during cell cycle arrest. Cell Cycle ;9: 4323-4327.

[134] Rodier F, Campisi J (2011) Four faces of cellular senescence. J.Cell Biol.192: 547-556.

[135] Schmitt CA (2007) Cellular senescence and cancer treatment. Biochim.Biophys.Acta ;1775: 5-20.

[136] Fimia GM, Stoykova A, Romagnoli A, Giunta L, Di BS, Nardacci R, et al. (2007) Ambra1 regulates autophagy and development of the nervous system. Nature ;447: 1121-1125.

[137] Liang C, Feng P, Ku B, Dotan I, Canaani D, Oh BH, et al. (2006) Autophagic and tumour suppressor activity of a novel Beclin1-binding protein UVRAG. Nat.Cell Biol.8: 688-699.

[138] Mehrpour M, Esclatine A, Beau I, Codogno P (2010) Overview of macroautophagy regulation in mammalian cells. Cell Res.20: 748-762.

[139] Ku B, Woo JS, Liang C, Lee KH, Jung JU, Oh BH (2008) An insight into the mechanistic role of Beclin 1 and its inhibition by prosurvival Bcl-2 family proteins. Autophagy.4: 519-520.

[140] Liang XH, Jackson S, Seaman M, Brown K, Kempkes B, Hibshoosh H, et al. (1999) Induction of autophagy and inhibition of tumorigenesis by beclin 1. Nature ;402: 672-676.

[141] Yue Z, Jin S, Yang C, Levine AJ, Heintz N (2003) Beclin 1, an autophagy gene essential for early embryonic development, is a haploinsufficient tumor suppressor. Proc.Natl.Acad.Sci.U.S.A ;100: 15077-15082.

[142] Qu X, Yu J, Bhagat G, Furuya N, Hibshoosh H, Troxel A, et al. (2003) Promotion of tumorigenesis by heterozygous disruption of the beclin 1 autophagy gene. J.Clin.Invest ;112: 1809-1820.

[143] Matsui Y, Takagi H, Qu X, Abdellatif M, Sakoda H, Asano T, et al. (2007) Distinct roles of autophagy in the heart during ischemia and reperfusion: roles of AMP-activated protein kinase and Beclin 1 in mediating autophagy. Circ.Res.100: 914-922.

[144] Takahashi Y, Coppola D, Matsushita N, Cualing HD, Sun M, Sato Y, et al. (2007) Bif-1 interacts with Beclin 1 through UVRAG and regulates autophagy and tumorigenesis. Nat.Cell Biol.9: 1142-1151.

[145] Ionov Y, Nowak N, Perucho M, Markowitz S, Cowell JK (2004) Manipulation of nonsense mediated decay identifies gene mutations in colon cancer Cells with microsatellite instability. Oncogene ;23: 639-645.

[146] Kim MS, Jeong EG, Ahn CH, Kim SS, Lee SH, Yoo NJ (2008) Frameshift mutation of UVRAG, an autophagy-related gene, in gastric carcinomas with microsatellite instability. Hum.Pathol.39: 1059-1063.

[147] Bekri S, Adelaide J, Merscher S, Grosgeorge J, Caroli-Bosc F, Perucca-Lostanlen D, et al. (1997) Detailed map of a region commonly amplified at 11q13-->q14 in human breast carcinoma. Cytogenet.Cell Genet.79: 125-131.

[148] Lee JW, Jeong EG, Soung YH, Nam SW, Lee JY, Yoo NJ, et al. (2006) Decreased expression of tumour suppressor Bax-interacting factor-1 (Bif-1), a Bax activator, in gastric carcinomas. Pathology ;38: 312-315.

[149] Coppola D, Khalil F, Eschrich SA, Boulware D, Yeatman T, Wang HG (2008) Down-regulation of Bax-interacting factor-1 in colorectal adenocarcinoma. Cancer ;113: 2665-2670.

[150] Cuddeback SM, Yamaguchi H, Komatsu K, Miyashita T, Yamada M, Wu C, et al. (2001) Molecular cloning and characterization of Bif-1. A novel Src homology 3 domain-containing protein that associates with Bax. J.Biol.Chem.276: 20559-20565.

[151] Yang YP, Liang ZQ, Gu ZL, Qin ZH (2005) Molecular mechanism and regulation of autophagy. Acta Pharmacol.Sin.26: 1421-1434.

[152] Gozuacik D, Kimchi A (2004) Autophagy as a cell death and tumor suppressor mechanism. Oncogene ;23: 2891-2906.

[153] Samuels Y, Velculescu VE (2004) Oncogenic mutations of PIK3CA in human cancers. Cell Cycle ;3: 1221-1224.

[154] Saeki K, Yuo A, Okuma E, Yazaki Y, Susin SA, Kroemer G, et al. (2000) Bcl-2 down-regulation causes autophagy in a caspase-independent manner in human leukemic HL60 cells. Cell Death.Differ.7: 1263-1269.

[155] Pattingre S, Tassa A, Qu X, Garuti R, Liang XH, Mizushima N, et al. (2005) Bcl-2 antiapoptotic proteins inhibit Beclin 1-dependent autophagy. Cell ;122: 927-939.

[156] Maiuri MC, Le TG, Criollo A, Rain JC, Gautier F, Juin P, et al. (2007) Functional and physical interaction between Bcl-X(L) and a BH3-like domain in Beclin-1. EMBO J.26: 2527-2539.

[157] Mathew R, Kongara S, Beaudoin B, Karp CM, Bray K, Degenhardt K, et al. (2007) Autophagy suppresses tumor progression by limiting chromosomal instability. Genes Dev.21: 1367-1381.

[158] Karantza-Wadsworth V, Patel S, Kravchuk O, Chen G, Mathew R, Jin S, et al. (2007) Autophagy mitigates metabolic stress and genome damage in mammary tumorigenesis. Genes Dev.21: 1621-1635.

[159] Tasdemir E, Maiuri MC, Galluzzi L, Vitale I, Djavaheri-Mergny M, D'Amelio M, et al. (2008) Regulation of autophagy by cytoplasmic p53. Nat.Cell Biol.10: 676-687.

[160] Zong WX, Thompson CB (2006) Necrotic death as a cell fate. Genes Dev.20: 1-15.

[161] Balkwill F, Charles KA, Mantovani A (2005) Smoldering and polarized inflammation in the initiation and promotion of malignant disease. Cancer Cell ;7: 211-217.

[162] Vakkila J, Lotze MT (2004) Inflammation and necrosis promote tumour growth. Nat.Rev.Immunol.4: 641-648.

[163] Degenhardt K, Mathew R, Beaudoin B, Bray K, Anderson D, Chen G, et al. (2006) Autophagy promotes tumor cell survival and restricts necrosis, inflammation, and tumorigenesis. Cancer Cell ;10: 51-64.

[164] Boya P, Gonzalez-Polo RA, Casares N, Perfettini JL, Dessen P, Larochette N, et al. (2005) Inhibition of macroautophagy triggers apoptosis. Mol.Cell Biol.25: 1025-1040.

[165] Majmundar AJ, Wong WJ, Simon MC (2010) Hypoxia-inducible factors and the response to hypoxic stress. Mol.Cell ;40: 294-309.

[166] Mazure NM, Pouyssegur J (2010) Hypoxia-induced autophagy: cell death or cell survival? Curr.Opin.Cell Biol.22: 177-180.

[167] Bellot G, Garcia-Medina R, Gounon P, Chiche J, Roux D, Pouyssegur J, et al. (2009) Hypoxia-induced autophagy is mediated through hypoxia-inducible factor induction of BNIP3 and BNIP3L via their BH3 domains. Mol.Cell Biol.29: 2570-2581.

[168] Rabinowitz JD, White E (2010) Autophagy and metabolism. Science ;330: 1344-1348.

[169] Fung C, Lock R, Gao S, Salas E, Debnath J (2008) Induction of autophagy during extracellular matrix detachment promotes cell survival. Mol.Biol.Cell ;19: 797-806.

[170] Kenific CM, Thorburn A, Debnath J (2010) Autophagy and metastasis: another double-edged sword. Curr.Opin.Cell Biol.22: 241-245.

[171] Debnath J (2008) Detachment-induced autophagy during anoikis and lumen formation in epithelial acini. Autophagy.4: 351-353.

[172] Yerbes R, Palacios C, Lopez-Rivas A (2011) The therapeutic potential of TRAIL receptor signalling in cancer cells. Clin.Transl.Oncol.13: 839-847.

[173] Ashkenazi A, Dixit VM (1998) Death receptors: signaling and modulation. Science ;281: 1305-1308.

[174] Nguyen T, Zhang XD, Hersey P (2001) Relative resistance of fresh isolates of melanoma to tumor necrosis factor-related apoptosis-inducing ligand (TRAIL)-induced apoptosis. Clin.Cancer Res.7: 966s-973s.

[175] Pavet V, Portal MM, Moulin JC, Herbrecht R, Gronemeyer H (2011) Towards novel paradigms for cancer therapy. Oncogene ;30: 1-20.

[176] Hetschko H, Voss V, Seifert V, Prehn JH, Kogel D (2008) Upregulation of DR5 by proteasome inhibitors potently sensitizes glioma cells to TRAIL-induced apoptosis. FEBS J.275: 1925-1936.

[177] Johnstone RW, Frew AJ, Smyth MJ (2008) The TRAIL apoptotic pathway in cancer onset, progression and therapy. Nat.Rev.Cancer ;8: 782-798.

[178] Speirs CK, Hwang M, Kim S, Li W, Chang S, Varki V, et al. (2011) Harnessing the cell death pathway for targeted cancer treatment. Am.J.Cancer Res.1: 43-61.

[179] Martinez-Lostao L, Marzo I, Anel A, Naval J (2012) Targeting the Apo2L/TRAIL system for the therapy of autoimmune diseases and cancer. Biochem.Pharmacol.83: 1475-1483.

[180] Aggarwal BB (2003) Signalling pathways of the TNF superfamily: a double-edged sword. Nat.Rev.Immunol.3: 745-756.

[181] Holler N, Zaru R, Micheau O, Thome M, Attinger A, Valitutti S, et al. (2000) Fas triggers an alternative, caspase-8-independent cell death pathway using the kinase RIP as effector molecule. Nat.Immunol.1: 489-495.

[182] Vanden Berghe T, van LG, Saelens X, Van GM, Brouckaert G, Kalai M, et al. (2004) Differential signaling to apoptotic and necrotic cell death by Fas-associated death domain protein FADD. J.Biol.Chem.279: 7925-7933.

[183] Tarhini AA, Kirkwood JM (2007) Oblimersen in the treatment of metastatic melanoma. Future.Oncol.3: 263-271.

[184] Heidari N, Hicks MA, Harada H (2010) GX15-070 (obatoclax) overcomes glucocorticoid resistance in acute lymphoblastic leukemia through induction of apoptosis and autophagy. Cell Death.Dis.1: e76.

[185] McCoy F, Hurwitz J, McTavish N, Paul I, Barnes C, O'Hagan B, et al. (2010) Obatoclax induces Atg7-dependent autophagy independent of beclin-1 and BAX/BAK. Cell Death.Dis.1: e108.

[186] Oltersdorf T, Elmore SW, Shoemaker AR, Armstrong RC, Augeri DJ, Belli BA, et al. (2005) An inhibitor of Bcl-2 family proteins induces regression of solid tumours. Nature ;435: 677-681.

[187] van Delft MF, Wei AH, Mason KD, Vandenberg CJ, Chen L, Czabotar PE, et al. (2006) The BH3 mimetic ABT-737 targets selective Bcl-2 proteins and efficiently induces apoptosis via Bak/Bax if Mcl-1 is neutralized. Cancer Cell ;10: 389-399.

[188] Cragg MS, Harris C, Strasser A, Scott CL (2009) Unleashing the power of inhibitors of oncogenic kinases through BH3 mimetics. Nat.Rev.Cancer ;9: 321-326.

[189] Carrington EM, Vikstrom IB, Light A, Sutherland RM, Londrigan SL, Mason KD, et al. (2010) BH3 mimetics antagonizing restricted prosurvival Bcl-2 proteins represent another class of selective immune modulatory drugs. Proc.Natl.Acad.Sci.U.S.A ;107: 10967-10971.

[190] Chen DJ, Huerta S (2009) Smac mimetics as new cancer therapeutics. Anticancer Drugs ;20: 646-658.

[191] Straub CS (2011) Targeting IAPs as an approach to anti-cancer therapy. Curr.Top.Med.Chem.11: 291-316.

[192] Gyrd-Hansen M, Meier P (2010) IAPs: from caspase inhibitors to modulators of NF-kappaB, inflammation and cancer. Nat.Rev.Cancer ;10: 561-574.

[193] Kaelin WG, Jr. (2009) Synthetic lethality: a framework for the development of wiser cancer therapeutics. Genome Med.1: 99.

[194] Chan DA, Giaccia AJ (2011) Harnessing synthetic lethal interactions in anticancer drug discovery. Nat.Rev.Drug Discov.10: 351-364.

[195] Bolderson E, Richard DJ, Zhou BB, Khanna KK (2009) Recent advances in cancer therapy targeting proteins involved in DNA double-strand break repair. Clin.Cancer Res.15: 6314-6320.

[196] Dai Y, Grant S (2010) New insights into checkpoint kinase 1 in the DNA damage response signaling network. Clin.Cancer Res.16: 376-383.

[197] Calvert H, Azzariti A (2011) The clinical development of inhibitors of poly(ADP-ribose) polymerase. Ann.Oncol.22 Suppl 1: i53-i59.

[198] Javle M, Curtin NJ (2011) The potential for poly (ADP-ribose) polymerase inhibitors in cancer therapy. Ther.Adv.Med.Oncol.3: 257-267.

[199] Imre G, Larisch S, Rajalingam K (2011) Ripoptosome: a novel IAP-regulated cell death-signalling platform. J.Mol.Cell Biol.3: 324-326.

[200] Safa AR, Pollok KE (2011) Targeting the Anti-Apoptotic Protein c-FLIP for Cancer Therapy. Cancers.(Basel) ;3: 1639-1671.

[201] Maiuri MC, Zalckvar E, Kimchi A, Kroemer G (2007) Self-eating and self-killing: crosstalk between autophagy and apoptosis. Nat.Rev.Mol.Cell Biol.8: 741-752.

[202] Chen N, Karantza-Wadsworth V (2009) Role and regulation of autophagy in cancer. Biochim.Biophys.Acta ;1793: 1516-1523.

[203] Carew JS, Nawrocki ST, Kahue CN, Zhang H, Yang C, Chung L, et al. (2007) Targeting autophagy augments the anticancer activity of the histone deacetylase inhibitor SAHA to overcome Bcr-Abl-mediated drug resistance. Blood ;110: 313-322.

[204] Apel A, Herr I, Schwarz H, Rodemann HP, Mayer A (2008) Blocked autophagy sensitizes resistant carcinoma cells to radiation therapy. Cancer Res.68: 1485-1494.

[205] Nahta R, Yuan LX, Zhang B, Kobayashi R, Esteva FJ (2005) Insulin-like growth factor-I receptor/human epidermal growth factor receptor 2 heterodimerization contributes to trastuzumab resistance of breast cancer cells. Cancer Res.65: 11118-11128.

[206] Esparis-Ogando A, Ocana A, Rodriguez-Barrueco R, Ferreira L, Borges J, Pandiella A (2008) Synergic antitumoral effect of an IGF-IR inhibitor and trastuzumab on HER2-overexpressing breast cancer cells. Ann.Oncol.19: 1860-1869.

[207] Ropero S, Menendez JA, Vazquez-Martin A, Montero S, Cortes-Funes H, Colomer R (2004) Trastuzumab plus tamoxifen: anti-proliferative and molecular interactions in breast carcinoma. Breast Cancer Res.Treat.86: 125-137.

[208] Yang S, Wang X, Contino G, Liesa M, Sahin E, Ying H, et al. (2011) Pancreatic cancers require autophagy for tumor growth. Genes Dev.25: 717-729.

[209] Guo JY, Chen HY, Mathew R, Fan J, Strohecker AM, Karsli-Uzunbas G, et al. (2011) Activated Ras requires autophagy to maintain oxidative metabolism and tumorigenesis. Genes Dev.25: 460-470.

[210] Abe A, Yamada H, Moriya S, Miyazawa K (2011) The beta-carboline alkaloid harmol induces cell death via autophagy but not apoptosis in human non-small cell lung cancer A549 cells. Biol.Pharm.Bull.34: 1264-1272.

[211] Yu L, Alva A, Su H, Dutt P, Freundt E, Welsh S, et al. (2004) Regulation of an ATG7-beclin 1 program of autophagic cell death by caspase-8. Science ;304: 1500-1502.

[212] Yu L, Wan F, Dutta S, Welsh S, Liu Z, Freundt E, et al. (2006) Autophagic programmed cell death by selective catalase degradation. Proc.Natl.Acad.Sci.U.S.A ;103: 4952-4957.

[213] Lum JJ, DeBerardinis RJ, Thompson CB (2005) Autophagy in metazoans: cell survival in the land of plenty. Nat.Rev.Mol.Cell Biol.6: 439-448.

[214] Wu YT, Tan HL, Huang Q, Kim YS, Pan N, Ong WY, et al. (2008) Autophagy plays a protective role during zVAD-induced necrotic cell death. Autophagy.4: 457-466.

[215] Amaravadi RK, Thompson CB (2007) The roles of therapy-induced autophagy and necrosis in cancer treatment. Clin.Cancer Res.13: 7271-7279.

[216] Lieberthal W, Triaca V, Levine J (1996) Mechanisms of death induced by cisplatin in proximal tubular epithelial cells: apoptosis vs. necrosis. Am.J.Physiol ;270: F700-F708.

[217] Healy E, Dempsey M, Lally C, Ryan MP (1998) Apoptosis and necrosis: mechanisms of cell death induced by cyclosporine A in a renal proximal tubular cell line. Kidney Int.54: 1955-1966.

[218] Chan FK, Shisler J, Bixby JG, Felices M, Zheng L, Appel M, et al. (2003) A role for tumor necrosis factor receptor-2 and receptor-interacting protein in programmed necrosis and antiviral responses. J.Biol.Chem.278: 51613-51621.

[219] Cho DH, Jo YK, Hwang JJ, Lee YM, Roh SA, Kim JC (2009) Caspase-mediated cleavage of ATG6/Beclin-1 links apoptosis to autophagy in HeLa cells. Cancer Lett.274: 95-100.

[220] Djavaheri-Mergny M, Maiuri MC, Kroemer G (2010) Cross talk between apoptosis and autophagy by caspase-mediated cleavage of Beclin 1. Oncogene ;29: 1717-1719.

[221] Hou W, Han J, Lu C, Goldstein LA, Rabinowich H (2010) Autophagic degradation of active caspase-8: a crosstalk mechanism between autophagy and apoptosis. Autophagy.6: 891-900.

Chapter 6

Cellular Caspases: New Targets for the Action of Pharmacological Agents

Tatyana O. Volkova and Alexander N. Poltorak

Additional information is available at the end of the chapter

1. Introduction

The number of scientific papers devoted to the study of caspases has lately being constantly growing. Caspases are a family of cysteine-dependent aspartate specific proteases. Caspases play an essential role in the apoptosis, necrosis and inflammation processes. Apoptosis is asynchronous programmed cell death, which helps maintain the physiological balance and genetic stability of the organism through self-destruction of genetically modified defect cells. Apoptosis may happen also in normal cells, e.g., during embryogenesis. During apoptosis, activated endogenous nucleases cleave DNA into fragments, but cell membranes and the intracellular matter remain intact, nor is there tissue damage or leukocytic infiltration. In contrast to apoptosis, necrosis is a pathological form of cell death caused by acute damage, rupture of the membrane, release of the cytoplasm content, and the inflammatory process induced thereby [1, 2].

As of now, at lease 14 caspases have been described from mammals: 8 caspases are involved in apoptosis, 5 activate anti-inflammatory cytokines, and one acts in keratinocyte differentiation. This division is, however, rather arbitrary – we know from a number of papers that some apoptotic caspases may, depending on the conditions, participate in other cell life processes, such as proliferation [3], differentiation [4, 5], modification of susceptibility to leukocyte lysis [2, 6].

During apoptosis, different caspases perform different functions. Depending on the phase at which those proteins enter the apoptotic cascade one distinguishes initiator (apical) and effector (executioner) caspases. E.g., caspase-2, -8, -9, -10, and -12 are initiator ones; caspase-3, -6, -7 – effector ones. All caspases are originally inactive, but activated when needed by cleavage of a small fragment by initiator caspases. Initiator caspase activation is more sophisticated – by special protein complexes: apoptosomes, PIDDosomes, DISC (*death-*

inducing signalling complexes) [7, 8]. Under certain conditions effector caspases may act as initiator ones to accelerate apoptotic reactions.

Research into the phenomenon of programmed cell death started in the late 1960s. One of the pioneers in the sphere was John Kerr, who studied the death of hepatocytes in rats with acute liver failure [9]. In 1972, a team of British scientists headed by Kerr first coined the term apoptosis to denote programmed cell death. The authors described two phases of the process (formation of apoptotic bodies, and their phagocytosis by other cells), and stressed that apoptosis is an active and controlled process. Yet, nothing was known at the time about the factors and mechanisms of this type of cell death. Studies of the structure of caspases began only in 1994 [10, 11]. In 1996, it was found that cytochrome c together with ATP promote activation of caspase-3 [12]. Caspases are now investigated in various model systems both *in vitro* and *in vivo*. The most popular systems in use are tumor cell lines of various histogenesis. Most studies employ hemopoietic tumor cells (lines *K562, HL-60, Raji, HEL*, etc.), cells derived from solid tumors (lines *HeLa, RPTC, HCT 116*, etc.), as well as stem cells. Through this approach one can not just study the structure, activation mechanisms and basic functions of caspases in normal cells, but also identify the biological role of these enzymes in emerging and progressing pathological processes, first of all, oncologic and immuno-inflammatory ones. Besides, the chemical reagents used in those model systems usually activate or inhibit cell processes, including apoptosis. These facts open up immense opportunities in the analysis of the activity of the substances considered for applicability as components of new targeted drugs. The clue to the effectiveness of the chemical reagents for the treated cells will be the changes in the susceptibility of those cells to cytotoxic lysis of blood leukocytes, namely natural killer cells and cytotoxic T-lymphocytes. Since proteases play an important part in cytotoxic lysis, caspases in this case can also be viewed as candidate molecules whose activity can be modulated to accelerate or damp apoptosis, as well as modify the susceptibility of pathological cells to leukocytolysis.

Hence, the primary objective of the chapter is to demonstrate by means of the experimental data the crucial role of caspases in the induction and progress of tumor cell apoptosis, as well as in the modification of tumor cell susceptibility to the cytotoxic lysis of natural killer cells upon treatment with chemical compounds, part of which have been newly synthesized. The model system used in our study was the human erythromyeloid leukemic cell line *K562* (Russian Cell Culture Collection, Institute of Cytology, Russian Academy of Sciences, St. Petersburg, Russia), as well as *K562/2-DQO* and *K562/4-NQO* subline cells. To induce apoptosis and modulate cell susceptibility to the lytic action of blood leukocytes we used a group of N-bearing heterocyclic reagents – derivatives of quinoline and pyridine: quinoline – *Q*, 2-methylquinoline – *2-MeQ*, quinoline-1-oxide – *QO*, 2-methylquinoline-1-oxide – *2-MeQO*, 4-nitroquinoline-1-oxide – *4-NQO*, 2-methyl-4-nitroquinoline-1-oxide – *2-Me-4-NQO*, 2-(4'-dimethylaminostyryl)quinoline-1-oxide – *2-DQO*, 4-(4'-dimethylaminostyryl)quinoline-1-oxide – *4-DQO*, 2-(4'-nitrostyryl)quinoline-1-oxide – *2-NSQO*, 4-(4'-nitrostyryl)quinoline-1-oxide – *4-NSQO*, and 4-(4'-dimethylaminostyryl)pyridine-1-oxide – *4-DPyO* (Figure 1).

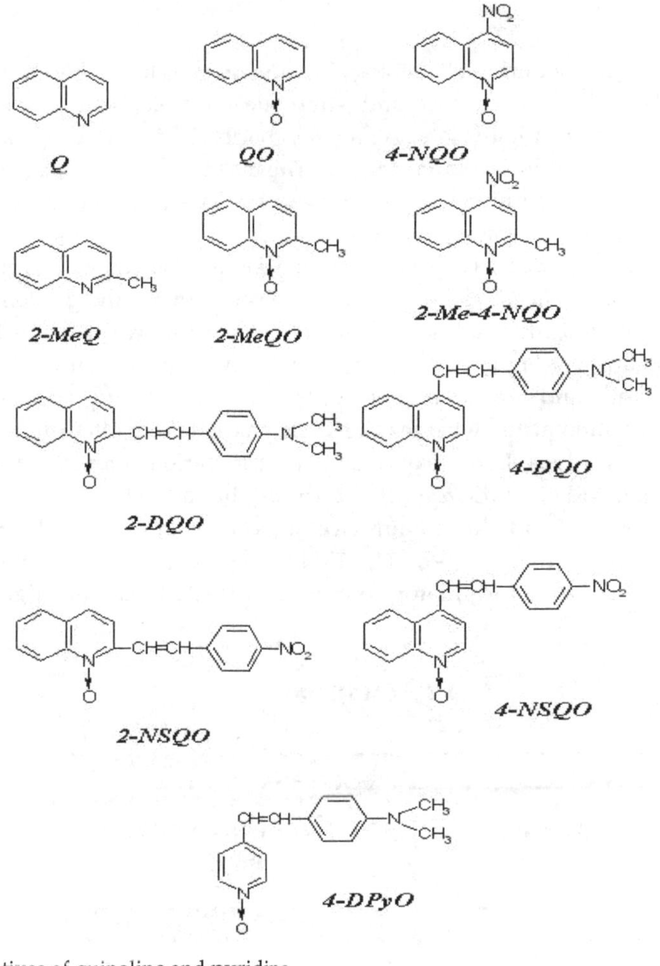

Figure 1. Derivatives of quinoline and pyridine

In current clinical practices, quinoline derivatives are used as drugs with high pharmacological potential – photoprotective, anti-inflammatory, immunomodulating, antioxidant, antiproliferative, antiaggregatory, hypolipidemic, and hypoglycemic. This group of chemical compounds and their intracellular metabolites can be highly competitive as structural analogs of key molecules of the cell (nitrogen bases, nucleosides, nucleotides, various co-enzymes, etc.), and take effect by activating or inhibiting a certain cell process. The compounds with such biological effects most frequently used in the oncology practice are 5-fluorouracil, cytarabine, methotrexate, vinblastine, dactinomycin, and others.

Derivatives of N-bearing heterocycles are very effective in treatment of tumors of various histogenesis and location, including leukemia, but many of them have limited applicability because of high toxicity. Finding the compounds that feature both high anti-tumor activity and low side effects is therefore an important task.

2. Body

A major problem in modern cell biology is the search for intracellular targets for pharmacologically active substances, and subsequent development of new generation medicines based on the resultant data. As a rule, molecules involved in basic processes of cell life are viewed as such intracellular targets. The molecules participating in two or more vital processes are of greatest interest. Such are caspases. They are synthesized in the cell as precursors (procaspases). Caspase precursors consist of a prodomain and two subunits: a small one and a large one. Some caspases contain a short linker sequence (around 10 amino acids) between the subunits. Caspases can activate one another: the prodomain is cleaved from the procaspase, and the small and large subunits of the two caspases form an active caspase – heterotetramer (Figure 2), which contains two active centres. Caspases of the initiator (pro-apoptotic and pro-inflammatory) type contain a long prodomain made up of approximately 100 amino acids, whereas the short prodomain of effector caspases contains only 30 amino acid residues. Long prodomains contain various motifs, mainly DED (*death effector domain*) and CARD (*caspase recruitment domain*), but also DID (*death inducing domain*). Each of the procaspases-8 and -10 contains two tandem copies of the DED, whereas CARD was found in caspases-1, -2, -4, -5, -9, -11, -12. Homotypic interactions between DED and CARD recruit procaspases into initiator complexes, activate them and trigger the caspase cascade [13].

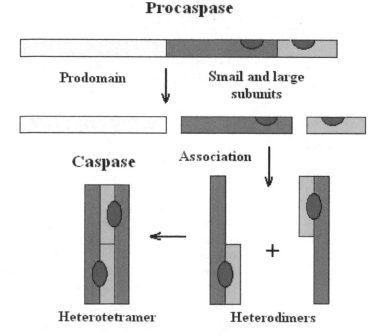

Figure 2. Proteolytic caspase activation (adopted from http://www.muldyr.ru/a/a/apoptoz_-_fazyi_apoptoza)

Apart from distinctions in the prodomain structure, caspases can be grouped by substrate specificity. E.g., caspases-6, -8 and -9 show preference for cleaving substrates with the valine/leucine–glutamate–threonine/histidine–aspartate   (V/L–E–T/H–D)   sequence, whereas caspases-3 and -7 selectively cleave the motif made up of the aspartate–glutamate–valine–aspartate (D–E–V–D) sequence. An optimal target for caspase-1 and its related caspases-4 and -5 are the sequences tyrosine–valine–alanine–aspartate (Y–V–A–D) or tryptophan/leucine–glutamate–histidine–aspartate (W/L–E–H–D) [14].

Various forms of post-translational modifications can influence the activity of caspases. Thus, phosphorylation of caspase-9 by the serine/threonine protein kinase Akt inhibits its activity. Such phosphorylation occurs away from the caspase active centre, and presumably hinders the clustering of the caspase subunits into the tetramer [13, 15]. Another variant of post-translational modification of caspases is S-nitrosylation. The NO radical group is tranfered to the cysteine of the caspase active centre and the R–S–NO group is formed [16, 17]. Furthermore, caspases are susceptible to oxidative modifications induced by active forms of oxygen, with disulphides formed in the process.

Apoptosis can be triggered by various mechanisms that activate caspases. Two major pathways are distinguished: activation involving cell receptors, and mitochondria-mediated activation.

Caspase activation involving cell receptors. The process of apoptosis often begins with ligation of specific extracellular ligands with cell death receptors on the membrane surface. Receptors of the apoptotic signal belong to the TNF (*tumor necrosis factor*) receptors superfamily [18]. The best studied death receptors, whose role in apoptosis has been identified and described, are CD95 (Fas or APO-1), and TNFR1 (p55 or CD120a) (Figure 3). Additional receptors are CARI, DR3 (*death receptor* 3), DR4, DR5, and DR6. All death receptors are transmembrane proteins which share a sequence of 80 amino acids at the cytoplasmic face termed the death domain (DD). It is required for apoptotic signalling [7, 18]. Extracellular regions of death receptors interact with ligand trimers (CD95L, TNF, Apo3L, Apo2L, etc.), and the latter trimerize the death receptors (crosslink 3 molecules of the receptor) [19, 20]. The thus activated receptor interacts with *a* corresponding intracellular adaptor(s). The adaptor for the CD95 (Fas/APO-1) receptor is FADD (*Fas-associated DD-protein*). The adaptor for the TNFR1 and DR3 receptors is TRADD (*TNFR1-associated DD-protein*) (Figure 3).

The adaptor associated with the death receptor interacts with caspases. The "ligand–receptor–adaptor–effector" interaction chain results in the formation of aggregates in which caspases are activated. These aggregates are called apoptosomes, apoptotic chaperones, or death-inducing signalling complexes (DISC). An example of apoptosome is the FasL–Fas–FADD–procaspase-8 complex, in which caspase-8 is activated (Figure 3) [7, 21]. Death receptors can mediate activation of caspases-2; -8, and -10 [22]. The activated initiator caspases then activate effector caspases.

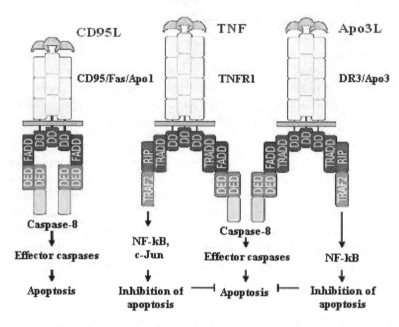

Figure 3. Receptors of the apoptotic signal (adopted from http://www.muldyr.ru/a/a/apoptoz_-_fazyi_apoptoza)

Mitochondria-mediated caspase activation. The mitochondrial apoptotic signalling pathway is realized through release of apoptogenic proteins from the intermembrane region into the cell cytoplasm. Presumably, there are two major pathways for the release of apoptogenic proteins: through formation of a giant pore followed by rupture of the mitochondrial membrane, or through opening of highly permeable channels on the mitochondrial outer membrane [23]. The apoptosome formation model can be represented as "Cytochrome c–Apaf-1–CARD–procaspase-9". Apaf-1 (*apoptosis protease activating factor-1*) undergoes conformational modifications induced by the reaction involving the loss of ATP energy. Procaspase-9 gets access to the CARD of Apaf-1. The thus activated caspase-9 recruits procaspase-3, which is, in turn, activated to form caspase-3 [24, 25].

In some apoptotic models the cytochrome is released by activation of the PTP (*permeability transition pore*) [26]. This pore is a compound complex made up of adenine nucleotide transporter (in the inner mitochondrial membrane), voltage-dependent anion channel, otherwise termed porin (in the outer mitochondrial membrane), and cyclophilin D (in the matrix of mitochondria). The 2.6–2.9 nm giant pore is non-specific, and molecules up to 1.5 Da can pass. Pore opening leads to mitochondrial swelling and rupture of the outer membrane. In addition to cytochrome c mitochondria emit other pro-apoptotic factors [23, 25]. Pore opening can be stimulated by inorganic phosphate, caspases, SH-reagents, cell exhaustion by reduced glutathione, formation of active forms of oxygen, uncoupling of oxidative phosphorylation, rise in Ca^{2+} content in the cytoplasm, effect of ceramide, depletion of the mitochondrial ATP pool, etc. [24, 27, 28].

A presumable alternative pathway for the release of apoptotic proteins from the mitochondrial intermembrane region is formation of a protein channel in the outer mitochondrial membrane. Whichever the pathway, the substances released into the cytoplasm are cytochrome c, procaspases-2, -3, -9, and AIF (*apoptosis inducing factor*) [14]. The release is promoted by Bcl-2 family proteins (Bax-protein). The flavoprotein AIF released from the mitochondrial intermembrane region is an apoptosis effector that would then act independently of caspases [24, 29].

The caspases activated through these pathways directly or indirectly promote the destruction of cell structures. Nuclear lamina proteins are hydrolysed, the cytoskeleton is disrupted, proteins regulating cytoadherence are degraded. Another essential function of effector caspases is inactivation of apoptosis-blocking proteins. To wit, they cleave the inhibitor DFF (*DNA fragmentation factor*), which prevents activation of the apoptotic CAD (*caspase-activated DNase*). Anti-apoptotic proteins of the Bcl-2 family are cleaved. Finally, the action of effector caspases results in dissociation of the regulatory and effector domains of the proteins involved in DNA replication and repair, mRNA splicing [30, 31].

No matter which apoptotic mechanism the cell chooses, the process can be modulated. At present, the structure and action mechanisms of a great number of apoptosis modulating chemical compounds are known. Most of them influence DNA either directly or indirectly. Nonetheless, if apoptosis can be modulated, then modification of caspase expression is also possible (although there exists caspase-independent apoptosis). The tumor cell apoptosis modulators used in this study are derivatives of two heterocyclic compounds – quinoline and pyridine (Figure 1). This group of reagents, most of which are N-oxide derivatives, holds good promise for *in vitro* study of cell systems, because these compounds comprise several functional activity centres (at the heterocyclic nucleus, functional groups, and radicals). Most published data on the biological activity of quinoline N-oxide derivatives are focused on 4-nitroquinoline-1-oxide (*4-NQO*), whereas information on the activity of other structural analogs of the reagents is, unfortunately, almost totally missing.

4-NQO is a chemical carcinogen whose biological effects on cells are in many ways similar to the effect of UV light [32]. The action of *4-NQO* is studied using various model systems: prokaryotic cells *in vitro* [33], eukaryotic cells *in vitro* [34], multicellular organisms [35]. The range of the investigated cell life processes and intracellular biomolecules is also diverse. For the neoplastic effect to take place *4-NQO* has to be metabolically activated in the cells to the proximate carcinogen 4-hydroxyaminoquinoline-1-oxide, which, after being acetylated, can covalently bind to DNA, namely adenine and guanine, to form stable monoadducts [36, 37]. Xanthine oxidase can metabolise *4-NQO* to a more reactive superoxide anion (5.5-dimethyl-1-pyrroline-N-oxide superoxide radical). Formation of substantial amounts of such structures first of all disturbs the processes of DNA replication and repair, thus inducing apoptosis (or necrosis). If, however, this does not happen, accumulation of multiple mutations in the genome would cause normal cells to transform into tumor cells. Some questions logically arise in this connection: how much similarity is there between the biological effects of the structural analogs of *4-NQO* and those of *4-NQO* itself? Would the application of these compounds enable activation of apoptosis in tumor cells by modulating the activity of caspases? Will the susceptibility of the reagent-treated cells lead to leukolysis change? It is these and other questions that we shall try to answer below.

3. Materials and methods

Cells of the human erythromyeloid leukemic cell line *K562* were incubated in the following medium: 89% RPMI 1640 (Institute of Poliomyelitis and Viral Encephalitides, Russian Academy of Medical Sciences, Moscow) supplemented with 11% fetal bovine serum (Institute of Epidemiology and Microbiology, Russian Academy of Medical Sciences, Moscow), 2 mM L-glutamine, 40 µg/ml gentamicin sulphate, 50 µM 2-mercaptoethanol (Ferak, Germany) in 95% O_2 and 5% CO_2 at 37°C. The cultures were seeded to 1 ml (3 ml) of the medium with the seeding density of 10^5 (10^6) cells/well, respectively. Quinoline derivatives in concentrations detailed in the text and figure captions were added to the medium. Cells were counted in Goryaev's chamber. EC_{50} cells incubated with the reagents were ascertained using the technique [38]. The viability was assessed using the test with 3-(4.5-dimethylthiasol-2-yl)-2.5-diphenyltetrazolium bromide (MTT) (Sigma, USA).

Assessment of viability (MTT-test). The technique was described in [39]. 100 µL aliquots of cell suspension (10^5 cells/well) were seeded to a 96-well flat bottom culture plate and treated with 100 µL of the chemical compound solution of the corresponding concentration. All assays were performed in three replicates. The plates were left to incubate for 24, 48, 72, and 96 hours. Three hours before the end of the incubation period the wells were stained with MTT solution (3-(4.5-dimethylthiasol-2-yl)-2.5-diphenyltetrazolium bromide (Sigma, USA)) to a final concentration of 0.25 mg/ml, and re-incubated in the dark in a humidified atmosphere at 37°C. Then, the supernatants were carefully removed and 200 µL aliquots of DMSO were added to each well. The residue was re-suspended and incubated for 15 min in the dark at room temperature. The optical density was measured at 540 nm with a Labsystems Multiscan Plus reader (LKB, Finland). In this case, staining of the cells incubated in the absence of chemical compounds was regarded as the control. Specific cell death was calculated by formula (1):

$$\text{Induced death, \%} = (1 - (D_{exp} - D_{env} / D_c - D_{env})) \times 100\ \% \tag{1}$$

where D_c is the optical density in control wells (cells without chemical reagent); D_{exp} – optical density in treated assays (cells with chemical reagent); D_{env} – optical density of the control environment.

Tumor cell clones resistant to 2-DQO and 4-NQO were obtained as described in [38]. Cell resistance to xenobiotics was induced by long-term (over a month) exposure in a culture medium containing *2-DQO* (10^{-9} M) or *4-NQO* (10^{-12} M). The concentration of the substances in the culture medium was increased every 14–21 days. The final doses for which resistant cell lines were obtained were 10^{-5} and 10^{-8} M, respectively.

Real-time PCR. Total RNA from peripheral blood leukocytes and tumor cells was extracted with the "YellowSolve" kit (Clonogen, Russia) following the manufacturer's guidelines. The extracted RNA template was treated with DNase (Sigma, USA). The concentration and purity of the RNA template was determined by spectrophotometry ("SmartSpec Plus", BioRad, USA). RNA nativity was determined by agarose gel electrophoresis. Complementary DNA was synthesized from 1 µg of total RNA using random hexaprimers and MMLV reverse transcriptase following the protocol proposed by the manufacturer (Sileks, Russia). RNA and cDNA samples were stored at -80°C. Gene expression was estimated by real-time PCR. The fluorophore for product detection was the intercalating SYBR Green I dye. Amplification was performed in an "iCycler Thermal Cycler" (BioRad, USA) with "iQ5 Optical System" V2.0 software (BioRad, USA) using real-time PCR assay kits in the presence of SYBR Green I. The PCR reaction mixture was prepared with the component volumes recommended by the manufacturer: we mixed 2.5 µl deoxynucleosidetriphosphates (2.5 mM), 2.5 µl 10-fold PCR buffer with SYBR Green I, 2.5 µl MgCl$_2$ (25 mM), 1 µl aliquots of forward and reverse primers (20 pmol/µl), 0.25 µl Taq-DNA-polymerase (5 U/µl), 2 µl template cDNA, deionized water – up to 25 µl per test tube. The PCR protocol was 15 sec at 95°C, 50 sec at 60°C (45 cycles). To determine the specificity of primer annealing PCR fragments were melted: for 1 min at 95°C, 1 min at 60°C, 10 sec at 60°C (80 cycles, the temperature raised by 0.5°C in each cycle). To exclude the possibility of the template cDNA being contaminated by the genomic DNA PCR was performed for each template under the same conditions with the RNA matrix. Primers for the nucleotide sequences of the investigated genes and the reference gene GAPDH were selected using the Primer Premier software ("Premier Biosoft", USA) or published sources (Table 1). Oligonucleotides were synthesized by the Syntol company (Russia). Gene expression was measured against the amount of GAPDH mRNA using the $2^{-\Delta\Delta Ct}$ method [40]. The resultant reaction products were separated in 8% polyacrilamide gel using the tris-borate buffer. PCR products were stained with 1% ethidium bromide solution and visualized in transmitted UV light using the low-molecular (501–567 bp) pUC19/Msp I fragment length marker (Syntol, Russia).

Gene	Gene Bank №	Sequence	Source
GAPDH F:	NM_	5'-GAAGGTGAAGGTCGGAGTC-3'	[41]
GAPDH R:	002046.3	5'-GAAGATGGTGATGGGATTTC-3'	
CASPASE 6 F:	NM_	5'-ACTGGCTTGTTCAAAGG-3'	[42]
CASPASE 6 R:	001226.3	5'-CAGCGTGTAAACGGAG-3'	
CASPASE 3 F:	NM_	5'-ATGGAAGCGAATCAATGGAC-3'	[43]
CASPASE 3 R:	004346.3	5'-ATCACGCATCAATTCCACAA-3'	
CASPASE 9 F:	NM_	5'-AACAGGCAAGCAGCAAAGTT-3'	[43]
CASPASE 9 R:	001229.2	5'-CACGGCAGAAGTTCACATTG-3'	

Table 1. Primers for the nucleotide sequences of the caspase genes under study and the reference gene GAPDH

The enzyme activity of caspases was determined by standard technique using specific substrates labeled with fluorescent marker (7-amino-4-trifluoromethylcumarin – AFC) (BioRad, USA), detected by variations in fluorescence or optical density [2]. 50 μl of lytic buffer prepared by mixing 920 μl of bidistilled H_2O, 40 μl of 25-fold reaction buffer and 10 μl of each of the four inhibitors: PMSF (phenylmethylsulfonyl fluoride) (35 mg/ml), pepstatin *A* (1 mg/ml), aprotinin (1 mg/ml), and leupeptin (1 mg/ml), was added to the tumor cells (10^6 cells). The 25-fold reaction buffer included the following components: 250 мM HEPES, pH 7.4, 50 mM EDTA, 2.5% CHAPS (3-((3-chloramidopropyl)dimethylammonio)-1-propanesulfonate), 125 mM dithiothreitol. After that, the cells were frozen three times in liquid nitrogen, the cell lysate then centrifuged in a microcentrifuge at 17 000 G (4^0 C) for 30 min, and the supernatant (template) collected. The activity of caspases-3, -6 and -9 was determined in the reaction buffer by mixing the template with the corresponding specific substrate. The substrate for caspase-3 was DEVD (Asp–Glu–Val–Asp), for caspase-6 – VEID (Val–Glu–Ile–Asp), for caspase-9 – LEHD (Leu–Glu–His–Asp). The amount of cleaved AFC was measured by spectrophotometry in FluoroMax ("Horiba-Scientific", Japan) at 395 nm 30, 60, 90, 120, 150, 180 min after the onset of the reaction. Then, the curve of caspase activity depending on the template and substrate incubation time was plotted. Plot ΔS versus Δt and calculate the slope ($\Delta S/\Delta t$).

$$\Delta S = [S(t_i) – B(t_i)] – [S(t_0) – B(t_0)], \qquad \Delta t = (t_i – t_0), \qquad (2)$$

S – sample signal at time t, and B – blank signal at time t; t_i – time of measurement, t_0 – time of initial measurement.

Cytochrome c reductase activity of microsomes was measured by spectrophotometry in FluoroMax ("Horiba-Scientific", Japan) at 25^0 C. The reaction medium contained microsomes (30 μg protein/ml), NADPH or NADH (50 μM), *2-NSQO* or *4-NSQO* (1–100 μM), and cytochrome c (20 μM).

Determination of the susceptibility of tumor cells to the cytotoxic lysis of human leukocytes. Lysis was studied in a homologous system. Human peripheral blood leukocytes were isolated from the blood of healthy males by a two-step procedure involving fractionation on a Ficoll-Hypaque gradient, followed by erythrocyte lysis by distilled water. The viability of leukocytes, estimated by the Trypan blue test, was at least 93–95 %.

The test for cytotoxicity of human leukocytes (effector cells) for *K562* cells (target cells) incubated with quinoline derivatives and labeled with ^{3}H-uridine followed the protocol [44]. Pre-assay incubation of the cultures with the reagents lasted 48 h and 96 h in all variants. The effector cell/target cell ratio was 50:1. The cytotoxic index (CI, %) was calculated by formula (3):

Cytotoxic index = 1 – (cpm) in experimental tests / (cpm) in control tests × 100 %,   (3)

where the control was *K562* cell cultures labeled with ^{3}H-uridine and free of effector-cells.

Quinoline derivatives were kindly provided by Prof. V.P. Andreev (St. Petersburg State University, Russia). The composition and structure of the resultant compounds were ascertained by elemental analysis, mass, IR-, NMR spectroscopy (^{1}H, ^{13}C, ^{15}N).

Reliability of the results was estimated using Student's t-test, and the non-parametric Mann–Whitney test.

4. Results

The first stage of the investigation of the biological activity of quinoline and pyridine derivatives was experiments to determine the effect of varying concentrations of Q, 2-MeQ, QO, 2-$MeQO$, 4-NQO, 2-Me-4-NQO, 2-DQO, 4-DQO, 2-$NSQO$, 4-$NSQO$, and 4-$DPyO$ on the viability of $K562$ cells. The resultant data were processed to calculate EC_{50} values of each compound. The results are detailed in Table 2 (the reagents are arranged in the order of decreasing toxicity). It follows from the results that the greatest toxic effect on tumor cells under the stated conditions was demonstrated by 2-Me-4-NQO and 4-NQO, and the lowest – by 2-$NSQO$ and 4-$NSQO$.

Reagent	EC_{50}, μM
2-Me-4-NQO	1.04
4-NQO	1.05
2-MeQ	1.26
Q	4.47
2-MeQO	23.44
4-DPyO	33.11
1. 2-DQO	2. 208.93
3. 4-DQO	4. 221.28
5. QO	6. 316.23
7. 2-NSQO	8. 570.10
9. 4-NSQO	10. 600.50

Table 2. Reagent concentrations resulting in 50% death of $K562$ cells (EC_{50})

The data presented in Table 2 indicate also that when the methyl radical ($Q \rightarrow 2$-MeQ, $QO \rightarrow 2$-$MeQO$, 4-$NQO \rightarrow 2$-Me-4-NQO) and the nitro group ($QO \rightarrow 4$-NQO, 2-$MeQO \rightarrow 2$-Me-4-NQO) were attached to the quinoline heterocycle, the toxicity of the compounds for $K562$ cells increased. For example, the EC_{50} of Q was 4.47 μM, 2-MeQ – 1.26 μM; QO – 316.23 μM, 2-$MeQO$ – 23.44 μM; 4-NQO – 1.05 μM, 2-Me-4-NQO – 1.04 μM, EC_{50} in the $QO/4$-NQO and 2-$MeQO/2$-Me-4-NQO pairs was 316.23 μM, 1.05 μM and 23.44 μM, 1.04 μM, respectively. When $K562$ cells were incubated with reagents comprising the N-oxide group (QO, 2-$MeQO$), the cell survival rate was, on the contrary, higher than for the cells incubated with Q

or *2-MeQ*, respectively (see EC$_{50}$). The toxic effect of the compounds was reduced also by addition of the *"styryl tail"* with the nitro group (*2-NSQO, 4-NSQO*) to the reagent, whereas substitution of the nitro group with the dimethylamino group (*2-DQO, 4-DQO*) resulted in a 2.72-fold rise in the toxicity (see EC$_{50}$). Translocation of the styryl group within the quinoline ring was also significant: if the group was located closer to the heteroatom (Fig. 1), the compound also became more toxic (EC$_{50}$ of *2-DQO* was 208.93 µM, that of *4-DQO* – 221.28 µM; EC$_{50}$ of *2-NSQO* – 570.10 µM, that of *4-NSQO* – 600.50 µM). When the quinoline cycle in the chemical compound was substituted with the pyridine cycle the reagent's toxicity for the cells increased 6.68-fold, e.g., *4-DQO* EC$_{50}$ was 221.28 µM, whereas *4-DPyO* EC$_{50}$ was 33.11 µM. This pattern was observed both on the 48 h and on the 96 h of incubation.

The results above suggest that the toxic effect of the investigated reagents on tumor cells depends on the presence/absence of certain type substituents (electron donors or electron acceptors) in the quinoline cycle, as well as on the direct bond of the named substituents to the quinoline cycle. Thus, attachment of the nitro group (electron acceptor) to the cycle – *QO* → *4-NQO, 2-MeQO* → *2-Me-4-NQO* – rendered the substance more toxic, whereas inclusion of the substituent in the *"styryl tail"* (*4-NQO* → *4-NSQO*) reduced the toxic effect of the reagent 571.90 fold (see EC$_{50}$).

Intracellular activation of aromatic nitrogen-bearing compounds, including *4-NQO* and other quinoline derivatives, is known to involve cytochrome P-450 and NADPH-dependent cytochrome P-450 reductase [45], glutathione-S-transferase, and quinone reductase [46]. As the result, there appears a set of metabolites most of which have a higher biological activity, and are potentially capable of interacting with high-molecular cell compounds (proteins, nucleic acids). Since an essential component part of research into the biological activity of chemical reagents *in vitro* is the study of their apoptosis inhibiting action, we shall now report the results on the effect of the investigated group of heterocycles on the expression (at the mRNA level) and enzymatic activity of cellular caspases.

Data on modifications of caspase-3, -6 and -9 mRNA expression in *K562* cells treated with *Q*, *QO* and *4-NQO* for 48 h are presented in Figure 4 (**a–c**). The reagent concentrations applied are detailed in the figure. The results suggest that cell incubation with *4-NQO* caused a rise in mRNA expression in all caspases; treatment with *Q* induced a rise in caspase-3 and -9 mRNA expression; whereas *QO* did not induce caspase expression under the given treatment conditions. A similar pattern was observed in the enzymatic activity (Figure 5 **a–c**). Cell incubation with *4-NQO* promoted the activity of all the caspases, whereas treatment with *Q* activated only caspases-3 and -9 (caspase-6 in this case was induced only on the 96 h). *QO* exhibited an activating effect on caspase-3 also on the 96 h of tumor cell treatment. Note that the activity of caspase-6 in the cells ($p<0.05$) (Figure 5 **b**) treated with *4-NQO* rose much less than that of caspases-3 and -9 ($p<0.01$) (Figure 5 **a, c**).

Caspase expression in *K562* cells incubated with *2-MeQ, 2-MeQO*, and *2-Me-4-NQO* did not differ from their expression in *Q, QO*, and *4-NQO* treatments, respectively.

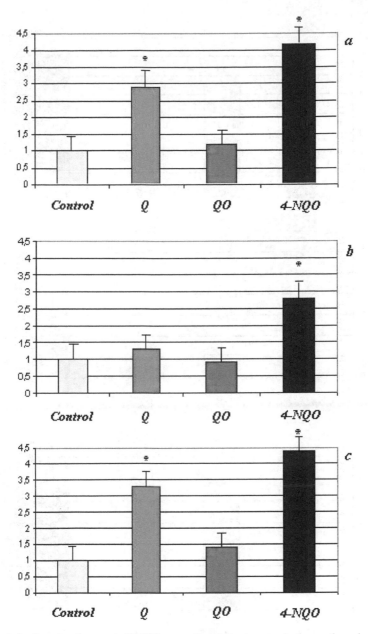

Results are presented as the ratio of caspase to GAPDH expression relative to a standard curve for each assay. Drug concentrations: Q – 0.3 µM, QO – 10 µM, $4\text{-}NQO$ – 0.001 µM. Incubation time – 48 h. Viability of cultured cells – 93–95%.

Y axis – relative expression, conventional unit; * – p<0.05

Figure 4. Caspase-3 (*a*), -6 (*b*) and -9 (*c*) mRNA levels were determined by quantitative real-time RT-PCR in *K562* cells treated with Q, QO and $4\text{-}NQO$ respectively.

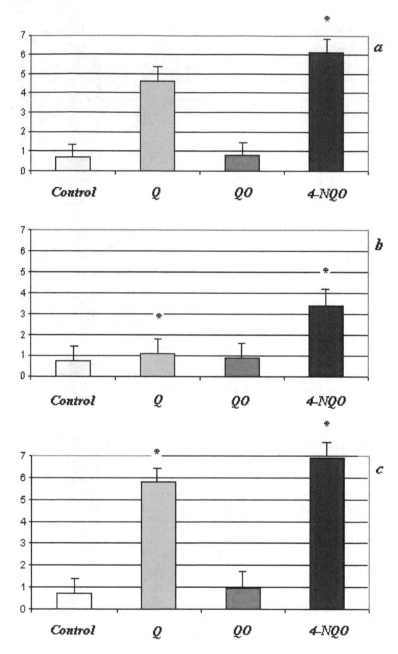

Drug concentrations: Q – 0.3 µM, QO – 10 µM, 4-NQO – 0.001 µM. Incubation time – 48 h.
Viability of cultured cells – 93–95%. Y axis – $\Delta S/\Delta t \times 10^4$; * – p<0.05

Figure 5. Caspase -3 (*a*), -6 (*b*) and -9 (*c*) activity in *K562* cells treated with *Q, QO* and *4-NQO* respectively.

Styryl derivatives also produced caspase inhibiting effects on *K562* cells. The results portrayed in Figures 6 and 7 evidence that treatment of tumor cells with *2-DQO, 4-DQO, 2-NSQO*, and *4-NSQO* promoted the activity of caspases-3 and -6, but with the latter two reagents the activity of the stated caspases was higher, and the rise in caspase-9 activity was recorded on the 24 h of application of the compounds to the incubation medium. Expression at the mRNA level correlated with data on the activity at the enzyme level.

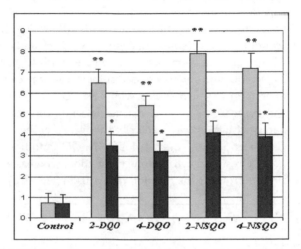

Drug concentrations – 1 µM. Incubation time – 48 h., Viability of cultured cells – 93–95%. Y axis – $\Delta S/\Delta t \times 10^{-4}$; * – $p<0.05$, ** $p<0.01$;
☐ caspase-3; ■ caspase-6

Figure 6. Caspase-3 and -6 activity in *K562* cells treated with *2-DQO, 4-DQO, 2-NSQO, 4-NSQO* respectively.

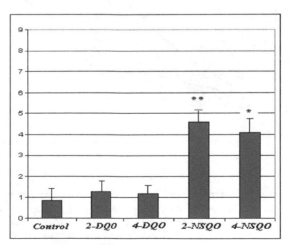

Drug concentrations – 1 µM. Incubation time – 24 h, Viability of cultured cells – 93–95%. Y axis – $\Delta S/\Delta t \times 10^{-4}$; * – $p<0.05$, ** $p<0.01$

Figure 7. Caspase-9 activity in *K562* cells treated with *2-DQO, 4-DQO, 2-NSQO, 4-NSQO* respectively

We have already mentioned that the release of cytochrome c from mitochondrial interior to the cell cytoplasm is a crucial stage in the activation of caspase cascades and triggering of apoptosis. In addition to involvement in the reactions of cellular caspase activation, "cytoplasmic" forms of cytochrome c also act as a substrate for microsomal NADPH-dependent cytochrome P-450 reductase [47]. This reductase facilitates electron transport to cytochrome P-450, which, in turn, participates in the metabolism of many heterocycles, including those investigated in the present paper. Figure 8 shows the kinetic curve of cytochrome c reduction by microsomes with no and with 10 μM 2-NSQO. Addition of 2-NSQO to the medium containing microsome suspension, NADPH, and cytochrome c causes a deceleration of cytochrome c reduction, which progresses with time. Depending on the reagent concentration, the reduction rate decreased at the greatest pace in the first 1–5 min of the reaction. Pre-incubation of microsomes with 2-NSQO for 5 min, followed by the addition of NADPH and cytochrome c to the reaction medium did not change the protein reduction rate as compared with the variant described above. After the NADPH-cytochrome c reductase activity had been totally inhibited by 2-NSQO, addition of NADH to the incubation medium partially restored cytochrome c reduction, whereas addition of NADPH and/or cytochrome c did not promote the reaction rate ($p > 0.05$) (Figure 8).

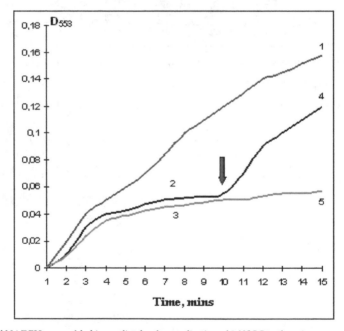

Cytochrome c and NADPH were added immediately after application of 2-NSQO to the microsome suspension (2) or after 5 min of pre-treatment of the microsomes with the reagent (3). NADH (50 μM) (4) or NADPH (50 μM) (5) were added to system 2 on the minute of the reading event. Light absorption by the system was measured in phosphate buffer on Labsystems Multiscan Plus reader (LKB, Finland) at 553 nm

Figure 8. Cytochrome c reduction by microsomes of K562 cells with no (1), and with 10 μM 2-NSQO (2–5).

Substitution of 2-*NSQO* with its structural analog (4-*NSQO*) yielded a similar pattern, but inhibition of the NADPH-cytochrome c reductase activity proceeded much slower, and addition of NADH to the incubation medium fully restored cytochrome c reduction. Dependence of the NADPH-cytochrome c reductase activity on the concentration of 2-*NSQO* and 4-*NSQO* is shown in Figure 9. The microsomal protein content being 30 µg/ml, 2-*NSQO* in a concentration of 10 µM inhibited the system's enzymatic activity by 50%, and at a concentration of 50 µM and higher the activity dropped by more than 80%. The 10 µM concentration of 4-*NSQO* inhibited the enzymatic activity of microsomes by 15%, and the 100 µM concentration – by 50%. The above results suggest that the apoptosis inducing effect of 2-*NSQO* and its structural analogs on the cells may be due to the irreversible inhibition of microsomal enzymes with NADPH-cytochrome c reductase activity. Where this effect happens, functioning of the whole electron transport chain in the cells is usually disrupted, there form large amounts of genotoxic products, and the cell eventually dies through apoptosis (or necroptosis).

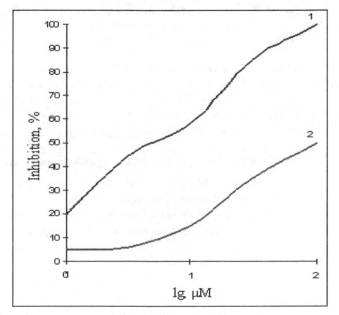

X axis– 2-*NSQO* and 4-*NSQO* concentrations, µM (logarithmic scale); Y axis – inhibition of the reaction rate (%) relative to the control. Modifications to the system were made in the first 5 min of the reaction

Figure 9. Effect of 2-*NSQO* (1) and 4-*NSQO* (2) on NADPH-cytochrome c reductase activity of the microsomes of *K562* cells.

Another important fact is that when some quinoline derivatives interacted with DNA the greatest hypochromic shift in DNA UV spectra (which evidences the formation of complexes) was observed in treatments with 4-*NQO*, 4-*NSQO*, 4-*DQO*, 4-*DPyO*, whereas mixing of *Q* or 2-*MeQ* with DNA did not cause changes in the absorption spectra (Table 3). This fact supports the assumption that for such biological effect to appear the latter two reagents must first undergo metabolic activation in the cell, and form electrophilic centres within the molecule. Other N-bearing heterocyclic compounds are potentially able to bind to DNA without prior intracellular activation (e.g., through intercalation), so that their biological effects would appear sooner.

Reagent	Percent of the reagents-related hypochromic effect of DNA
4-NQO	10.7±1.2
4-NSQO	9.7±1.3
4-DQO	9.4±1.8
4-DPyO	7.0±1.4
QO	2.4±0.6
Q	0
2-MeQ	0

Table 3. Hypochromic effect of DNA upon binding with quinoline and pyridine derivatives. DNA from chicken erythrocytes was used in the experiment. The spectra were obtained in phosphate buffer on a Labsystems Multiscan Plus reader (LKB, Finland) at λ 260 nm.

Many chemical compounds, including anticancer drugs, can induce tumor cell apoptosis both *in vivo* and *in vitro* [48–50]. However, a major cause of failure in the application of cytostatic agents in clinical practice is the multiple drug resistance (MDR) phenotype induced in the tumor cells. Multidrug resistance is a condition when tumor cells are unsusceptible to a variety of chemotherapeutic agents differing in chemical structure and the mechanisms through which they affect the cell. MDR is a serious hindrance to success in the treatment of malignant tumors, including leukosis. Latest studies have demonstrated that the molecular mechanisms of MDR are multifarious, and the drug resistance of a cell can depend on various mechanisms triggered at different stages of the toxic impact of the agent on the cell – from restricted accumulation of the drug in the cell to cancellation of the substance-induced cell death programme. Interplay of several protective mechanisms is not uncommon, but one mechanism usually prevails. The best studied mechanisms, whose clinical significance for certain neoplastic forms (such as chronic myeloid leukemia or chronic lymphatic leukemia) has been ascertained, are: activation of transmembrane transport proteins that remove various substances from the cell (namely, P-glycoprotein – Pgp); activation of the glutathione system enzymes that detoxify the drugs; modifications in the genes and proteins that control apoptosis and cell survival [51].

We have therefore obtained *K562* cell sublines resistant to *2-DQO* and *4-NQO* (*K562/2-DQO*, *K562/4-NQO*), and characterized their capacity to undergo induced apoptosis under the effect of structurally different chemical reagents as compared with the parental line cells.

The MDR development mechanisms related to inhibition of cytostatic-induced apoptosis have lately been actively investigated. Of greatest interest among the cytostatic agents are DNA-tropic compounds (adriamycin, actinomycin D), antimetabolites (5-fluorouracil, methotrexate), reagents interacting with the mitotic spindle microtubules (vincristine, vinblastine, taxol), and others. *K562/4-NQO* and *K562/2-DQO* cells were tested for susceptibility to the following chemical compounds: DNA intercalating agent ethidium bromide, microtubule polymerization inhibitor colchicine; and quinoline derivatives – 2-DQO and 4-NQO, respectively. The results are shown in Table 4 and Figure 10.

Cell line	EC$_{50}$, µM			
	Ethidium bromide	Colchicine	4-NQO	2-DQO
K562	2.5	0.002	1.06	79.7
K562/4-NQO	39.8	0.0004	—	398.2
K562/2-DQO	7.9	0.120	79.0	—

Table 4. Susceptibility of *K562/4-NQO* and *K562/2-DQO* cells to cytostatic agents. Clones resistant to 4-NQO and 2-DQO were obtained after [38]. Cell resistance to the xenobiotics was induced by long-term (over 1 month) exposure to 10^{-12} M 4-NQO or 10^{-9} M 2-DQO in the culture medium. The concentration of the substances was then increased every 14th–21st days. The final concentrations of the reagents for which resistant cell lines were obtained were 10^{-8} M for 4-NQO, and 10^{-5} M for 2-DQO.

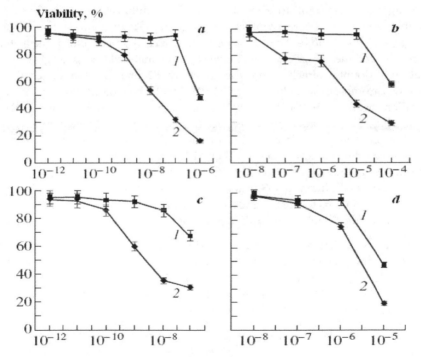

a – cells were treated with 4-NQO: 1 – Viability of *K562/2-DQO* cells, 2 – Viability of *K562* cells;
b – cells were treated with 2-DQO: 1 – Viability of *K562/4-NQO* cells, 2 – Viability of *K562* cells;
c – cells were treated with colchicines: 1 – Viability of *K562/2-DQO* cells, 2 – Viability of *K562* cells;
d – cells were treated with ethidium bromide: 1 – Viability of *K562/2-DQO* cells, 2 – Viability of *K562* cells.
X axis – drug concentration (M). Incubation time – 48 h.

Figure 10. Viability of *K562*, *K562/2-DQO* and *K562/4-NQO* cell lines treated with different xenobiotics.

Measurements of the viability of the cells of the parental line and the derived sublines, incubated for 48 h in the presence of various concentrations of the reagents, revealed substantial differences in the toxic effect of the xenobiotics on the tumor cells. *K562/4-NQO* cells were highly resistant to ethidium bromide and *2-DQO*, but susceptible to colchicine (Table 4, Figure 10). Contrastingly, *K562/2-DQO* cells demonstrated quite high resistance to colchicine and *4-NQO*, but lower resistance to ethidium bromide (Table 4, Figure 10). One should mention that the resistance of *K562/2-DQO* cells to ethidium bromide was 5.03 times lower than that of *K562/4-NQO* cells.

It is now solid knowledge that when subjected to stepped selection for resistance to some cytotoxic agents (plant alkaloids, antibiotics), cells of tumor lines in vitro become cross-resistant to quite a number of cytostatics, which differ both in the structure and genesis, and in the effect on different cellular targets [52–54]. The spectra of the drugs to which the cells develop cross-resistance, as well as the mechanisms behind it may vary depending on the selection agent. Grinchuk et al. [55] isolated three adriamycin-resistant clones *C9*, *B2*, and *B3*, through multi-step selection from the *K562* cell population. The cells of the adriamycin-resistant clones displayed cross-resistance to colchicine, actinomycin D, and ethidium bromide – agents of the multidrug resistance group. Amplification of the gene *mdr1* was detected in the cells of the resistant clones at DNA hybridization with Southern blot. Karyological analysis of the resistant cells performed at early stages of their incubation with adriamycin showed the genome contained extra genetic material (morphological markers of amplification – double minichromosomes in clones *B2* and *B3*, and uniformly stained regions in the chromosomes of clone *C9* cells). The karyotype modifications are attributed to destabilization of the *K562*/adriamycin cell genome as the cells acquired the MDR phenotype. Thus, distinctions in the resistance to the toxic effect of xenobiotics among cells of different sublines may be related to specific characteristics of the expression of *mdr* genes and then P glycoprotein, expression of other MDR proteins, genome destabilization, and, not least, changes in the activity of cellular enzymes involved in the first (e.g., cytochrome P-450) and second (e.g., glutathione-S-transferase) phases of the reagent metabolism in the cells. These and other factors can, under certain conditions, play a decisive role in the modulation of the functional activity of apoptosis inducing agents, including caspases.

Figure 11*a* shows the results on relative expression of caspase-3, -6, and -9 genes in the cells of the *K562* parental line and the derivative sublines (data from RT-PCR analysis). One can see that the mRNA expression of the named caspases in *K562/4-NQO* and *K562/2-DQO* cells is significantly lower ($p<0.05$) than in the parental *K562*, and that is a potential explanation of the resistance of the tumor cells to *4-NQO* and *2-DQO*.

In the treatment with colchicine mRNA expression of caspases-3, -6, and -9 was significantly promoted in the parental line *K562* and *K562/4-NQO* – $p<0.05$ (Figure 11 *b*). In *K562/2-DQO* cells only caspase-6 expression was activated ($p<0.05$). Incubation of tumor cells with ethidium bromide promoted the mRNA expression of the three investigated caspases in the parent cells; caspase-3 expression was promoted in *K562/2-DQO* cells; no significant changes in caspase mRNA expression was observed in the *K562/4-NQO* line. Changes in the enzymatic activity correlated with mRNA expression.

One of the preconditions for *in vivo* development and progress of tumors is that the tumor cells gain another type of resistance – resistance to factors of anti-tumor immunity. Such factors are, first of all, natural killer (NK) cells and cytotoxic T lymphocytes. The mechanisms of identification of *de novo* tumor cells in the organism by natural killers have not been fully elucidated, but the mechanisms of the lysis of such cells are quite well studied. NK cells and antigen-specific cytotoxic T lymphocytes, as well as these IL-2 activated cells are known to lyse target cells both by perforin-dependent cytolysis, and by inducing Fas-dependent apoptosis [56, 57]. In this case, the target cells can modulate the rate of release of the cytolytic granule content, including perforin and granzymes, or suppress killer activation by reducing the frequency and affinity of the formation of target cell-effector cell conjugates, the latter being a quite frequent phenomenon in the system [58]. Perforin is structurally homologous to the *C9* complement component, and has a similar action mechanism (formation of polyperforin pores made up of 12–18 monomers, with the inner diameter of 10–20 nm) [56]. On the other hand, no DNA fragmentation occurs when the complement acts on the cells, wherefore we presume that the action of perforin involves also other factors (namely serine proteases of the granzyme family [58], which can permeate the target cell cytosol via perforin-generated pores in the plasma membrane and trigger apoptotic events, for instance by means of caspase activation.

Some authors have demonstrated in their papers that treatment of tumor cells with chemical reagents modifies their susceptibility to the cytotoxic lysis of homologous effector cells, and the modification may be either for amplification or for reduction of the effect depending on the cell line and the treatment settings. E.g., treatment of *K562* cells with sodium butyrate [59], fagaronine, adriamycin, aclacinomycin A [60] makes the named cells less susceptible to natural killer lysis. Quite a number of lymphoma cell lines (*Raji* and *Daudi* lymphoma cell lines, human lymphoblastoid cell line *NAD-7*, human T-lymphoblastoid line *Molt-4*) treated with PMA, sodium butyrate, retinoic acid demonstrate higher susceptibility to the cytotoxic activity of natural killers [61–63]. An interesting effect was observed in the case of *CB-1 B* cell lines. Clone *26CB-1* is more resistant to NK cells than clone *13CB-1*. After they had been treated with sodium butyrate, iodinated deoxyuridine, 5-azacytidine, tunicamycin, the susceptibility of clone *26CB-1* cells to cytotoxic lysis increased, whereas the susceptibility of clone *13CB-1* cells remained nearly unmodified [64].

Distinctions in the chemical inducers' structure and mechanisms of action on cells may have different effects on modulation of the expression of positive and negative apoptosis regulators, which quantitative ratios may contribute to establishment of both the initial susceptibility of the cells to natural killer lysis and the post-treatment susceptibility. This may as well be valid for the cell sublines with a more or less differentiated phenotype.

Data in Figure 12 (**a, b**) show that the treatment of tumor cells with *4-NQO, 2-NSQO,* and *4-NSQO* for 48 h caused a significant rise (p<0.05) in susceptibility to the lysis of human leukocytes containing NK cells. Caspase expression and cell apoptosis induction were also promoted in this case. In treatments with other quinoline derivatives the cytotoxic index in the cultures did not change. A similar pattern was observed also when incubation lasted 96 h.

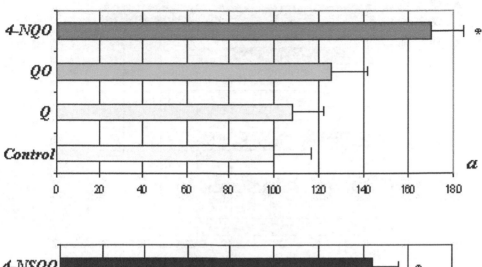

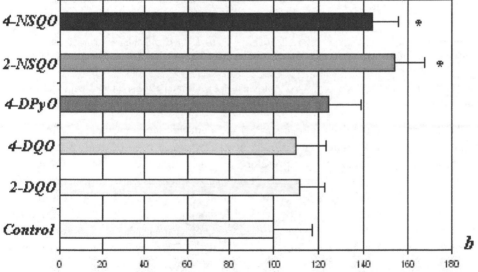

Drug concentrations: Q – 0.3 μM, QO – 10 μM, 4-NQO – 0.001 μM, styryl derivatives – 1 μM. X-axis: cytotoxic index ratio of effector cells with treated cell lines to control cell lines without treatment, % (cytotoxic index value of control cells is 100 %). Effector cells to target cells ratio is 50:1. Incubation time – 48 h; * $p<0.05$.

Figure 12. The *K562* cell line sensitivity to cytotoxic lysis by human peripheral blood white cells after treatment with quinoline and pyridine derivatives

When cells of sublines *K562/4-NQO* and *K562/2-DQO* were used as the targets, changes in susceptibility to the leukocyte lysis effect were recorded only in the *K562/4-NQO*–colchicine system (Figure 13).

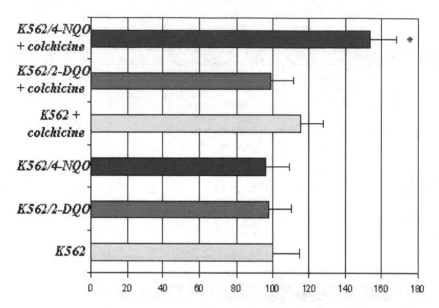

Drug concentrations – 0.1 μM for *K562/2-DQO* cells, 0.001 μM for *K562* cells, 0.001 μM for *K562/4-NQO* cells. X-axis: cytotoxic index ratio of effector cells with treated cell lines to control cell lines without treatment, % (cytotoxic index value of control cells is 100 %). Effector cells to target cells ratio is 50:1. Incubation time – 48 h; * p<0.05.

Figure 13. The *K562* cell line sensitivity to cytotoxic lysis by human peripheral blood white cells after treatment with colchicines.

Thus, treatment of tumor cells with chemical compounds may simultaneously induce cell apoptosis and modulate their susceptibility to the cytotoxic lysis of blood leukocytes. In these treatment conditions the expression and/or activity of intracellular apoptosis regulators, namely caspases, changes, and this change may tell on the effectiveness of the leukocyte lysis effect, which is based on induction of target cell apoptosis. It is essential that the susceptibility of tumor cells to the lysis of leukocytes containing NK cells changed in those variants where the expression and activity of caspases-3, -6, -9 were promoted (cultures treated with *4-NQO, 2-NSQO,* and *4-NSQO*). On the other hand, the equivocal nature of the results proves that the process involves extra intracellular factors. One can therefore assume that similar patterns of the response of tumor cells to treatment with chemical compounds can occur *in vivo* in humans and animals treated with the anticancer drugs stimulating cell apoptosis. This effect can persist when nontoxic or low-toxicity doses are applied, and caspases, being a cell's major apoptosis inducing factor, can be viewed as potential targets for pharmacologically active agents.

5. Conclusions

Over 40 years have passed since the first publication devoted to apoptosis. Since then, the problem of apoptosis regulation mechanisms has become central to pharmacology, genetics, virology, cytology, immunology, biochemistry, embryology, and other areas of modern science at the cellular and molecular levels. It is obvious that the problem addressed in this chapter is of high applied value, for the key aim of the effort is to enhance the effectiveness of anticancer therapy. The industry of rational drug design is becoming more and more widespread in practical medicine. The principal concepts in drug design are the target and the drug. The target is a low-molecular biological structure presumably linked to a certain function which disruption would lead to disease, and to which a certain impact should be applied. The most common targets are receptors, enzymes, hormones. The drug is most often a chemical compound (usually a low-molecular one) that specifically interacts with the target, modifying the cell response in one way or another. One of the earliest and most seminal stages of drug design is accurate identification of the target by influencing which one can specifically regulate certain biochemical processes, leaving others unaffected as much as possible. This is not however always feasible: by no means are all diseases caused by dysfunction of just one protein or gene. Preference should therefore be given to the processes' principal biomolecules. In terms of apoptosis, caspases clearly are such biomolecules. Hence, the choice of methods for experimental validation of caspases as potential specific targets for drugs is a topical challenge for the nearest future.

6. Abbreviations:

Quinoline – *Q*, 2-methylquinoline – *2-MeQ*, quinoline-1-oxide – *QO*, 2-methylquinoline-1-oxide – *2-MeQO*, 4-nitroquinoline-1-oxide – *4-NQO*, 2-methyl-4-nitroquinoline-1-oxide – *2-Me-4-NQO*, 2-(4'-dimethylaminostyryl)quinoline-1-oxide – *2-DQO*, 4-(4'-dimethylaminostyryl)-quinoline-1-oxide – *4-DQO*, 2-(4'-nitrostyryl)quinoline-1-oxide – *2-NSQO*, 4-(4'-nitrostyryl)-quinoline-1-oxide – *4-NSQO*, and 4-(4'-dimethylaminostyryl)pyridine-1-oxide – *4-DPyO*; death effector domain – DED, caspase recruitment domain – CARD, death inducing domain – DID, p53-induced protein with a death domain – PIDD, death-inducing signalling complex – DISC, DNA fragmentation factor – DFF, caspase-activated DNase – CAD, permeability transition pore – PTP, apoptosis inducing factor – AIF, tumor necrosis factor receptor – TNFR, death domain – DD, death receptor – DR, Fas-associated DD-protein – FADD, TNFR1-associated DD-protein – TRADD.

Author details

Tatyana O. Volkova
Department of Biochemistry, School of Medicine,
Petrozavodsk State University, Petrozavodsk, Russia

Alexander N. Poltorak
Department of Pathology, School of Medicine,
Tufts University, Boston, USA

Acknowledgement

We are sincerely grateful to the Chair of the Petrozavodsk State University Molecular Biology Department (Russia), DSc, Professor N. N. Nemova for her invaluable help in the preparation of the manuscript. Special thanks are due also to Professor V. P. Andreev of the St. Petersburg State University Organic Chemistry Department (Russia) for the chemical compounds he has provided.

The study was supported by the Government of the Russian Federation grant № 11.G34.31.0052 (Ordinance 220), and RF Presidential grant for leading scientific schools № 1642.2012.4.

7. References

[1] Kopnin BP (2000) Targets of oncogenes and tumor suppressors: key for understanding basic mechanisms of carcinogenesis. Biochemistry (Moscow) 65 (1): 2-27.

[2] Volkova TO, Nemova NN (2006) Molecular Mechanisms of Apoptosis of Leukemic Cells. Moscow: Nauka, 208 p.

[3] Xu XM, Yuan GJ, Deng JJ, Guo HT, Xiang M, Yang F, Ge W, Chen SY (2012) Inhibition of 12-lipoxygenase reduces proliferation and induces apoptosis of hepatocellular carcinoma cells in vitro and in vivo. Hepatobiliary Pancreat Dis Int. 11 (2): 193-202.

[4] Hu WF, Gong L, Cao Z, Ma H, Ji W, Deng M, Liu M, Hu XH, Chen P, Yan Q, Chen HG, Liu J, Sun S, Zhang L, Liu JP, Wawrousek E, Li DW (2012) αA- and αB-crystallins interact with caspase-3 and Bax to guard mouse lens development. Curr Mol Med. 12 (2): 177-87.

[5] Volkova TO, Nemova NN (2008) Functional redistribution of caspase activities in K562 cells, induced for differentiation and apoptosis by thiazophosphol derivatives. Biomed Khim. 54 (6): 643-8.

[6] Seeger JM, Schmidt P, Brinkmann K, Hombach AA, Coutelle O, Zigrino P, Wagner-Stippich D, Mauch C, Abken H, Krönke M, Kashkar H (2010) The proteasome inhibitor bortezomib sensitizes melanoma cells toward adoptive CTL attack. Cancer Res. 70 (5): 1825-34.

[7] Festjens N, Cornelis S, Lamkanfi M, Vandenabeele P (2006) Caspase-containing complexes in the regulation of cell death and inflammation. Biol Chem. 387 (8): 1005-16.

[8] Kersse K, Verspurten J, Vanden Berghe T, Vandenabeele P (2011) The death-fold superfamily of homotypic interaction motifs. Trends Biochem Sci. 36 (10): 541-52.

[9] Kerr J F R, Wyllie A H, Currie A R (1972) Apoptosis: a basic biological phenomenon with wide-ranging implications in tissue kinetics. Br. J. Cancer. 26: 239-257.

[10] Kumar S, Kinoshita M, Noda M, Copeland NG, Jenkins NA (1994) Induction of apoptosis by the mouse Nedd2 gene, which encodes a protein similar to the product of the Caenorhabditis elegans cell death gene ced-3 and the mammalian IL-1 beta-converting enzyme. Genes Dev. 8 (14): 1613-26.

[11] Wang L, Miura M, Bergeron L, Zhu H, Yuan J (1994) Ich-1, an Ice/ced-3-related gene, encodes both positive and negative regulators of programmed cell death. Cell. 78 (5): 739-50.

[12] Liu X, Kim CN, Yang J, Jemmerson R, Wang X (1996) Induction of apoptotic program in cell-free extracts: requirement for dATP and cytochrome c. Cell. 86 (1): 147-57.

[13] Chang HY, Yang X (2000) Proteases for cell suicide: functions and regulation of caspases. Microbiol Mol Biol Rev. 64 (4): 821-46.

[14] Haunstetter A, Izumo S (1998) Apoptosis: basic mechanisms and implications for cardiovascular disease. Circ Res. 82 (11): 1111-29.

[15] Wang Y, Zhao Y, Liu Y, Tian L, Jin D (2011) Chamaejasmine inactivates Akt to trigger apoptosis in human HEp-2 larynx carcinoma cells. Molecules. 16 (10): 8152-64.

[16] Chandra J, Samali A, Orrenius S (2000) Triggering and modulation of apoptosis by oxidative stress. Free Radic Biol Med. 29 (3-4): 323-33.

[17] Slomiany BL, Slomiany A (2010) Constitutive nitric oxide synthase-mediated caspase-3 S-nitrosylation in ghrelin protection against Porphyromonas gingivalis-induced salivary gland acinar cell apoptosis. Inflammopharmacology. 18 (3): 119-25.

[18] Schmitz I, Kirchhoff S, Krammer PH (2000) Regulation of death receptor-mediated apoptosis pathways. Int J Biochem Cell Biol. 32 (11-12): 1123-36.

[19] Ferrao R, Wu H (2012) Helical assembly in the death domain (DD) superfamily. Curr Opin Struct Biol. 22 (2): 241-7.

[20] Esposito D, Sankar A, Morgner N, Robinson CV, Rittinger K, Driscoll PC (2010) Solution NMR investigation of the CD95/FADD homotypic death domain complex suggests lack of engagement of the CD95 C terminus. Structure. 18 (10): 1378-90.

[21] Lavrik IN, Krammer PH (2012) Regulation of CD95/Fas signaling at the DISC. Cell Death Differ. 19 (1): 36-41.

[22] Pennarun B, Meijer A, de Vries EG, Kleibeuker JH, Kruyt F, de Jong S (2010) Playing the DISC: turning on TRAIL death receptor-mediated apoptosis in cancer. Biochim Biophys Acta. 1805 (2): 123-40.

[23] Shang YC, Chong ZZ, Hou J, Maiese K (2009) FoxO3a governs early microglial proliferation and employs mitochondrial depolarization with caspase 3, 8, and 9 cleavage during oxidant induced apoptosis. Curr Neurovasc Res. 6 (4): 223-38.

[24] Twiddy D, Brown DG, Adrain C, Jukes R, Martin SJ, Cohen GM, MacFarlane M, Cain K (2004) Pro-apoptotic proteins released from the mitochondria regulate the protein composition and caspase-processing activity of the native Apaf-1/caspase-9 apoptosome complex. J Biol Chem. 279 (19): 19665-82.

[25] Yuan S, Yu X, Topf M, Ludtke SJ, Wang X, Akey CW (2010) Structure of an apoptosome-procaspase-9 CARD complex. Structure. 18 (5): 571-83.

[26] Kowaltowski AJ, Castilho RF, Vercesi AE (2001) Mitochondrial permeability transition and oxidative stress. FEBS Lett. 495 (1-2): 12-5

[27] Hu Y, Benedict MA, Ding L, Núñez G (1999) Role of cytochrome c and dATP/ATP hydrolysis in Apaf-1-mediated caspase-9 activation and apoptosis. EMBO J. 18 (13): 3586-95.

[28] Bajgar R, Seetharaman S, Kowaltowski AJ, Garlid KD, Paucek P (2001) Identification and properties of a novel intracellular (mitochondrial) ATP-sensitive potassium channel in brain. J Biol Chem. 276 (36): 33369-74.

[29] Chiang JH, Yang JS, Ma CY, Yang MD, Huang HY, Hsia TC, Kuo HM, Wu PP, Lee TH, Chung JG (2011) Danthron, an anthraquinone derivative, induces DNA damage and caspase cascades-mediated apoptosis in SNU-1 human gastric cancer cells through mitochondrial permeability transition pores and Bax-triggered pathways. Chem Res Toxicol. 24 (1): 20-9.

[30] Hsu PC, Huang YT, Tsai ML, Wang YJ, Lin JK, Pan MH (2004) Induction of apoptosis by shikonin through coordinative modulation of the Bcl-2 family, p27, and p53, release of cytochrome c, and sequential activation of caspases in human colorectal carcinoma cells. J Agric Food Chem. 52 (20): 6330-7.

[31] Widlak P, Garrard WT Roles of the major apoptotic nuclease-DNA fragmentation factor-in biology and disease (2009) Cell Mol Life Sci. 66 (2): 263-74.

[32] Héron-Milhavet L, Karas M, Goldsmith CM, Baum BJ, LeRoith D (2001) Insulin-like growth factor-I (IGF-I) receptor activation rescues UV-damaged cells through a p38 signaling pathway. Potential role of the IGF-I receptor in DNA repair. J Biol Chem. 276 (21): 18185-92.

[33] Zhang M, Qiao X, Zhao L, Jiang L, Ren F (2011) Lactobacillus salivarius REN counteracted unfavorable 4-nitroquinoline-1-oxide-induced changes in colonic microflora of rats. J Microbiol. 49 (6): 877-83.

[34] Zhang S, Han Z, Kong Q, Wang J, Sun B, Wang G, Mu L, Wang D, Liu Y, Li H (2010) Malignant transformation of rat bone marrow-derived mesenchymal stem cells treated with 4-nitroquinoline 1-oxide. Chem Biol Interact. 188 (1): 119-26.

[35] Chu M, Su YX, Wang L, Zhang TH, Liang YJ, Liang LZ, Liao GQ (2012) Myeloid-derived suppressor cells contribute to oral cancer progression in 4NQO-treated mice. Oral Dis. 18 (1): 67-73.

[36] Fann YC, Metosh-Dickey CA, Winston GW, Sygula A, Rao DN, Kadiiska MB, Mason RP (1999) Enzymatic and nonenzymatic production of free radicals from the carcinogens 4-nitroquinoline N-oxide and 4-hydroxylaminoquinoline N-oxide. Chem Res Toxicol. 12 (5): 450-8.

[37] Benson AM (1993) Conversion of 4-nitroquinoline 1-oxide (4NQO) to 4-hydroxyaminoquinoline 1-oxide by a dicumarol-resistant hepatic 4NQO nitroreductase in rats and mice. Biochem Pharmacol. 46 (7): 1217-21.

[38] Tsuruo T, Iida H, Tsukagoshi S, Sakurai Y (1981) Overcoming of vincristine resistance in P388 leukemia in vivo and in vitro through enhanced cytotoxicity of vincristine and vinblastine by verapamil. Cancer Res. 41 (5): 1967-72.

[39] Liu D, Wang Y, Wang B (1997) The proliferation inhibition and differentiation inducing effects of all-trans retinoic acid on human pancreatic adenocarcinoma cell line JF-305. Zhonghua Wai Ke Za Zhi. 35 (3): 153-5.

[40] Livak KJ, Schmittgen TD (2001) Analysis of relative gene expression data using real-time quantitative PCR and the 2(-Delta Delta C(T)). Method Methods. 25 (4): 402-8.

[41] Kolomeichuk SN, Terrano DT, Lyle CS, Sabapathy K, Chambers TC (2008) Distinct signaling pathways of microtubule inhibitors – vinblastine and Taxol induce JNK-dependent cell death but through AP-1-dependent and AP-1-independent mechanisms, respectively. FEBS J. 275 (8): 1889-99.

[42] Bozec A, Ruffion A, Decaussin M, Andre J, Devonec M, Benahmed M, Mauduit C (2005) Activation of caspases-3, -6, and -9 during finasteride treatment of benign prostatic hyperplasia. J Clin Endocrinol Metab. 90 (1): 17-25.

[43] Mrass P, Rendl M, Mildner M, Gruber F, Lengauer B, Ballaun C, Eckhart L, Tschachler E (2004) Retinoic acid increases the expression of p53 and proapoptotic caspases and sensitizes keratinocytes to apoptosis: a possible explanation for tumor preventive action of retinoids. Cancer Res. 64 (18): 6542-8.

[44] Anisimov AG, Bolotnikov IA, Volkova TO (2000) Changes in the K562 cell sensitivity to nonspecific lysis by human and rat leukocytes under the influence of sodium butyrate, dimethyl sulfoxide and phorbol-12-myristate-13-acetate. Ontogenez. 31 (1): 47-52.

[45] Josephy PD, Evans DH, Williamson V, Henry T, Guengerich FP (1999) Plasmid-mediated expression of the UmuDC mutagenesis proteins in an Escherichia coli strain engineered for human cytochrome P450 1A2-catalyzed activation of aromatic amines. Mutat Res. 429 (2): 199-208.

[46] Kawabata K, Tanaka T, Honjo S, Kakumoto M, Hara A, Makita H, Tatematsu N, Ushida J, Tsuda H, Mori H (1999) Chemopreventive effect of dietary flavonoid morin on chemically induced rat tongue carcinogenesis. Int J Cancer. 83 (3): 381-6.

[47] Ponnamperuma K, Croteau R (1996) Purification and characterization of an NADPH-cytochrome P450 (cytochrome c) reductase from spearmint (Mentha spicata) glandular trichomes. Arch Biochem Biophys. 329 (1): 9-16.

[48] Brunelle JK, Zhang B (2010) Apoptosis assays for quantifying the bioactivity of anticancer drug products. Drug Resist Updat. 13 (6): 172-9.

[49] Lu D, Choi MY, Yu J, Castro JE, Kipps TJ, Carson DA (2011) Salinomycin inhibits Wnt signaling and selectively induces apoptosis in chronic lymphocytic leukemia cells. Proc Natl Acad Sci U S A. 108 (32): 13253-7.

[50] Du BY, Song W, Bai L, Shen Y, Miao SY, Wang LF (2012) Synergistic effects of combination treatment with bortezomib and Doxorubicin in human neuroblastoma cell lines. Chemotherapy. 58 (1): 44-51.

[51] Stavrovskaya AA (2000) Cellular mechanisms of multidrug resistance of tumor cells. Biochemistry (Moscow). 65 (1): 95-106.

[52] Donenko FV, Sitdikova SM, Kabieva AO (1993) The cross resistance to cytostatics of leukemia P388 cells with induced resistance to doxorubicin. Biull Eksp Biol Med. 116 (9): 309-11.

[53] Theile D, Ketabi-Kiyanvash N, Herold-Mende C, Dyckhoff G, Efferth T, Bertholet V, Haefeli WE, Weiss J (2011) Evaluation of drug transporters' significance for multidrug resistance in head and neck squamous cell carcinoma. Head Neck. 33 (7): 959-68.

[54] Shah PS, Pham NP, Schaffer DV (2012) HIV Develops Indirect Cross-resistance to Combinatorial RNAi Targeting Two Distinct and Spatially Distant Sites. Mol Ther. 20 (4): 840-8.

[55] Grinchuk TM, Pavlenko MA, Lipskaia LA, Sorokina EA, Tarunina MV, Berezkina EV, Kovaleva ZV, Ignatova TN (1998) Resistance to adriamycin in human chronic promyeloleukemia line K562 correlates with directed genome destabilization--amplification of MDR1 gene and nonrandom changes in karyotype structure. Tsitologiia. 40 (7): 652-60.

[56] Voskoboinik I, Dunstone MA, Baran K, Whisstock JC, Trapani JA (2010) Perforin: structure, function, and role in human immunopathology. Immunol Rev. 235 (1): 35-54.

[57] Liu Z, Liu R, Qiu J, Yin P, Luo F, Su J, Li W, Chen C, Fan X, Zhang J, Zhuang G (2009) Combination of human Fas (CD95/Apo-1) ligand with adriamycin significantly enhances the efficacy of antitumor response. Cell Mol Immunol. 6 (3): 167-74.

[58] Rousalova I, Krepela E (2010) Granzyme B-induced apoptosis in cancer cells and its regulation (review). Int J Oncol. 37 (6): 1361-78.

[59] Laskay T, Kiessling R (1986) Interferon and butyrate treatment leads to a decreased sensitivity of NK target cells to lysis by homologous but not by heterologous effector cells. Nat Immun Cell Growth Regul. 5 (4): 211-20.

[60] Benoist H, Comoe L, Joly P, Carpentier Y, Desplaces A, Dufer J (1989) Comparative effects of fagaronine, adriamycin and aclacinomycin on K562 cell sensitivity to natural-killer-mediated lysis. Lack of agreement between alteration of transferrin receptor and CD15 antigen expressions and induction of resistance to natural killer. Cancer Immunol Immunother.30(5):289-94.

[61] Lazarus AH, Baines MG (1985) Studies on the mechanism of specificity of human natural killer cells for tumor cells: correlation between target cell transferrin receptor expression and competitive activity. Cell Immunol. 96 (2): 255-66.

[62] Joshi SS, Sinangil F, Sharp JG, Mathews NB, Volsky DJ, Brunson KW (1988) Effects of differentiation inducing chemicals on in vivo malignancy and NK susceptibility of metastatic lymphoma cells. Cancer Detect Prev. 11 (3-6): 405-17.

[63] Zanyk MJ, Banerjee D, McFarlane DL (1988) Flow cytometric analysis of the phenotypic changes in tumour cell lines following TPA induction. Cytometry. 9 (4): 374-9.

[64] Clark EA, Sturge JC, Falk LA Jr (1981) Induction of target antigens and conversion to susceptible phenotype of NK-cell-resistant lymphoid cell line. Int J Cancer. 28 (5): 647-54.

Chapter 7

Some Findings on Apoptosis in Hepatocytes

Mayumi Tsuji and Katsuji Oguchi

Additional information is available at the end of the chapter

1. Introduction

For development of novel strategy for the treatment of hepatic diseases, in which changes in apoptosis are targeted, it is most important to understand the molecular mechanisms by which apoptotic process is regulated. Rapid and efficient removal of unwanted cells such as aged, damaged, genetically mutated, or virally-infected ones is indispensable in the maintenance of the health of the liver. Means for removal of such cells are provided by nature as apoptosis, a well-controlled and programmed cell death process, which is especially important in the liver where cells are exposed to various toxins and viruses [1]. In a healthy adult body, the number of cells removed by apoptotic process in a certain period of time is comparable to the number of cells that proliferate by mitosis. Proper homeostasis of organs is thus maintained. However, under certain pathophysiological conditions, the balance between the growth and the loss of the cells is often upset, due to the onset of some liver diseases resulting in the loss of tissue homeostasis. Insufficient apoptosis will promote the growth of hepatocarcinoma or biliary carcinoma by failing the removal of cells growing uncontrollably in genetically mutated cells or in chronic or persistent inflammations [2-5]. Persistent stimulation causing apoptosis may be a factor to promote high rate regeneration of cancer cells in the tissue, elevating the risk of errors in mitosis. In contrast, excessive and/or persistent apoptosis may lead to acute damages such as fulminant hepatitis or reperfusion injury [6, 7], or chronic persistent damages such as alcoholic liver diseases, cholestatic liver disease, or viral hepatitis [8-11]. Inhibition of apoptosis in the liver injury or selective killing of malignant cells in tumors will provide strategies for treatment of hepatic diseases. In fact, new drug development is ongoing targeting apoptosis through the understanding of molecular process and pathophysiological role of apoptosis, and such substances are now tested in clinical trial or used as new options for certain human diseases. In this review, we focus the subject on the role of apoptosis in cholestatic liver diseases or alcoholic liver injury in which we carried out some investigations [12-15].

2. Cholestasis and hydrophilic or hydrophobic bile acids

It has been demonstrated that hydrophobic bile acids damage cellular functions by affecting intracellular organelle and signaling system at the concentrations 100-500 μM, which are lower than those at which they show cytotoxic or detergent actions. Combettes and his co-workers reported that lithocholic acid (LCA) and taurolithocholic acid (TLCA) induce release of calcium ion (Ca^{2+}) from the endoplasmic reticulum (ER) and the increased level of Ca^{2+} within the cell mediates cytotoxicity due to these hydrophobic bile acids [16]. Combettes and his colleagues speculated that this increase in intracellular Ca^{2+} levels occurs because LCA activates the inositol (1,4,5)-triphosphate (IP$_3$) receptor, independent from IP$_3$ itself, resulting in the release of Ca^{2+} from the intracellular organelle, ER [17]. On the other hand, there is a report that the increase in intracellular Ca^{2+} level induced by hydrophobic bile acid depends on extracellular Ca^{2+} level [18]. Spivey and his colleagues reported that at 250 μM, glycochenodeoxycholic acid (GCDCA) induces the impairment of mitochondrial function and cellular ATP depletion, followed by a sustained rise in cytosolic Ca^{2+} resulting from an influx of extracellular Ca^{2+} leading to the death of hepatocytes, and that this cytotoxicity decreases in the order of GCDCA>CDCA>tauro-CDCA [19].

GCDCA is also reported to enhance the mitochondrial membrane permeability and induce cytotoxicity, while ursodeoxycholic acid (UDCA) shows suppression of GCDCA-induced enhancement of mitochondrial membrane permeability and cytotoxicity [20, 21]. Bile acid-induced enhancement of mitochondrial membrane permeability was also shown in the in vivo study of cholestasis [22]. Considering the data that increase in reactive oxygen species (ROS) generation in hepatocyte and generation of H$_2$O$_2$ stimulated by tauro-CDCA (TCDCA) in mitochondria preceded TCDCA-induced hepatocyte necrosis, it was speculated that generation of ROS by hydrophobic bile acid constitutes one of the causative factors of hepatic injury in cholestasis [23].

In cholestatic liver disease, loss of hepatocytes or appearance of apoptotic body in hepatocytes was morphologically observed, and involvement of apoptosis was suggested in the hepatocyte injury in cholestasis [24]. Patel and his colleagues reported for the first time that low concentration of glycodeoxycholic acid induced apoptosis in hepatocytes, and pointed out that bile acids may induce necrosis at higher concentrations and apoptosis at lower concentrations in hepatocytes [25, 26]. Furthermore, Sokol and his colleagues reported that hydrophobic bile acid induced lipid peroxidation and mitochondrial dysfunction via enhancement of mitochondrial membrane permeability [27], and we ourselves also reported mitochondria-mediated time- and concentration-dependent apoptotic cell death and endoplasmic reticulum (ER) stress-mediated apoptosis of hepatocytes induced by GCDCA [28, 29].

Ursodeoxycholic acid (UDCA), a hydrophilic bile acid, has been widely used as a therapeutic agent for primary biliary cirrhosis (PBC) or cholestasis [30]. Suggested mechanism of action of UDCA includes promotion of bile secretion, detoxification metabolism in the liver, and antioxidant stress response, but its molecular mechanism is still not clarified in detail. In vivo studies showed that UDCA is protecting hepatocytes from hydrophobic bile acid-induced apoptosis [31]. Furthermore, tauro-conjugated form of UDCA, tauro-UDCA (TUDCA), was shown to protect hepatocytes in ischemia-reperfusion injury in rats [32], and ethanol-fed rats [33]. When toxic bile acid is given to rats, apoptosis is induced in the liver, but when UDCA is given in combination, apoptosis is suppressed by inhibition of translocation of pro-apoptotic protein Bax from cytosol to mitochondria and ROS generation. This suppression is observed in cells other than hepatocytes, and UDCA was shown to act on classic mitochondrial-mediated pathway in different types of cells [34]. Taurourusodeoxycholic acid (TUDCA), a substance in which UDCA is conjugated with taurine, was shown to play an important role in various disorders including some liver diseases, type-II diabetes and metabolic syndrome [35, 36] possibly by its actions shown in isolated mitochondria to stabilize mitochondrial membrane directly through affecting channel formation by Bax [37], and bringing changes in ER stress-mediated pathways by decreasing caspase-12 activity and decrease in Ca^{2+} releases.

However, in a clinical trial for reevaluation of effectiveness in PBC patients, the effectiveness of UDCA was not acknowledged [38]. In rat isolated primary cultured hepatocytes, UDCA, given in combination with hydrophobic bile acids, was shown to be cytotoxic [39] and activates the pro-apoptotic pathway in some condition [40].

3. Apoptosis induced by hydrophobic bile acids in rat hepatocytes

Bile acids are synthesized from cholesterol in the liver, and act as surfactants that help digestion and absorption of lipids, and lipid-soluble vitamins. Major bile acids found in human bile are cholic and chenodeoxycholic acids, and they are secreted into bile as conjugates with taurine or glycine, via amide-bond. Most of bile acids secreted into the duodenum are reabsorbed by active transport in the terminal ileum, and returned to the liver. Bile acids are not so potently toxic as to injure hepatocytes in healthy subjects, but if the bile acid levels in the liver are too high or the ratio of hydrophobic bile acids to hydrophilic bile acids increases, as in the cases when there is some abnormality in bile acid synthesis, they induce apoptosis or necrosis [41]. The potency of their hepatotoxicity is, in the decreasing order, LCA > deoxycholic acid (DCA) > CDCA > CA > UDCA > dehydrocholic acid. The total bile acid in the liver tissue in normal subjects is not more than 10 μM when determined as serum bile acid level, but in patients with cholestasis, CDCA level in the liver tissue increases to about 20 times higher than in the normal case, and the serum bile acid level elevated to 10 to 30 times (100-300 μM) higher than normal, of which hydrophobic bile acid accounted for about 50-60 % [42, 19]. Cholestatic liver diseases are associated with bile duct obstruction by the formation of biliary stones, genetic defects, hepatotoxicity, hepatobiliary tumors [43]. Acute and chronic cholestasis induces hepatocelluar injury, biliary dilatation, hepatic fibrosis, cirrhosis, and ultimately hepatic failure [44]. Decrease in bile flow or total obstruction of bile duct upon cholestasis is induced by the stasis of metabolized products such as cholesterol and bile acids in the liver which are normally eliminated into bile. Especially some hydrophobic bile acids induce cytotoxicity in cultured hepatocytes [45]. In a major hypothesis on the mechanisms of hepatocyte injury, bile acids accumulated in the hepatocytes are regarded to be the major cause of cell death [23]. While bile acids such as taurolithocholic acid (TLCA), deoxycholic acid (DCA), glycodeoxycholic acid (GDCA), glycochenodeoxycholic acid (GCDCA) are known to induce necrosis of hepatocytes via increase in oxidative stress mediated by hydroperoxide generation by mitochondria [23], attention is recently focused on apoptotic cell death associated with these bile acids [45, 46].

The mechanism and pathways of bile acid-induced apoptosis in hepatocytes are reported to include death receptor-mediated pathway [9] and increase in ROS generation [31]. We ourselves reported that, in primary cultured hepatocytes isolated from healthy rats, apoptosis is induced by a hydrophobic bile acid in a concentration- and time-dependent manner [28]. When hepatocyte was cultured with 200 μM of GCDCA for 6 hours, ssDNA and caspase-3 activity which are measures of apoptosis increased about 10 times higher than in the untreated cultured hepatocytes (Fig. 1). Also, in hepatocytes treated with 200 μM GCDCA for 4hours, the level of mRNA of Fas and activities of caspase-8 and caspase-3 were significantly increased compared with those in untreated hepatocytes. Further, mitochondrial membrane potential difference of 200 μM GCDCA-treated hepatocytes was decreased by 60% compared with the same in untreated hepatocytes. GDCDA, a hydrophobic bile acid, was demonstrated to induce an increase in Fas death receptors located in cell membrane to activate apoptotic pathway.

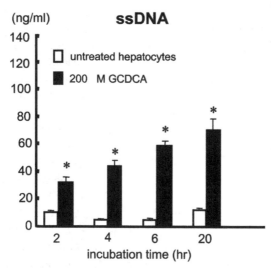

Figure 1. The effect of GCDCA on detection of apoptotic cells. Isolated hepatocytes (1x10[6] cells/well) were treated with GCDCA at 37 °C. Formamide-denaturable DNA was detected in apoptotic hepatocytes. The ssDNA was stained with the primary antibody (anti-ssDNA-mAb) and peroxidase-labelled secondary antibody for color development. Each value represents the mean+SEM of 6-12 samples. Statistically significant changes are indicated as *p<0.05 compared with the untreated hepatocytes.

On the other hand, oriental traditional pharmacognosy has utilized bear bile as a medicine for improving gastrointestinal symptoms. The effective ingredient of the bear bile is UDCA, which has been used widely as a choleretic or calculolytic drug for cholesterol gallstone. UDCA is now used for a wide range of liver disease treatment, and the evidence for its effectiveness has been reported one after another. The indications of UDCA include primary biliary sclerosis (PBC), calculolysis, prevention of gallstone formation and UDCA is demonstrated to be effective in most of these. However, there has been a report that in a clinical trial for reevaluation of effectiveness in PBC patients, the effectiveness of UDCA was not acknowledged [47]. In rat isolated primary cultured hepatocytes, when given in combination with hydrophobic bile acids, UDCA was reported to be cytotoxic [48] and activates the proapoptotic pathway [49]. We ourselves examined the action of UDCA on GCDCA-induced apoptosis in rat primary cultured hepatocytes [28]. When hepatocytes were incubated in the co-presence of UDCA and GCDCA, UDCA significantly inhibited GCDCA-induced apoptosis in a short incubation (4 hours), but a prolonged incubation (20-hour incubation) potentiated the apoptosis (Fig. 2). In the study of decrease in mitochondrial membrane potential difference, a further decrease in the potential was observed in co-incubation of hepatocytes with UDCA and GCDCA even in a short incubation. It was suggested that when cholestatic condition is severe and hepatotoxicity of bile acids is more potent, UDCA might potentiate the toxicity and we pointed out the need for attention in the clinical use [28].

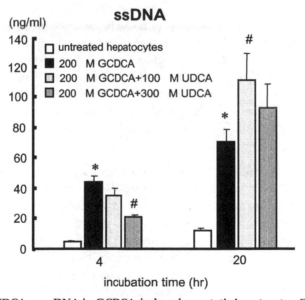

Figure 2. Effect of UDCA on ssDNA in GCDCA-induced apoptotic hepatocytes. Each value represents the mean+SEM of 5-15 samples. Other culture conditions were the same as in Fig. 1. *p < 0.05: significant difference from untreated cells, # p < 0.05: significant difference from GCDCA-treated cells.

4. Endoplasmic reticulum involvement in hydrophobic bile acid-induced apoptosis in rat hepatocytes

In addition to the major two intracellular apoptotic pathways, that is, death receptor-mediated and mitochondria-mediated pathways, attention has recently been focused on the ER stress-mediated pathway. The response against the accumulation of unfolded proteins in ER is called unfolded protein response (UPR) and this is featured by activation of three different signaling pathways, inositol-requiring (IRE)-1⊚, protein kinase RNA-activated (PKR)-like ER kinase (PERK), activating transcription factor (ATF)-6⊚. Changes in ER functions are induced by various stimulations, extrinsic chemicals administered pharmacologically, or increase in physiologically secreted proteins, and these upsets are called ER stress, which can be detected by the UPR transducer activation. ER stress is observed in many liver diseases, and UPR activity is correlated with hepatic resistance against insulin in obesity and fatty liver. Chronic viral B and C hepatitis, alcohol-induced liver injury, ischemic re-perfusion damages, and cholestatic liver injury are also correlated with UPR activity. Prolonged or potent ER stress induces apoptosis.

Just like an anti-apoptotic protein Bcl-2, pro-apoptotic proteins Bax and Bak are localized in membranes of ER, and regulate homeostasis of Ca^{2+} within the cell. Release of Ca^{2+} from ER activates calpain, which in turn activates caspase-12 (in human cells, caspase-4), and initiates apoptosis. Ca^{2+} intake by mitochondria leads to enhancement of mitochondrial membrane permeability, and cytochrome c is released from mitochondria. Membranes of mitochondria and ER are connected through protein junctions [50]. These junctions seem to facilitate transports of Ca^{2+} and phospholipids, and are possibly associated with apoptosis. However, the exact signaling pathway that mediates ER stress-induced apoptosis in hepatocytes is not clarified yet. UPR is activated for restoration of ER homeostasis. The upset of ER homeostasis induced by damages or activations of UPR sensors is observed in some liver diseases, and ER stress is observed in various liver diseases. Drugs that are hepatotoxic activate several intracellular stress responses (for example, lysosome disorder, increased permeability of mitochondrial membrane, oxidative stress, inflammation and so on) in hepatocytes.

Correlations have been observed between these responses and ER dysfunction. Steatosis occurs in the liver after an acute ER stress mediated by a transcription factor for lipid regulation, sterol regulatory element-binding protein (SREBP)-1c and SREBP-2c. Activation of nuclear factor kappa B (NF-κB) occurs in the downstream of ER upset in alpha-1 antitrypsin (AAT) deficiency disease. Chronic viral hepatitis (hepatitis C virus HCV, hepatitis B virus HBV) accompanies ER dysfunction. Prolonged ER stress leads to apoptosis by activation of C/enhancer binding protein (EBP) homologous protein (CHOP), change in Ca2+ homeostasis and premature resuming of mRNA translation.

In cholestasis, toxic hydrophobic bile acid/salts are retained in the liver due to impaired biliary excretion. Sodium deoxycholate (DC) induced expressions of UPR genes BIP and

CHOP in vitro [51]. When hepatocytes of CHOP-null mice were incubated with toxic GCDCA, cell death was decreased [52]. In a hereditary model of intrahepatic cholestasis, accumulation of bile acids in the liver was correlated with ER stress [53]. We ourselves studied the effects of GCDCA treatment in hepatocytes, and found that when the culture medium contained Ca^{2+}, persistent increase in intracellular Ca^{2+} was observed while only transient increase in intracellular Ca2+ was observed when cells were cultured in Ca^{2+} free medium, showing that GCDCA promotes Ca^{2+} influx from extracellular matrix and release of Ca^{2+} from ER. Activations of calpain and caspase-12 due to the increase in intracellular Ca^{2+} were also observed, and we reported that GCDCA induces apoptosis mediated by ER stress [29]. Furthermore, we reported the correlation between ER stress associated with GCDCA and caspase-8 activation. GCDCA activates caspase-8 via Fas death receptor, but when hepatocytes were pretreated with a caspase-8 inhibitor, z-IETD-FMK, expressions of BIP, an ER chaperone molecule, and CHOP, an ER stress response transcription factor, were suppressed. From these, we speculated that caspase-8 activated by GCDCA regulates ER stress [54].

5. Detailed mechanism of hydrophobic bile acid-induced apoptosis in HepG2 cells

A hydrophobic bile acid GCDCA activates caspase-8 in hepatocytes via Fas death receptor in the cell membrane. Activated caspase-8 enhances mitochondrial membrane permeability to facilitate the release of cytochrome c, a pro-apoptotic protein, activates caspase-9 and caspase-3 to induce apoptosis. Caspase-8 activated by GCDCA cleaves BAP31 protein in ER membrane and may possibly be associated with Ca^{2+} release from mitochondria from ER. BAP31 is cleaved to be BAP20 which is an activated form. As shown in Fig. 3, intact (uncleaved) BAP31 protein is observed in ER in untreated hepatic cells, but in GCDCA-treated hepatic cells, the content of intact BAP31 protein is decreased. The decrease in intact BAP31 observed in GCDCA-treated hepatocytes was suppressed by pretreatment with z-IETD-FMK. When recombinant active caspase-8 was added to hepatocytes, decrease in intact BAP31 was observed. GCDCA-induced increased intracellular Ca^{2+} interacted with mitochondria and caused its dysfunction, followed by increase in mitochondrial membrane permeability and release of pro-apoptotic factors from mitochondria. Also, calpain was activated by the increase in Ca^{2+} release followed by activation of caspases. Furthermore, ER chaperone BIP is usually bound to an ER stress sensor located within the lumen of ER, but GCDCA treatment causes the increase in unfolded protein, and promotes the release of BIP from the stress sensor and increases the expression of CHOP, an ER stress-related transcription factor that works in the downstream of PERK, an ER stress sensor (Fig. 4). CHOP induces the transcription of Bim, a pro-apoptotic BH3 only protein [55]. Membranes of mitochondria and ER are connected by complexes composed of tethering proteins [50].

These complexes facilitate the transport of calcium and phospholipids and possibly are associated with apoptosis. GCDCA is known to induce apoptosis via mitochondria- and ER stress-mediated pathways via death receptor, but mitochondria and ER themselves seem to be interrelated to each other.

6. Some new findings with effects of ethanol-induced oxidative stress on apoptosis in SK-Hep1 cells

Overconsumption of alcohol is associated with deaths of about 2 million people per year throughout the world. In its early stage, fatty liver is induced, which may progress to liver disorders accompanied with hepatocyte death, inflammation, and fibrosis, further to cirrhosis and hepatocarcinoma [56, 57]. Rate of incidences of acute alcohol intoxication and heavy alcohol drinking indifferent to one's health is globally increasing and acute liver injury caused by alcohol is attracting attention [58]. Suggested molecular mechanisms for liver disorders induced by alcohol include increase of ROS and changes in various signaling pathways, but the molecular mechanism for hepatocyte death was not clarified yet. In experimental studies using model animals of ethanol-induced liver injury, it was emphasized that apoptosis is playing an important role in the pathogenesis of alcoholic

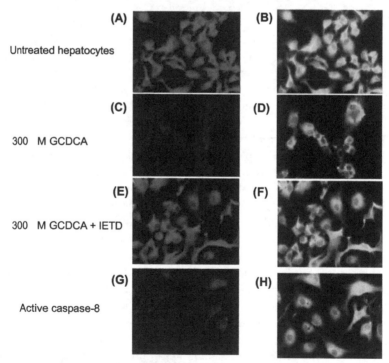

Figure 3. The effect of GCDCA on BAP31 in HepG2 cells.HepG2 cells treated with GCDCA (300 μM for 24 hours) were double-labeled for BAP31 with rat anti-BAP31 antibody (A, C, E, G) and for ER with Alexa Fluor 488 (B, D, F, H). Original magnification × 200. (untreated hepatocytes (A, B), hepatocytes treated with 300 μM GCDCA (C, D), 300 μM GCDCA + Z-IETD-FMK (E, F), recombinant active caspase-8 (G, H)

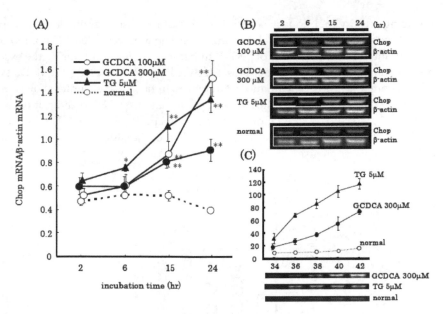

Figure 4. The effect of GCDCA on the CHOP mRNA expression in isolated heaptocytes. (A) CHOP mRNA expression was determined by semiquantitative PCR. Line graphs representation of the observed fold induction normalized to β-actin for each time point. (B) RT-PCR analysis of CHOP and β-actin expression at each incubation time after treatment with GCDCA. (C) Reaction cycles-PCR product yield curves of each reaction mixture were plotted. The intensity of fluorescence was fitted to the data in the linear portion of curves. The resulting CHOP mRNA / β-actin mRNA ratio is represented as the mean+SEM of 8-15 samples. *p<0.05 and **p<0.01: significant difference from untreated hepatocytes (control).

hepatitis or alcoholic liver cirrhosis [3, 59]. Clinical studies suggested similar findings. Various factors including cytotoxicity by alcohol and its metabolites, changes in metabolizing enzymes, invasion of inflammatory cells, reactive oxygen species, cytokines, hepatic microcirculation, nutritional factors, etc. are involved in the onset of apoptosis. Among others, correlation between ROS or oxidative stress and hepatocyte apoptosis is attracting attention. Orally fed ethanol is absorbed in the upper digestive organs, mainly small intestine, and 90% of it is metabolized in the liver. Upon metabolization, ROS is generated in the liver by alcohol dehydrogenase, microsome-ethanol oxydizing system (MEOS) via cytochrome P450, and NADPH oxidase (NOX) in the cell membrane, which is a center for ROS generation (Fig. 5). In general, oxidative stress inflicts cytotoxicity when excessive ROS is generated in the cells. Antioxidant was shown to reduce hepatocyte apoptosis in acute ethanol-addicted rats [60]. On the other hand, several death receptors and their ligands (especially Fas/FasL) are over-expressed in the liver cells of alcoholic hepatitis patients compared with those in healthy subjects. The levels of Fas and FasL are increased in serious alcoholic hepatitis patients, but there are still many uncertainties about the details of their mediators and biological importance [61]. The increase in FasL might be mediated by ROS or increase in NF-κB which increases the transcription of Fas and FasL genes based on

TNF-α inducing activity [62]. In fact, serum TNF-α level increases also in alcoholic hepatitis patients, and plays an important role in inflicting hepatotoxicity [63]. Chronic ethanol administration increases expressions of TNF-R in hepatocytes [64], and during the exposure to ethanol, hepatocytes underwent apoptosis induced by TNF-α. TNF-α/ TNF-R1 system seems to be required in the cell death mediated by Fas. In fact, recent studies showed that in TNF-R1/TNF-R2 double knockout mice, apoptosis mediated by TNF-α did not occur, and the mice showed resistance against induced fulminant hepatic failure [65]. The activation of TNF-α/TNF-R1 complex may be co-working with the signaling conveyed by Fas for inducing hepatocyte apoptosis.

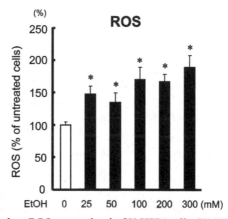

Figure 5. The effect of ethanol on ROS generation in SK-HEP1 cells. SK-HEP1 cells (1×10⁶ cells/ well) were treated with ethanol (25-300 mM) for 5 hours at 37°C. The generation of ROS in SK-HEP1 treated with ethanol was analyzed spectrofluorometrically. Each value represents the mean ± S.E.M. of 6-12 samples. *:p<0.05: significant difference from untreated SK-HEP-1 cells.

Histologically, nick-end label positive apoptotic cells were abundantly found around Mallory body in biopsied tissue from patients with alcoholic liver injury [66]. In in vivo animal studies, hepatocyte apoptosis is increased in animals fed with alcohol-containing feed, and long-term alcohol administration is suggested to induce hepatocyte death [67]. Apoptotic cells are observed even in normal cells in some part around the central vein, but in rats fed with alcohol for a long term, many apoptotic cells were observed in various parts around the central vein.

We ourselves reported the ethanol-induced apoptosis in cultured hepatocytes, and showed that apoptosis is induced in vitro in the presence of alcohol. We also found that at ethanol concentrations lower than those that induce apoptosis, significant increase in ROS generation was observed. It was suggested that in the process of apoptosis induced by ethanol, ROS generation by NOX was important as inducer of apoptosis [68]. As shown in Fig. 6, in the presence of 200 μM ethanol, the mRNA expression of p22phox, which is a protein constitutively bound to NOX to enhance the action of NOX4, was found significantly increased and the increase of the expression was suppressed by the pretreatment with N-acetyl-L-cysteine (NAC), a precursor of glutathione and an antioxidant.

The results above prove that the cell sublines we have obtained have become cross-resistant to chemical reagents of the MDR group, and that such resistance may be due, i.a., to modifications in caspase expression.

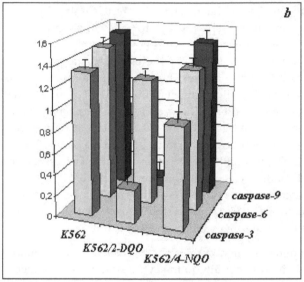

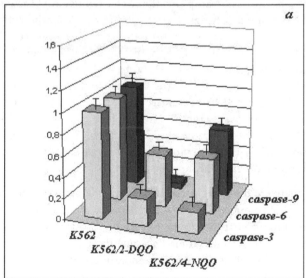

Results are presented as the ratio of caspase to GAPDH expression relative to a standard curve for each assay. Drug concentrations: 0.1 µM for *K562/2-DQO* cells, 0.001 µM for *K562* cells, 0.001 µM for *K562/4-NQO* cells. Incubation time – 48 h. Viability of cultured cells – 93–95%. Z axis – relative expression, conventional unit.

Figure 11. Caspase-3, -6 and -9 mRNA levels were determined by quantitative real-time RT-PCR in *K562, K562/2-DQO* and *K562/4-NQO* cell lines (*a*) treated with colchicine (*b*) respectively.

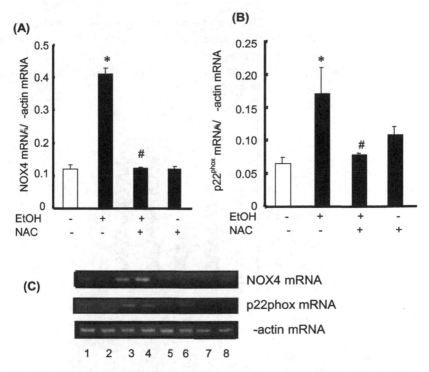

Figure 6. The effect of ethanol on NOX-4 and p22phox mRNA in SK-HEP-1 cells. (A): The expression of NOX-4 mRNA in SK-HEP1 treated with ethanol (200 mM) and NAC (10 mM). (B): The expression of p22phox mRNA in SK-HEP1 treated with ethanol and NAC. The resulting NOX-4 or p22phox mRNA/β-actin mRNA ratio is represented as the mean ± S.E.M. of 8-12 samples. (C): RT-PCR analysis of NOX-4 or p22phox and β-actin expression after treatment with ethanol or ethanol+NAC. Lane1, 2: untreated SK-HEP1 cells, Lane3, 4: 200 mM ethanol, Lane5, 6: ethanol+NAC, Lane7, 8: 10 mM NAC. *p < 0.05: significant difference from untreated cells, # p < 0.05: significant difference from 200 mM ethanol-treated cells.

In a recent study on the actions of ethanol on hepatocytes, the importance of mitogen-activated protein kinases (MAPKs) (mainly, ERK1/2, p38 and JNK1/2) and histone modification (acetylation, methylation, or phosphorylation) is emphasized. MAPK pathway is correlated with many signaling pathways including tyrosine, serine/threonine kinase, G protein and calcium signals [69]. MAPKs are a family of protein kinases of which main members are ERK1 and ERK2, p38MAPK, and c-Jun-N-terminal kinase/stress-activated protein kinase (JNK/SAPK). MAPKs regulate various biological processes including cell growth, proliferation, movement, inflammation, fatty degeneration, necrosis and apoptosis [70]. In a primary culture of rat hepatocytes, ethanol showed modest activation of ERK1/2 and notable activation of JNK [71]. In rat cultured hepatocytes, when ERK1/2 phosphorylation was inhibited by U-0126 (a MEK1/2 inhibitor), phosphorylation of JNK by ethanol was increased [72]. In previous studies on neurons, ethanol was found to activate MAPK cascade and increase ROS generation via p38MAPK pathway [73, 74].

We found that exposure of cultured hepatocytes to ethanol increase generation of ROS and MAPK (p38MAPK and JNK) phosphorylation activity. However, ROS generation was not significantly affected when hepatocytes were pretreated with MAPK inhibitors (SB202190 for p38MAPK, and SP600125 for JNK) [75]. These results suggest that ROS may be generated by the upstream effector of p38 MAPK (Fig. 7).

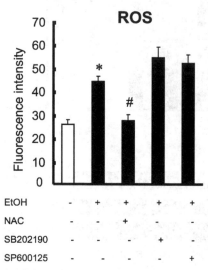

Figure 7. The effect of MAPK inhibitors on ROS generation in ethanol-induced apoptotic SK-Hep1 cells. Generation of ROS in SK-Hep1 cells treated with 200 mM ethanol or MAPK inhibitors (SB202190: p38 inhibitor, SP600125: JNK inhibitor) + 200 mM ethanol was analyzed spectrofluorometrically. *p < 0.05: significant difference from untreated cells, # p < 0.05: significant difference from 200 mM ethanol-treated cells.

Furthermore, as described above, overconsumption of alcohol induces various pathological stress responses and a part of them is endoplasmic reticulum (ER) stress response. ER stress is associated with alcoholic injury in such organs as the liver, pancreas, heart and brain. The possible mechanism for triggering alcoholic ER stress response is directly or indirectly correlated with alcohol metabolism, which, in turn, is correlated with toxic acetaldehyde and homocysteine, oxidative stress, upset of calcium or iron homeostasis, decrease in the ratio of S-adenosylmethionine/S-adenosyl- homocysteine and abnormal epigenetic modifications. Inhibition of triggering process of ER stress could hopefully be beneficial in the treatment of alcoholic diseases.

In the study of genetic expression in ethanol-fed mice, remarkable increase in caspase-12 mRNA and BIP, a ER chaperone, and CHOP mRNA [76]. When the protein levels were examined, BIP, CHOP and caspase-12 were increased. When CHOP null mice was fed with ethanol, apoptosis of hepatocyte was found to be dependent on CHOP [77], showing that ethanol-induced hepatic injury was associated with hepatocyte apoptosis mediated by CHOP. Furthermore, when micropigs were fed with ethanol, fatty liver and apoptosis are found to be correlated with the increase in mRNAs of CYP2E1, GRP78, SREBP-1c and also increase in protein levels of CYP2E1, GRP78, nuclear SREBP-1c and caspase-12 activity [78].

We found that when human hepatocarcinoma cell line, SK-Hep1, were exposed to ethanol, expressions of mRNAs of BIP, CHOP, and sXBP-1 were increased. The ethanol-induced increase in expressions were suppressed by NAC, an antioxidant, and we speculated the increase to be associated with oxidative stress induced by ethanol. We also observed a transient increase in intracellular Ca^{2+} and calpain activity, but they were not suppressed by NAC, and we believe these were independent from ethanol-induced oxidative stress. It was suggested, therefore, there are two pathways of ethanol-induced apoptosis mediated by ER stress, that is, ethanol-induced oxidative stress-dependent and independent pathways [79].

Possible approaches for the treatment of ER stress induced by ethanol include decreasing homocysteine and increase in SAM by betain or folic acid [80-82], improvement of protein folding using chemical chaperone, PBA(sodium phenylbutyrate) and TUDCA [81, 83, 84], inhibition of dephosphorylation of eukaryotic initiation factor-2α (eIF2α) using its inhibitor salubrinal [85], and reducing oxidative stress by decreasing ROS generation from oxidized protein using antioxidants. However, the results of clinical study cannot be obtained. Mechanism of ethanol-induced ER stress is too complex, and the approaches for treatment in human may not be so simple or universal. It may be necessary to employ properly combined therapy of every known beneficial medication.

7. Proposed conceptual diagram of apoptosis process in hepatocytes

Apoptosis is an indispensable process in line with cell proliferation for maintenance of tissue homeostasis and health by removal of injured and/or aged cells. This is especially important in the liver where cells are exposed to toxins and viruses [1]. Any loss of balance between cell death and proliferation due to excessive or insufficient apoptosis always leads to pathologic conditions due to unstable state. In the liver, massive hepatocyte apoptosis is observed in acute hepatic failure, and persistent hepatocyte apoptosis is associated with fibrilization, chronic dysfunction and cancerous transformation of the liver [86]. Apoptosis is induced by various intracellular and extracellular stimuli. In all types of hepatocytes, death receptors (especially Fas) are universally expressed [87], and hepatocyte apoptosis is usually transmitted by external pathways. Especially, activations of Fas and TNF-R1 are correlated with hepatocyte apoptosis in various liver diseases including viral hepatitis, fulminant hepatitis, cholestatic liver disease, alcoholic hepatitis, non-alcoholic fatty-liver disease (NAFLD) and non-alcoholic steatohepatitis (NASH), Wilson's disease, and ischemia-reperfusion injury [88]. In the study of cholestasis, because no liver injury was observed after ligation of the common bile duct (a model of extrahepatic cholestasis) in Fas-knockout mice, high levels of toxic intracellular bile salts were speculated to increase Fas in the cell membrane resulting in the activation of the receptor [89, 10]. Furthermore, apoptosis is induced via ER stress. Caspase-8 activated via death receptor may regulate ER stress mediated by BAP31 on ER [54]. In alcoholic liver injury, ROS generation is enhanced via NOX, and the excessive ROS interact with Fas death receptor [8] to induce mitochondria-mediated and ER stress-mediated apoptosis. Taken together, we propose a diagram showing the mechanism of hepatocyte apoptosis as seen in Fig. 8.

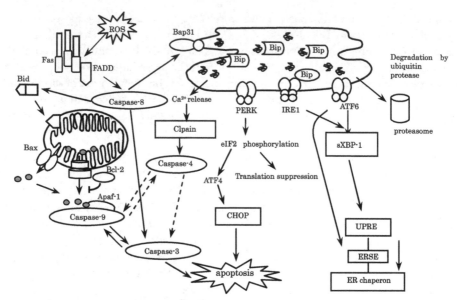

Figure 8. Schematic diagram of apoptotic pathway.

8. Conclusion

Apoptosis is involved in various diseases, and affects a wide variety of organs including the liver, kidney, central nervous system, and heart. This field of research has been increasingly active in both basic medical sciences and clinical levels, and in the future, when novel findings are obtained on how changes in the process of apoptosis lead to aggravation or improvement of diseases, an innovative strategy for more effective therapy will be designed. Progress of elucidation of apoptotic process is greatly sought after.

Author details

Mayumi Tsuji* and Katsuji Oguchi
Department of Pharmacology, Showa University, School of Medicine, Tokyo, Japan

Acknowledgement

This work has been supported by Special subsidy for promotion of higher level education & research in private universities from Japanese Ministry of Education,Culture,Sports,Science & Technology and Showa University Joint Research grant. The authors thank the graduate students and fellows who contributed to the studies.

* Corresponding Author

9. References

[1] Malhi H, Guicciardi ME, Gores GJ (2010) Hepatocyte death: a clear and present danger. Physiol. Rev. 90: 1165–1194.

[2] Oda T, Tsuda H, Scarpa A, Sakamoto M, Hirohashi S (1992) p53 gene mutation spectrum in hepatocellular carcinoma. Cancer Res. 52: 6358–6364.

[3] Que FG, Phan VA, Phan VH, Celli A, Batts K, LaRusso NF, Gores GJ (1999) Cholangiocarcinomas express Fas ligand and disable the Fas receptor. Hepatology 30: 1398–1404.

[4] Nzeako UC, Guicciardi ME, Yoon JH, Bronk SF, Gores GJ (2002) COX-2 inhibits Fas-mediated apoptosis in cholangiocarcinoma cells. Hepatology 35: 552-559.

[5] Pikarsky E, Porat RM, Stein I, Abramovitch R, Amit S, Kasem S, Gutkovich-Pyest E, Urieli-Shoval S, Galun E, Ben-Neriah Y (2004) NF-kappaB functions as a tumour promoter in inflammation-associated cancer. Nature 431: 461-466.

[6] Ogasawara J, Watanabe-Fukunaga R, Adachi M, Matsuzawa A, Kasugai T, Kitamura Y, Itoh N, Suda T, Nagata S (1993) Lethal effect of the anti-Fas antibody in mice. Nature 364: 806-809.

[7] Kohli V, Selzner M, Madden JF, Bentley RC, Clavien PA (1999) Endothelial cell and hepatocyte deaths occur by apoptosis after ischemia-reperfusion injury in the rat liver. Transplantation 67: 1099–1105.

[8] Natori S, Rust C, Stadheim LM, Burgart LJ, Gores GJ (2001) Hepatocyte apoptosis is a pathologic feature of human alcoholic hepatitis. J. Hepatol.34: 248-253.

[9] Faubion WA, Guicciardi ME, Miyoshi H, Bronk SF, Roberts PJ, Svingen PA, Kaufmann SH, Gores GJ (1999) Toxic bile salts induce rodent hepatocyte apoptosis via direct activation of Fas. J. Clin. Invest. 103: 137-145.

[10] Miyoshi H, Rust C, Roberts PJ, Burgart LJ, Gores GJ (1999) Hepatocyte apoptosis after bile duct ligation in the mouse involves Fas. Gastroenterology 117: 669-677.

[11] Canbay A, Higuchi H, Bronk SF, Taniai M, Sebo TJ, Gores GJ (2002) Fas enhances fibrogenesis in the bile duct ligated mouse: a link between apoptosis and fibrosis. Gastroenterology 123: 1323-1330.

[12] Patel T, Gores GJ. (1995) Apoptosis and hepatobiliary disease. Hepatology 21: 1725-1741.

[13] Patel T, Roberts LR, Jones BA, Gores GJ (1998) Dysregulation of apoptosis as a mechanism of liver disease: an overview. Semin. Liver Dis. 18: 105-114.

[14] Rust C, Gores GJ. (2000) Apoptosis and liver disease. Am. J.Med.108: 567-574.

[15] Yoon JH, Gores GJ. (2002) Death receptor-mediated apoptosis and the liver. J. Hepatol. 37: 400-410.

[16] Combettes L, Dumont M, Berthon B, Erlinger S, Claret M (1988) Release of calcium from the endoplasmic reticulum by bile acids in rat liver cells. J.Biol.Chem. 263: 2299-2303.

[17] Combettes L, Berthon B, Doucet E, Erlinger S, Claret M (1989) Characteristics of bile acid-mediated Ca^{2+} release from permeabilized liver cells and liver microsomes. J.Biol.Chem. 264: 157-167.

[18] Zimniak P, Little JM, Radominska A, Oelberg DG, Anwer MS, Lester R (1991) Taurine-conjugated bile acids act as Ca^{2+} ionophores. Biochemistry 30: 8598-8604.

[19] Spivey JR, Bronk SF, Gores GJ (1993) Glycochenodeoxycholate-induced lethal hepatocellular injury in rat hepatocytes. Role of ATP depletion and cytosolic free calcium. J.Clin.Invest. 92: 17-24.

[20] Botla R, Spivey JR, Aguilar H, Bronk SF, Gores GJ (1995) Ursodeoxycholate (UDCA) inhibits the mitochondrial membrane permeability transition induced by glycochenodeoxycholate: a mechanism of UDCA cytoprotection. J.Pharmacol.Exp.Ther. 272: 930-938.

[21] Tsuji M, Utanohara S, Takinishi Y, Yokochi A, Tanaka M, Oyamada H, Okazaki M, Oguchi K (2003) Tauroursodeoxycholic acid prevents glycochenodeoxycholic acid-induced apoptosis. Showa Univ.J.Med.Sci. 15: 265-277.

[22] Lieser MJ, Park J, Natori S, Jones BA, Bronk SF, Gores GJ (1998) Cholestasis confers resistance to the rat liver mitochondrial permeability transition. Gastroenterology 115: 693-701.

[23] Sokol RJ, Winklhofer-Roob BM, Devereaux MW, McKim JM Jr (1995) Generation of hydroperoxides in isolated rat hepatocytes and hepatic mitochondria exposed to hydrophobic bile acids. Gastroenterology 109: 1249-56.

[24] Searle J, Harmon BV, Bishop CJ, Kerr JFR (1987) The significance of cell death by apoptosis in hepatobiliary disease. J. Gastroenterol. Hepatol. 2: 77-96.

[25] Patel T, Bronk SF, Gores GJ (1994) Increases of intracellular magnesium promote glycodeoxycholate-induced apoptosis in rat hepatocytes. J. Clin. Invest. 94: 2183-2192.

[26] Zeid IM, Bronk SF, Fesmier PJ, Gores GJ (1997) Cytoprotection by fructose and other ketohexoses during bile salt-induced apoptosis of hepatocytes. Hepatology 25: 81-86.

[27] Sokol R J, Straka MS, Dahl R, Devereaux MW, Yerushalmi B, Gumpricht E, Elkins N, Everson G (2001) Role of oxidant stress in the permeability transition induced in rat hepatic mitochondria by hydrophobic bile acids. Pediatr. Res. 49: 519-531.

[28] Utanohara S, Tsuji M, Momma S, Morio Y, Oguchi K (2005) The effect of ursodeoxycholic acid on glycochenodeoxycholic acid-induced apoptosis in rat hepatocytes. Toxicology 214: 77-86.

[29] Tsuchiya S, Tsuji M, Morio Y, Oguchi K (2006) Involvement of endoplasmic reticulum in glycochenodeoxycholic acid-induced apoptosis in rat hepatocytes. Toxicol. Letters 166: 140-149.

[30] Lindor KD, Lacerda MA, Jorgensen RA, DeSotel CK, Batta AK, Salen G, Dickson ER, Rossi SS, Hofmann AF (1998) Relationship between biliary and serum bile acids and response to ursodeoxycholic acid in patients with primary biliary cirrhosis. Am. J. Gastroenterol. 93: 1498-1504.

[31] Rodrigues CM, Fan G, Ma X, Kren BT, Steer CJ(1998) A novel role for ursodeoxycholic acid in inhibiting apoptosis by modulating mitochondrial membrane perturbation. J. Clin. Invest. 101: 2790-2799.

[32] Ishigami F, Naka S, Takeshita K, Kurumi Y, Hanasawa K, Tani T (2001) Bile salt tauroursodeoxycholic acid modulation of Bax translocation to mitochondria protects

the liver from warm ischemia-reperfusion injury in the rat. Transplantation 72: 1803-1807.

[33] Colell A, Coll O, García-Ruiz C, París R, Tiribelli C, Kaplowitz N, Fernández-Checa JC (2001) Tauroursodeoxycholic acid protects hepatocytes from ethanol-fed rats against tumor necrosis factor-induced cell death by replenishing mitochondrial glutathione.Hepatology 34: 964-971.

[34] Castro R, Solá S, Steer C, Rodrigues C (2007) Bile acids as modulators of apoptosis. In Hepatotoxicity: From Genomics to in- Vitro and in-Vivo Models. Sahu S, editor. John Wiley & Sons, West Sussex, UK. 39-419.

[35] Xie Q, Khaoustov VI, Chung CC, Sohn J, Krishnan B, Lewis DE, Yoffe B (2002) Effect of tauroursodeoxycholic acid on endoplasmic reticulum stress-induced caspase-12 activation. Hepatology 36: 592–601.

[36] Ozcan U, Yilmaz E, Ozcan L, Furuhashi M, Vaillancourt E, Smith RO, Gorgun CZ, Hotamisligil GS (2006) Chemical chaperones reduce ER stress and restore glucose homeostasis in a mouse model of type 2 diabetes. Science 313: 1137–1140.

[37] Rodrigues CM, Sola S, Sharpe JC, Moura JJ, Steer CJ (2003) Tauroursodeoxycholic acid prevents Bax-induced membrane perturbation and cytochrome C release in isolated mitochondria. Biochemistry 42: 3070-3080.

[38] Goulis J, Leandro G, Burroughs AK (1999) Randomised controlled trials of ursodeoxycholic-acid therapy for primary biliary cirrhosis: a meta-analysis. Lancet 354: 1053-1060.

[39] Rolo AP, Palmeira CM, Wallace KB (2002) Interactions of combined bile acids on hepatocyte viability: cytoprotection or synergism. Toxicol. Letters 126: 197-203.

[40] Qiao L., Yacoub A., Studer E., Gupta S., Pei XY, Grant S, Hylemon PB, Dent P (2002) Inhibition of the MAPK and PI3K pathways enhances UDCA-induced apoptosis in primary rodent hepatocytes. Hepatology 35: 779-789.

[41] Attili AF, Angelico M, Cantafora A, Alvero A, Copocaccia L (1986) Bile acid-induced liver toxicity; Relation to the hydrophobic-hydrophilic balance of bile acids. Med. Hypothesis 19: 57-68.

[42] Crosignani A, Podda M, Battezzati PM, Bertolini E, Zuin M, Watson D, Setchell KD (1991) Changes in bile acid composition in patients with primary biliary cirrhosis induced by ursodeoxycholic acid administration. Hepatology 14: 1000-1007.

[43] Hirschfield GM, Heathcote EJ, Gershwin ME (2010) Pathogenesis of cholestatic liver disease and therapeutic approaches. Gastroenterology 139: 1481-96.

[44] Sellinger M, Boyer JL (1990) Physiology of bile secretion and cholestasis. Prog. Liver Dis. 9: 237-59.

[45] Perez MJ, Briz O (2009) Bile-acid-induced cell injury and protection. World J. Gastroenterol. 15: 1677-89.

[46] Guicciardi ME, Gores GJ (2002) Bile acid-mediated hepatocyte apoptosis and cholestatic liver disease. Dig. Liver Dis. 34: 387–92.

[47] Goulis J, Leandro G, Burroughs AK (1999) Randomised controlled trials of ursodeoxycholic-acid therapy for primary biliary cirrhosis: a meta-analysis. Lancet 354: 1053-1060.

[48] Rolo AP, Palmeira CM, Wallace KB (2002) Interactions of combined bile acids on hepatocyte viability: cytoprotection or synergism. Toxicol. Letters 126: 197-203.

[49] Qiao L., Yacoub A., Studer E., Gupta S., Pei XY, Grant S, Hylemon PB, Dent P (2002) Inhibition of the MAPK and PI3K pathways enhances UDCA-induced apoptosis in primary rodent hepatocytes. Hepatology 35: 779-789.

[50] Kornmann B, Currie E, Collins SR, Schuldiner M, Nunnari J, Weissman JS, Waiter P (2009) An ER-mitochondria tethering complex revealed by a synthetic biology screen. Science 325: 477-481.

[51] Bernstein H, Payne CM, Bernstein C, Schneider J, Beard SE, Crowley CL (1999) Activation of the promoters of genes associated with DNA damage, oxidative stress, ER stress and protein malfolding by the bile salt, deoxycholate. Toxicol. Lett. 108: 37-46.

[52] Tamaki N, Hatano E, Taura K, Tada M, Kodama Y,Nitta T, Iwaisako K, Seo S, Nakajima A, Ikai I, Uemoto S (2008) CHOP deficiency attenuates cholestasis-induced liver fibrosis by reduction of hepatocyte injury. Am. J. Physiol. Gastrointest. Liver Physiol. 294: G498-G505.

[53] Bertolotti A, Zhang Y, Hendershot LM, Harding HP, Ron D (2000) Dynamic interaction of BiP and ER stress transducers in the unfolded-protein response. Nat. Cell Biol. 2: 326–332.

[54] Iizuka T, Tsuji M, Oyamada H, Morio Y, Oguchi K (2007) Interaction between caspase-8 activation and endoplasmic reticulum stress in glycochenodeoxycholic acid-induced apoptotic HepG2 cells. Toxicology 241: 146-156.

[55] Puthalakath H, O'Reilly LA, Gunn P, Lee L, Kelly PN, Huntington ND, Hughes PD, Michalak EM, McKimm-Breschkin J, Motoyama N, Gotoh T, Akira S, Bouillet P, Strasser A (2007) ER stress triggers apoptosis by activating BH3-only protein Bim. Cell 129: 1337-1349.

[56] Lucey MR, Mathurin P, Morgan TR (2009) Alcoholic hepatitis. N. Engl. J. Med. 360: 2758-2769.

[57] Purohit V, Gao B, Song BJ (2009) Molecular mechanisms of alcoholic fatty liver. Alcohol Clin. Exp. Res. 33: 191-205.

[58] Mathurin P,Deltenre P (2009) Effect of binge drinking on the liver: an alarming public health issue? Gut 58: 613-617.

[59] Benedetti A, Brunelli E, Risicato R, Cilluffo T, Jézéquel AM, Orlandi F R (1998) Subcellular changes and apoptosis induced by ethanol in rat liver. J. Hepatol. 6: 137-143.

[60] Kurose I, Higuchi H, Miura S, Saito H, Watanabe N, Hokari R, Hirokawa M, Takaishi M, Zeki S, Nakamura T, Ebinuma H, Kato S, Ishii H (1997) Oxidative stress-mediated apoptosis of hepatocytes exposed to acute ethanol intoxication. Hepatology 25: 368-378.

[61] Taieb J, Mathurin P, Poynard T, Gougerot-Pocidalo MA, Chollet-Martin S (1998) Raised plasma soluble Fas and Fasligand in alcoholic liver disease. Lancet 351: 1930-1931.

[62] Hug H, Strand S, Grambihler A, Galle J, Hack V, Stremmel W, Krammer PH, Galle PR (1997) Reactive oxygen intermediates are involved in the induction of CD95 ligand mRNA expression by cytostatic drugs in hepatoma cells. J. Biol. Chem. 272: 28191-28193.

[63] McClain C, Hill D, Schmidt J, Diehl AM (1993) Cytokines and alcoholic liver disease. Semin. Liver Dis. 13: 170-182.

[64] Deaciuc IV, D'Souza NB, Spitzer JJ (1995) Tumor necrosis factor-alpha cell-surface receptors of liver parenchymal and nonparenchymal cells during acute and chronic alcohol administration to rats. Alcohol Clin. Exp. Res. 19: 332-338.

[65] Costelli P, Aoki P, Zingaro B, Carbó N, Reffo P, Lopez-Soriano FJ, Bonelli G, Argilés JM, Baccino FM (2003) Mice lacking TNFa receptors 1 and 2 are resistant to death and fulminant liver injury induced by agonistic anti-Fas antibody. Cell Death Differ. 10: 997-1004.

[66] Kawahara H, Matsuda Y, Takase S (1994) Is apoptosis involved in alcoholic hepatitis? Alcohol 29: 113-118.

[67] Goldin RD, Hunt NC, Clark J, Wickramsinghe SN (1993) Apoptotic bodies in a murine model of alcoholic liver disease: Reversibility of ethanol induced changes. J. Pathol. 171: 73-76.

[68] Mochizuki Y, Tsuji M, Nakajima A, Inagaki M, Hayashi T, Seino T, Oguchi K (2012) Ethanol increases NADPH oxidase-derived oxidative stress and induces apoptosis in human liver adenocarcinoma cell (SK-HEP-1). Showa Univ. J. Med. Sci. 24: in press.

[69] Roux PP, Blenis J (2004) ERK and p38 MAPK-activated protein kinases: a family of protein kinases with diverse biological functions. Microbiol. Mol. Biol. Rev. 68: 320-344.

[70] Boutros T, Chevet E, Metrakos P (2008) Mitogen-activated protein (MAP) kinase /MAP kinase phosphatase regulation: roles in cell growth, death, and cancer. Pharmacol. Rev. 60: 261-310.

[71] Lee YJ, Aroor AR, Shukla SD (2002) Temporal activation of p42/44 mitogen-activated protein kinase and c-Jun N-terminal kinase by acetaldehyde in rat hepatocytes and its loss after chronic ethanol exposure. J. Pharmacol. Exp. Ther. 301: 908–914.

[72] Lee YJ, Shukla SD (2005) Pro- and anti-apoptotic roles of c-Jun N-terminal kinase (JNK) in ethanol and acetaldehyde exposed rat hepatocytes. Eur. J. Pharmacol. 508: 31-45.

[73] Aroor AR, Shukla DS (2004) MAP kinase signaling in diverse effects of ethanol, Life. Sci. 74: 2339-2364.

[74] Lee CS, Kim YJ, Ko HH, Han ES (2005) Synergistic effects of hydrogen peroxide and ethanol on cell viability loss in PC12 cells by increase in mitochondrial permeability transition. Biochem. Pharmacol. 70: 317-325.

[75] Morio Y, Tsuji M, Inagaki M, Nakagawa M, Asaka Y, Oyamada H, Furuya K, Oguchi K (2012) Ethanol induced oxidative stress in human liver adenocarcinoma cells (SK-Hep1) leads to Fas-mediated apoptosis: Role of JNK and p38MAPK. In preparation.

[76] Ji C, Kaplowitz N (2003) Betaine decreases hyperhomocysteinemia, endoplasmic reticulum stress, and liver injury in alcohol-fed mice. Gastroenterology 124: 1488-1499.

[77] Ji C, Mehrian-Shai R, Chan C, Hsu YH, Kaplowitz N (2005) Role of CHOP in hepatic apoptosis in the murine model of intragastric ethanol feeding. Alcohol Clin. Exp. Res. 29: 1496-1503.

[78] Esfandiari F, Villanueva JA, Wong DH, French SW, Halsted CH (2005) Chronic ethanol feeding and folate deficiency activate hepatic endoplasmic reticulum stress pathway in micropigs. Am. J Physiol. Gastrointest. Liver Physiol. 289: G54-G63.

[79] Ota M, Tsuji M, Mochizuki Y, Inagaki M, Murayama M, Emori H, Sambe T, Oguchi K (2011) Ethanol-induced stress leads to apoptosis via endoplasmic reticulum stress in SK-Hep1 cells. Showa Univ. J. Med. Sci. 23: 23-35.

[80] Halsted CH (2004) Nutrition and alcoholic liver disease. Semin. Liver Dis. 24: 289-304.

[81] Ji C (2008) Dissection of endoplasmic reticulum stress signaling in alcoholic and non-alcoholic liver injury. J. Gastroenterol. Hepatol. 23: S16-S24.

[82] Dara L, Ji C, Kaplowitz N (2011) The contribution of endoplasmic reticulum stress to liver diseases. Hepatology 53: 1752-1763.

[83] Malhi H, Kaufman RJ (2011) Endoplasmic reticulum stress in liver disease. J. Hepatol. 54: 795-809.

[84] Ji C, Kaplowitz N, Lau MY, Kao E, Petrovic LM, Lee AS (2011) Liver-specific loss of glucose-regulated protein 78 perturbs the unfolded protein response andexacerbates a spectrum of liver diseases in mice. Hepatology 54: 229-239.

[85] Boyce M, Bryant KF, Jousse C, Long K, Harding HP, Scheuner D, Kaufman RJ, Ma D, Coen DM, Ron D, Yuan J (2005) A selective inhibitor of eIF2a dephosphorylation protects cells from ER stress. Science 307: 935-939.

[86] Malhi H, Gores GJ (2008) Cellular and molecular mechanisms of liver injury. Gastroenterology 134: 1641-1654.

[87] Gores GJ (1999) Death receptors in liver biology and pathobiology. Hepatology 29: 1-4.

[88] Akazawa Y, Gores GJ (2007) Death receptor-mediated liver injury. Semi. Liver Dis. 27: 327–338.

[89] Sodeman T, Bronk SF, Roberts PJ, Miyoshi H, Gores GJ (2000) Bile salts mediate hepatocyte apoptosis by increasing cell surface trafficking of Fas. Am. J. Physiol.Gastrointest. Liver Physiol. 278: G992-G999.

Chapter 8

Cytocidal Effects of Polyphenolic Compounds, Alone or in Combination with, Anticancer Drugs Against Cancer Cells: Potential Future Application of the Combinatory Therapy

Bo Yuan, Masahiko Imai, Hidetomo Kikuchi, Shin Fukushima, Shingo Hazama, Takenori Akaike, Yuta Yoshino, Kunio Ohyama, Xiaomei Hu, Xiaohua Pei and Hiroo Toyoda

Additional information is available at the end of the chapter

1. Introduction

A growing body of clinical and experimental evidence has revealed a strong impact of drug resistance on clinical outcomes, especially in cancer therapy, since carcinogenesis is a multi-step, multi-pathway and multi-focal process, which involves a series of epigenetic and genetic alterations [1-3]. In order to solve the serious issue facing clinical treatment, combination therapy is now widely advocated for clinical use and has been shown to have a beneficial effect on patient satisfaction [3, 4]. For instance, 5-fluorouracil (5-FU) and leucovorin with either irinotecan or oxaliplatin have been widely used for the treatment of patients with colorectal cancer [5, 6]. Furthermore, recently, various types of molecular target-based drugs, such as cetuximab and bevacizumab, are being used clinically. Although these continuous efforts to exploit potential combination therapies are ongoing, there is still a growing concern about treatment resistance, disease relapse and side effects of drugs clinically used. Of note, numerous components of edible plants, collectively termed phytochemicals that have beneficial effects for health, are increasingly being reported in the scientific literature and these compounds are now widely recognized as potential therapeutic compounds [1, 2, 4, 7, 8]. In fact, natural product derived substances, especially polyphenolic compounds with very little toxic effects on normal cells, have attracted great attention in the therapeutic arsenal in clinical oncology due to their chemopreventive, antitumoral, radiosensibilizing and chemosensibilizing activities against various types of aggressive and recurrent cancers [1, 8-10].

Apoptosis, or programmed cell death, plays a key role in the development and growth regulation of normal cells, and is often dysregulated in cancer cells [11, 12]. It has been accepted that the aim of anticancer therapy is generally focused on apoptosis induction in premalignant and malignant cells, although other multiple molecular mechanisms such as modulation of carcinogen metabolism, anti-angiogenesis and induction of differentiation are also known to be implicated in its anticancer activity [4, 13]. So far, two principal signal pathways of apoptosis have been identified [11, 12]. The intrinsic mechanism of apoptosis involves a mitochondrial pathway. Apoptosis stimuli destruct mitochondrial membrane structure under the control of Bcl-2 (B-cell leukemia/lymphoma) family, resulting in the release of mitochondrial proteins including cytochrome c. Once cytochrome c is released it activates caspase-9 (initiator caspase) through the interaction with Apaf-1 (apoptotic protease activating factor-1) and dATP [14, 15], which ultimately leads to caspase-3 and -7 (effector caspases) activation [16]. On the other hand, the extrinsic pathway induced by death receptors, such as tumor necrosis factor receptor (TNFR) and Fas, is responsible for the activation of caspase-8 and caspase-10 (initiator caspase) accompanied by the activation of caspase-3 and -7 [16]. The effector caspases are the final mediators in the intrinsic and extrinsic pathways that cleave substrates and lead to cell death. Moreover, a third pathway involving endoplasmic reticulum (ER) stress and caspase-12 has been reported to be associated in apoptosis [4, 13]. A number of markers have been utilized to reveal apoptosis status including cell viability, cytochrome c release, caspase-3 activation, poly (ADP-ribose) polymerase (PARP) cleavage, and DNA fragmentation [17].

Reactive oxygen species (ROS) have been widely believed to play a pivotal role in a wide variety of cellular functions, including cell proliferation and differentiation [3, 11]. Further-more, oxidative stress, as a result of alterations of redox homeostasis due to an imbalance between ROS production and elimination, is known to be involved in many diseases such as hypoxic injury [11, 18]. Therefore, maintaining ROS homeostasis is crucial for normal cell growth and survival. Generally, cancer cells appear to generate more ROS than do normal cells due to its increased aerobic glycolysis. Furthermore, cancer cells exhibit increased ROS production and altered redox status. Recent studies suggest that these biochemical charac-teristics of cancer cells can be exploited for therapeutic benefits [3, 18]. Especially, tumors in advanced stage frequently exhibit multiple genetic alterations and high oxidative stress, suggesting that it is possible to preferentially eliminate these cells by pharmacological ROS insults. However, the upregulation of antioxidant capacity in adaptation to intrinsic oxida-tive stress in cancer cells can confer drug resistance [3, 18]. Thus, abrogation of such drug-resistant mechanisms by redox modulation could have significant therapeutic implications [2, 3, 18, 19]. Indeed, it has been known that altered redox status is closely associated with apoptosis induction in various cancer cells [2, 3, 18, 19]. Collectively, manipulating ROS levels by redox modulation is a way to selectively kill cancer cells without causing signifi-cant toxicity to normal cells.

Polyphenolic compounds such as flavonoids and curcumin have been shown to induce apoptosis in various malignant cells including solid tumors and hematologic malignant cells [1, 2, 7-9, 20-22]. Interestingly, the mechanisms underlying the apoptosis induction, associated with their antitumoral, chemopreventive and chemotherapeutic activities, have been shown to be implicated in alteration of redox status, since polyphenolic compounds are well known to possess both antioxidant and prooxidant activity [1, 2, 7, 8, 10, 20, 22-26].

In this chapter, we will highlight the recent advances on the cancer preventative activities of the polyphenolic compounds, including flavonoids such as anthocyanins, and *Vitex agnus-castus* fruit extract (Vitex) in which flavonoids are one of major components, as well as curcumin based on the most recent results from in vitro cell culture and in vivo animal model tumor systems. We will further summarize the detailed mechanisms underlying their cytocidal effects focusing on apoptosis induction. We will also provide detailed insight into potential future clinical application of these promising candidates endowed with potent antitumor activities, alone or in combination with other anticancer clinical drugs based on preclinical and clinical trial results.

2. Cancer preventative activities of the polyphenolic compounds, anthocyanins, Vitex, and curcumin

2.1. Resource and chemistry of anthocyanins, Vitex, and curcumin

The most abundant flavonoid constituents of fruits and vegetables are anthocyans (i.e. anthocyanins (glycosides), and their aglycones, anthocyanidins) that confer bright red or blue coloration on berries and other fruits and vegetables [8, 20]. Anthocyanins are especially interesting with respect to other flavonoids because they occur in the diet at relatively high concentrations. The daily intake of anthocyanins in the US diet has been suggested to be 180-255 mg/day, in contrast, the daily intake of most other dietary flavonoids, including genistein, quercetin and apigenin, is estimated to be only 20-25 mg/day [27]. Anthocyanidins are a diphenylpropane-based polyphenolic ring structure, and are limited to a few structure variants including cyanidin, delphinidin, malvidin, pelargonidin, peonidin and petunidin (Figure 1), with a distribution in nature of 50%, 12%, 12%, 12%, 7%, and 7%, respectively, and they present almost exclusively as glycosides, anthocyanins [28]. Furthermore, epidemiological evidence has demonstrated that consumption of fruits and berries has been associated with decreased risk of developing cancer [29].

Vitex agnus-castus is a shrub of the Verbenaceae family and is found naturally in the Middle East and Southern Europe. Ripe fruit of *Vitex agnus-castus* has been used as a folk medicine for the treatment of various obstetric and gynecological disorders in Europe [30, 31]. Itokawa and colleagues have reported that Vitex (an extract from dried ripe *Vitex agnus-castus*) possesses cytocidal effects on P388, a mouse leukemia cells, suggesting its antitumor activity, and that flavonoids such as luteolin (Figure 2) are one of its major constituents [32, 33].

Curcumin, a hydrophobic polyphenol, also known as turmeric, is a major bioactive ingredient extracted from the rhizome of the plant *Curcuma longa* [34, 35]. Curcumin has been used as a dietary supplement as well as therapeutic agent in Chinese medicine and

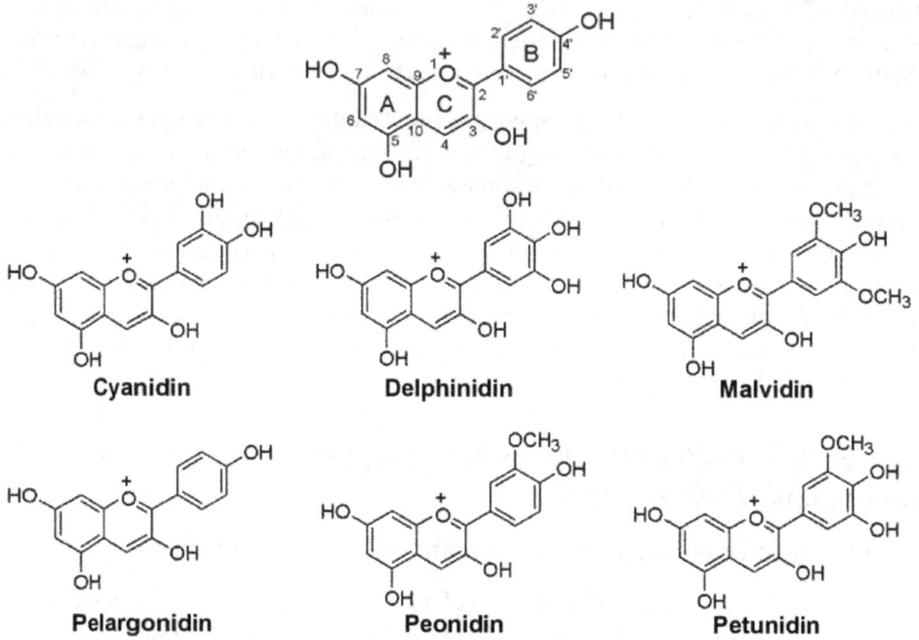

Figure 1. Chemical structure of anthocyanidins

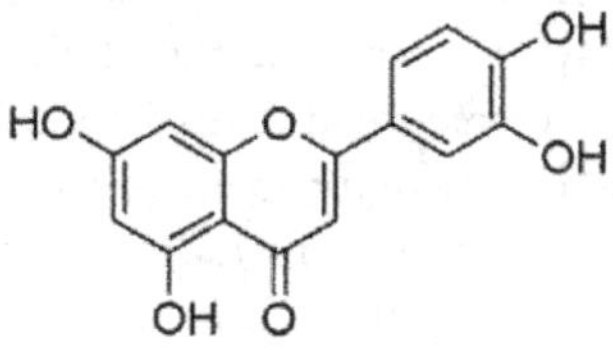

Figure 2. Chemical structure of luteolin

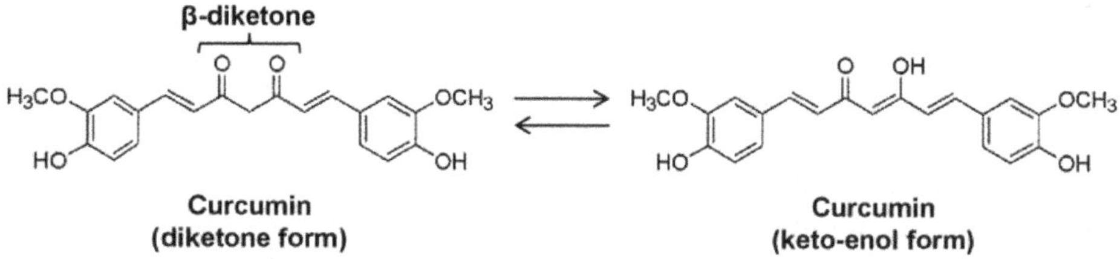

Figure 3. Chemical structures of naturally occurring curcumin. The scheme shows the diketone and keto-enol forms of curcumin. Curcumin exists as an equilibrium mixture of two tautomeric forms in solution. The enol structure of curcumin, which is stabilized by intramolecular H-bonding, is the most energetically stabilized and favored form.

other Asian medicines for centuries due to its safety and tolerance [34, 35]. Curcumin has a unique conjugated structure including two methylated phenols linked by the enol form of a heptadiene-3,5-diketone that gives the compound a bright yellow color (Figure 3) [7, 35].

2.2. Involvement of altered redox status in apoptosis induction triggered by polyphenolic compounds

2.2.1. Anthocyans

It has been demonstrated that anthocyanin-rich extracts from berries and grapes, and several pure anthocyanins and anthocyanidins, exhibit pro-apoptotic effects in multiple cell types such as colon [23, 36], breast [37, 38], prostate [39, 40], and leukemia cancer cells [10, 41]. They induce apoptosis through both intrinsic (mitochondrial) and extrinsic (Fas) pathways. In the intrinsic pathway, the treatment of cancer cells with anthocyanin results in destabilization of the mitochondrial membrane, cytochrome c release and activation of caspase-9, and -3 as well as pro-apoptotic protein such as apoptosis inducing factor [10, 37, 40]. In the extrinsic pathway, anthocyanins modulate the expression of Fas and FasL (Fas ligand) in cancer cells, which result in the activation of caspase-8, then cleaves Bid to tBid, and ultimately stimulates cytochrome c release [41]. Of note, several lines of evidence have indicated that oxidative stress resulted from stimulation of ROS production and/or insufficient ROS elimination is implicated in anthocyanins-triggered apoptosis induction in cancer cells, although broad biological activities including antimutagenesis and anticarcinogenesis of anthocyanins are generally attributed to their antioxidant activity [2, 7, 10, 22-25]. Indeed, it has been demonstrated that the most common type of anthocyanins, cyanidin-3-rutinoside, induced apoptosis in a human myeloid leukemia cell line, HL-60, in a dose- and time-dependent manner accompanied by accumulation of peroxides [10]. Cyanidin-3-rutinoside treatment resulted in ROS-dependent activation of p38 mitogen-activated protein kinase (MAPK) and c-Jun N-terminal kinase (JNK), which contributed to cell death by activating the mitochondrial pathway mediated by Bim, one of proapoptotic gene of Bcl-2 family [10]. More importantly, cyanidin-3-rutinoside treatment did not lead to increased ROS accumulation in normal human peripheral blood mononuclear cells (PBMNC) without inducing cytotoxic effects on these cells [10] as indicated by our preliminary data concerning treatment of HL-60 and PBMNC with anthocyanidin (unpublished data). These results suggest that cyanidin-3-rutinoside could be used in leukemia therapy with the advantages of being widely available and selective against tumors. More recently, delphinidin and cyanidin were reported to induce apoptosis in a human metastatic colorectal cancer cell line, LoVo and LoVo/ADR, a doxorubicin-resistant LoVo, but not in cells originating from a primary tumor site, *i.e.* Caco-2 [23]. Furthermore, LoVo/ADR was more sensitive to anthocyanins than LoVo cells [23]. It has been reported that the rate of lactate production is significantly higher in LoVo/ADR than in LoVo cells [42]. Therefore, the differences in changes of cellular energy metabolism associated with neoplastic transformation has been suggested to contribute the differential sensitivities to these anthocyanins [43]. Moreover, ROS accumulation, inhibition of glutathione reductase, and depletion of GSH were observed in the apoptosis triggered by anthocyanidins in these metastatic colorectal cancer cell lines [23]. These experimental results are in good agreement with previous reports on a possible role of flavonoids, as well as other phytochemicals, in modulating the glutathione (GSH) antioxidant activity, including regulation of intracellular GSH levels through targeting its synthesis, induction of multiple resistant protein 1 mediated GSH efflux, or inhibition of glutathione peroxidase enzyme activity observed in hematopoietic malignant and solid cancer cells in vitro and in vivo [19, 24, 44]. Furthermore, an in vitro study showed that anthocyanidins inhibit glutathione reductase (GR) in an oxygen-dependent manner, presumably via the effect of superoxide [45]. Therefore, these results suggest that anthocyanidins may be used as sensitizing agents through modulating intracellular redox status in various cancer therapy.

Intriguingly, a good correlation has been found between anthocyanin chemical structure and chemoprotective activity. Indeed, several lines of evidence have shown that the number of hydroxyl groups on the B-ring of anthocyanidins is associated with the potency of prooxidative [45-47], apoptotic induction [48], anti-transformation [49], as well as anti-oxidative activities [1, 8]. For instance, delphinidin and cyanidin that possess *ortho*-dihydroxyphenyl structure on the B-ring, showed stronger apoptotic induction in human leukemia cells [48] and inhibitory effect on 12-O-tetradecanoylphorbol-13-acetate (TPA)-induced cell transformation [49]. Furthermore, a similar trend of structure-activity relationship was also observed in the suppression of transcription factors closely associated with intracellular redox status, such as activator protein-1 (AP-1), nuclear factor-κB (NF-κB), and CCAAT/enhancer-binding protein (C/EBPδ) [20, 49, 50]. Structure-activity studies also suggested that the potency as inhibitors of epidermal growth factor receptor (EGFR), a target of an expanding class of anticancer therapies [51], might be positively correlated with the presence of hydroxyl functions in positions 3′ and 5′ of ring B of the anthocyanidinos molecule, and inversely with the presence of methoxy groups in these positions. All of these findings provide important molecular basis for the antitumor properties of anthocyanidins.

2.2.2. Vitex

We have been interesting the effects of naturally derived flavonoids on the growth of various types of cancer cells. Of those, we have demonstrated that Vitex exhibits cytotoxic activities against various types of solid tumor cells, such as KATO-III (a human gastric signet ring carcinoma cell line), COLO 201 (a human colon adenocarcinoma cell line), MCF-7 (a human breast carcinoma cell line) [52]. More interestingly, no apparent cytotoxicity was observed in non-tumor cells, such as human uterine cervical canal fibroblast (HCF) and human embryo fibroblast (HE-21) when treated with concentrations showing significant cytotoxicity in tumor cells, suggesting a selective cytotoxic activity against tumor cells [52]. We further demonstrated that Vitex induced apoptosis accompanied by an accumulation of intracellular ROS along with the decrease in the levels of intracellular GSH in KATO-III cells [22]. At the same time, our experimental data demonstrated a decrease in the amount of Bcl-2, Bcl-xL and Bid proteins; an increase in Bad protein; activation of caspase-8, -9 and -3; a leakage of cytochrome *c* from mitochondria in the cells [22]. Furthermore, the addition of an antioxidant, N-acetyl-L-cysteine (NAC), or exogenous GSH significantly abrogated the effects of Vitex [22]. Together, our results suggest that a crosstalk between intrinsic and extrinsic pathway via Bid activation as a result of oxidative stress plays a critical role in Vitex-induced apoptosis in KATO-III cells. We also demonstrated that apoptosis induction was observed in Vitex-treated COLO 201, concomitantly with a significant increase in heme oxygenase-1 (HO-1) gene expression as observed in KATO-III cells [53]. On the other hand, unlike KATO-III, apoptosis induction was not abrogated in the presence of antioxidants, such as NAC [53]. We further demonstrated that after treatment with Vitex, the upregulation of ER stress-related genes, such as glucose-regulated protein 78 (GRP78) and C/EBP-homologous protein (CHOP) along with the activation of caspase-9 and -3 were observed in COLO 201 [21]. However, an inhibitor for JNK significantly suppressed the apoptosis induction associated with caspase-3 activation [21]. These results thus suggest that the activation of JNK, and caspase-9 and -3 resulted from ER stress contributed to the apoptosis induction in Vitex-treated COLO 201 cells. Taken together, it seems that either oxidative stress-dependent or ER stress-dependent apoptosis would be triggered in cancer cells treated with Vitex depending on different cell types. Most importantly, our in vivo experimental data revealed that the administration of Vitex significantly suppressed tumor growth in COLO 201 xenograft mice, although more studies must be conducted to understand detailed in vivo pharmacological characterization of Vitex treatment [21]. In addition, as shown in Figure 4, we recently demonstrated a significant dose-dependent cytocidal effect in both a well-differentiated hepatocellular carcinoma (HCC) cell line, HepG2 and an undifferentiated HCC cell line, HLE, although the levels of cytotoxic activities of Vitex varied between two cells. Similarly, a significant dose-dependent cytocidal effect was also observed in both cells when treated with as high as 20 μg/ml luteolin, one of major constituents of Vitex. However, there was a trend to increase cell proliferation in both cells when treated with a relatively lower concentration of luteolin. On the other hand, 5-FU induced a dose-dependent cytocidal effect on HLE, but not in HepG2. These results indicated that while the cytocidal effect of 5-FU was more selective against undifferentiated hepatocellular carcinoma HLE, Vitex and luteolin exhibited significant cytocidal effects on both well-differentiated and undifferentiated hepatocellular carcinoma cells, suggesting a possible broad usefulness of these compounds to hepatocellular carcinoma therapy. Of note, Vitex has been used to treat patients with various obstetric and gynecological disorders in Europe [30, 31]. Moreover, it is interesting to note that Vitexins, which are isolated from the seed of Chinese herb *Vitex Negundo* and bear a basic flavonoid structure, show cytotoxic and antitumor effects against breast, prostate and ovarian cancer cells through apoptosis induction via an intrinsic pathway based on in vitro and in vivo xenograft tumor models [54]. Therefore, our results provide a new insight into the clinical use of Vitex for colon cancer and hepatocellular carcinoma besides those cancers mentioned above.

2.2.3. Curcumin

Curcumin has emerged worldwide as a potent therapeutic substance for treating diverse human diseases including various types of cancer, such as leukemia, colon cancer and pancreatic cancer [7, 25, 35, 55, 56]. Although the precise mode of action of this compound is

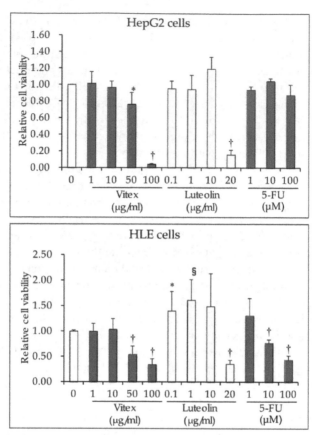

Figure 4. **Cytocidal effects of Vitex, luteolin and 5-FU on HepG2 and HLE cells.** Hepatocellular carcinoma (HCC) cell lines, HepG2 (well-differentiated) and HLE (undifferentiated) were kindly provided by Dr. Yamato Kikkawa (Laboratory of Clinical Biochemistry, School of Pharmacy, Tokyo University of Pharmacy and Life Sciences, Tokyo, Japan). After treatment with Vitex (final concentrations: 1, 10, 50 and 100 µg/ml), luteolin (final concentrations: 0.1, 1, 10 and 20 µg/ml) or 5-FU (1, 10 and 100 µM) for 48 h, cell viability was determined by XTT dye-reduction assay according to the method described previously [53]. Significant differences between treatment group and control (*: $p < 0.05$; §: $p < 0.01$; †: $p < 0.001$)

not yet elucidated, studies have shown that chemopreventive action of curcumin might be due to its ability to induce apoptosis through multiple signaling pathways, including intrinsic and extrinsic pathways as well as ER stress pathway [7, 57]. It has been suggested that curcumin-induced apoptosis is associated with ROS production and/or oxidative stress in cancer cells, in spite of its normal antioxidant capacity [7, 57]. Indeed, it has been demonstrated that curcumin can generate ROS as a prooxidant in the presence of copper in HL-60 cells, resulting in DNA damages and apoptotic cell death [58]. Furthermore, the prooxidant action of curcumin may be related to the conjugated β-diketone structure of this compound [58]. Kuo et al. also demonstrated that curcumin induced a dose- and time-dependent apoptotic cell death in the same cells, concomitant with a decrease of Bcl-2

accumulated in red blood cells during the consecutive administration of ATO to APL patients [66]. Furthermore, we have demonstrated for the first time that these arsenic metabolites also existed in cerebrospinal fluid [67], in which the concentrations of arsenic reached levels necessary for differentiation induction [68, 69]. We further investigated for the first time the arsenic speciation in plasma of bone marrow (BM), and demonstrated that speciation profiles of BM plasma were very similar to those of PB plasma, suggesting that speciation analysis of PB plasma could be predicative for BM speciation [70]. These findings on the pharmacokinetics of ATO in APL patients provide a new insight into clinical applications of ATO, and may contribute to better therapeutic protocols [4]. Recently, in order to understand the clinical side effects of arsenite, we also investigated the effects of arsenite on human-derived normal cells, since the clinical side effects of drugs are always found as a harmful and undesired effect on normal cells and/or tissues. Based on a study using a unique in vitro cell culture system comprising the primary culture chorion and amnion cells established in our laboratory [71-73], we demonstrated for the first time that transporter genes, such as aquaporin 9 and multidrug resistance associated protein 2 are involved in controlling intracellular arsenic accumulation in these primary cultured normal cells, which then contribute to differential sensitivity to arsenite cytotoxicity among these cells [74].

The successful clinical efficacy of ATO in the treatment of APL patients has led to investigations on exploring potential treatment applications for other malignancies, including ATO-resistant hematopoietic cancer and solid tumors [75, 76]. In order to further extend our previous study and promote the clinical application of arsenite, we have been seeking to explore potential candidate agents, which are expected to not only potentiate the efficacy of ATO but also possibly reduce its dosages [4]. In this regard, using HL-60 cells which are reported to show resistance to arsenite, we found that delphinidin showed selective cytotoxic effects on the cells, but minimal effects on PBMNC, and sensitized the cells to arsenite, resulting in the enhancement of arsenite cytotoxicity (Yuan et al. manuscript in preparation). Therefore, our experimental data suggest that sensitization of HL-60 cells to arsenite achieved by the combination with delphinidin could benefit a reduced dosages of arsenite in clinical application, contributing to minimize side effect. The clinical trial planning is now underway.

Of note, it is well known that oxidative stress is involved in the mechanisms underlying the therapeutic efficacy of arsenite and plays a major role in the toxicity of arsenite [4, 13, 77]. In fact, in order to maximally exploit effective ROS-mediated cell death without causing significant toxicity to normal cells, redox-based drug combination strategies have been proposed [3, 11, 18]. Based on the strategies, ROS-generating reagents including natural products derived substances, especially phenolic compounds, have received much attention due to their cytocidal effects on tumor cells but little on normal cells. In this regard, like the cytocidal effects of cyanidin-3-rutinoside on leukemia cells [10], delphinidin 3-sambubioside has also been demonstrated to induce apoptosis in HL-60 through ROS-mediated mitochondria pathway [78], suggesting these anthocyanins are good candidates for ROS-generating reagents, thereby possibly potentiate the action of

ATO. Indeed, quercetin and/or genistein are flavonoid with multiple biochemical effects such as downregulation of phosphoinositide 3-kinase/Akt signaling pathway and NF-κB transcription factor activity [79, 80], tyrosine kinase inhibition [81, 82]. Furthermore, both quercetin and genistein have been reported to selectively potentiate ATO-induced apoptosis via ROS generation resulted from intracellular GSH depletion, and activation of intrinsic and extrinsic apoptotic pathway in human leukemia cell lines such as HL-60, U937 and THP-1, but not in phytohemagglutinin-stimulated non-tumor peripheral blood lymphocytes [24, 83]. These results thus suggest that these flavonoids might be used to increase the clinical efficacy of ATO. Similar to flavonoids, subcytotoxic concentrations of curcumin also has been demonstrated to stimulate ROS production and potentiate apoptosis induction by ATO in leukemia cell lines via deregulation of Akt phosphorylation [25].

With these considerations in mind, a lot of preclinical and clinical trials have been carried out. For instance, we recently reported for the first time that 5-FU in combination with Vitex achieved an enhanced cytocidal effect on COLO 201 cells, but lesser cytotoxic effect on human PBMNC [9]. It has also been demonstrated that delphinidin induces apoptosis and inhibited NF-κB signaling in prostate tumor cells in vitro and in a human prostate tumor xenograft in nude mice in vivo [39]. Furthermore, twenty-five colon cancer patients without receiving prior therapy and surgery consumed 60 g/day (20 g/3x/day) of black raspberry powder daily for 2–4 weeks. Biopsies of normal-appearing and tumor tissues were taken from these patients before and after berry treatment. Intake of berries reduced proliferation rates and increased apoptosis in colon tumors but not in normal-appearing crypts [8]. On the other hand, supplementation of anthocyanins in the diet of cancer patients receiving chemotherapy did not result in increased inhibition of tumor development when compared to chemotherapy alone [84]. These conflicting findings suggest that a large scale of clinical trial is needed.

A phase I clinical trial has shown that quercetin, another one of flavonoids, can be safely administered to patients with ovarian cancer or hepatoma by intravenous injection of bolus at a dose of 1400 mg/m² [82]. Moreover, the evidence of antitumor activity was seen in the clinical trial based on sustained fall in serum CA 125 levels, which is proposed for the use as a surrogate marker of response [85]. Furthermore, similar to our previous report [9], CDF (a difluorinated analog of curcumin), alone or in combination with 5-FU and oxaliplatin, was more potent than curcumin alone in reducing the number of chemoresistant HCT-116 and HT-29 colon cancer cells expressing CD44 and CD166 stem cell-like markers [56]. Concomitantly, cell growth inhibition, apoptosis induction and disintegration of colonospheres in these colon cancer cells were also observed in the study [56]. Moreover, clinical trials have confirmed the safety and feasibility to use curcumin in combination therapy with current chemotherapeutic treatments [7]. More recent results from a phaseI/II study on 21 advanced pancreatic cancer patients with gemcitabine-based chemotherapy have indicated that the median overall survival time of the patients after a treatment with curcumin plus gemcitabine or gemcitabine/S-1 combination was 161 days and 1-year survival rate of 19% (95% confidence interval) [86]. Among eighteen evaluable patients, no patients experienced a partial or complete response and five patients (28%) demonstrated stable disease according to Response Evaluation Criteria in Solid Tumors (RECIST) [86].

Although many encouraging results of in vitro and in vivo studies suggest polyphenolic compounds as a promising candidate for cancer therapy, either alone or in combination with current anticancer drugs, the therapeutic applications of these compounds in humans are limited by their high metabolic instability as well as poor absorption and bioavailability [1, 7, 8, 20, 87]. In this regard, the selective delivery of nanotechnology-based formulations of these polyphenolic compounds to tumors, alone or in combination with other anticancer drugs, has been of great interest [7, 26, 88]. For instance, pegylated liposomal quercetin was shown to significantly improve its solubility and bioavailability and be a potential application in the treatment of tumor based on a study using CT26 (a mouse colorectal carcinoma cell line), LL/2 (Lewis lung cancer cell line) and H22 (a hepatoma cell line) xenograft mice [88]. Furthermore, diverse curcumin formulations have been developed with different nanotechnology consisting of its encapsulation or conjugation with nanoparticles, polymeric micelles or liposomes to improve its stability, bioavailability and specific and sustained delivery into cancer cells and, consequently, its anticarcinogenic effects [7]. In particular, the systemic administration of gemcitabine plus polymeric micelle-encapsulated curcumin formulation enhanced greater bioavailability in plasma and tissues as compared to that of free curcumin in xenograft models of human pancreatic cancer established in athymic mice [89]. In consequence, the combinatory administration efficiently block tumor growth and metastases in this animal model of pancreatic cancer. Furthermore, an inhibition of NF-κB and its targeted genes are implicated in the tumor growth inhibition [89]. Therefore, the use of nanotechnology-based formulations of polyphenolic compounds and their novel chemical analogs probably represents a potential alternative strategy of great clinical interest for overcoming the high metabolic instability and poor bioavailability of these compounds, which are among the principal factors limiting their therapeutic applications.

4. Conclusion

A striking global research on substances derived from natural products including polyphenolic compounds is being explored to understand the detailed mechanisms of their chemopreventive, antitumoral and chemosensibilizing activities against various types of aggressive and recurrent cancers. Besides the involvement of altered redox status in apoptosis induction triggered by these compounds, anti-inflammatory effects, anti-angiogenesis, anti-invasiveness and induction of differentiation are well known to be implicated in their broad biological functions. It is worthy of note that flavonoids have been revealed to inhibit the function of ATP-binding cassette transporters such as multidrug resistance-associated proteins as well as P-glycoprotein [90], similar to our recent study [91]. On the other hand, a recent study demonstrated that berry anthocyanins, such as cyanidin-3-galactosidee, and cyanidin-3-glucoside as well as peonidin-3-glucosid, exhibit affinities for the efflux transporters breast cancer resistance protein (BCRP) and consequently may be actively transported out of intestinal tissues and endothelia [92]. However, the same report also demonstrated that some berry anthocyanins and anthocyanidins, such as delphinidin, cyanidin and cyanidin-3-rutinoside, act as BCRP inhibitor, while some of them, such as malvidin, malvidin-3-galactoside and petunidin, exhibited bimodal activities serving as BCRP substrates and inhibitors at low concentrations and high concentrations, respectively [92]. These findings suggest that a variety of biological activities of anthocyanins and anthocyanidins may be attributed in part to their inhibitory effects on those drug transporters, paradoxically, may be abolished as a result of efflux through those transporters. These findings also raised a pharmacological and pharmaceutical concern about formulatability of the dietary constituent, and warn us against the casual use of herbs and/or other botanicals in cancer patient care.

Author details

Bo Yuan, Masahiko Imai, Hidetomo Kikuchi, Shin Fukushima, Shingo Hazama, Takenori Akaike, Yuta Yoshino, Kunio Ohyama and Hiroo Toyoda
Department of Clinical Molecular Genetics, School of Pharmacy, Tokyo University of Pharmacy & Life Sciences, Hachioji, Tokyo, Japan

Xiaomei Hu
National Therapeutic Center of Hematology of Traditional Chinese Medicine, XiYuan Hospital, China Academy of Traditional Chinese Medicine, Beijing, P.R. China

Xiaohua Pei
The Third Affiliated Hospital of Beijing University of Traditional Chinese Medicine, Beijing, P.R. China

Acknowledgement

This work was supported in part by grants from Japan China Medical Association to B.Y. This work was also supported in part by grants from the Ministry of Education, Culture, Sports, Science and Technology and by the Promotion and Mutual Aid Corporation for Private Schools of Japan. The authors thank Dr. Chieko Hirobe for encouraging suggestions and arranging sample supply, and Dr. Yamato Kikkawa for providing HCC cell lines for this study.

5. References

[1] Fimognari C, Lenzi M, Hrelia P (2008) Chemoprevention of cancer by isothiocyanates and anthocyanins: Mechanisms of action and structure-activity relationship. Curr. Med. Chem. 15:440-447.

[2] Reddy L, Odhav B, Bhoola KD (2003) Natural products for cancer prevention: A global perspective. Pharmacol. Ther. 99:1-13.

[3] Trachootham D, Alexandre J, Huang P (2009) Targeting cancer cells by ros-mediated mechanisms: A radical therapeutic approach? Nat. Rev. Drug Discov. 8:579-591.

[4] Yuan B, Yoshino Y, Kaise T, Toyoda H (2011) Application of arsenic trioxide therapy for patients with leukaemia. In: Sun HZ, editor. Biological Chemistry of As, Sb and Bi. John Wiley & Sons, New York, pp.263-292.

[5] Meyerhardt JA, Mayer RJ (2005) Systemic therapy for colorectal cancer. N. Engl. J. Med. 352:476-487.

[6] Hurwitz H, Fehrenbacher L, Novotny W, Cartwright T, Hainsworth J, Heim W, Berlin J, Baron A, Griffing S, Holmgren E, Ferrara N, Fyfe G, Rogers B, Ross R, Kabbinavar F (2004) Bevacizumab plus irinotecan, fluorouracil, and leucovorin for metastatic colorectal cancer. N. Engl. J. Med. 350:2335-2342.

[7] Mimeault M, Batra SK (2011) Potential applications of curcumin and its novel synthetic analogs and nanotechnology-based formulations in cancer prevention and therapy. Chin. Med. 6:31.

[8] Wang LS, Stoner GD (2008) Anthocyanins and their role in cancer prevention. Cancer Lett. 269:281-290.

[9] Imai M, Kikuchi H, Yuan B, Aihara Y, Mizokuchi A, Ohyama K, Hirobe C, Toyoda H (2011) Enhanced growth inhibitory effect of 5-fluorouracil in combination with vitex agnus-castus fruits extract against a human colon adenocarcinoma cell line, colo 201. J. Chin. Clin. Med. 6:14-19.

[10] Feng R, Ni HM, Wang SY, Tourkova IL, Shurin MR, Harada H, Yin XM (2007) Cyanidin-3-rutinoside, a natural polyphenol antioxidant, selectively kills leukemic cells by induction of oxidative stress. J. Biol. Chem. 282:13468-13476.

[11] Zhang Y, Du Y, Le W, Wang K, Kieffer N, Zhang J (2011) Redox control of the survival of healthy and diseased cells. Antioxid. Redox Signal. 15:2867-2908.

[12] Indran IR, Tufo G, Pervaiz S, Brenner C (2011) Recent advances in apoptosis, mitochondria and drug resistance in cancer cells. Biochim. Biophys. Acta. 1807:735-745.

[13] Wang ZY, Chen Z (2008) Acute promyelocytic leukemia: From highly fatal to highly curable. Blood 111:2505-2515.

[14] Wei MC, Zong WX, Cheng EH, Lindsten T, Panoutsakopoulou V, Ross AJ, Roth KA, MacGregor GR, Thompson CB, Korsmeyer SJ (2001) Proapoptotic bax and bak: A requisite gateway to mitochondrial dysfunction and death. Science 292:727-730.

[15] Shimizu S, Narita M, Tsujimoto Y (1999) Bcl-2 family proteins regulate the release of apoptogenic cytochrome c by the mitochondrial channel vdac. Nature 399:483-487.

[16] Earnshaw WC, Martins LM, Kaufmann SH: Mammalian caspases (1999) Structure, activation, substrates, and functions during apoptosis. Annu. Rev. Biochem. 68:383-424.

[17] McCarthy NJ, Evan GI (1998) Methods for detecting and quantifying apoptosis. Curr. Top. Dev. Biol. 36:259-278.

[18] Schumacker PT (2006) Reactive oxygen species in cancer cells: Live by the sword, die by the sword. Cancer cell 10:175-176.

[19] Trachootham D, Zhou Y, Zhang H, Demizu Y, Chen Z, Pelicano H, Chiao PJ, Achanta G, Arlinghaus RB, Liu J, Huang P (2006) Selective killing of oncogenically transformed cells through a ros-mediated mechanism by beta-phenylethyl isothiocyanate. Cancer cell 10:241-252.

[20] Cooke D, Steward WP, Gescher AJ, Marczylo T (2005) Anthocyans from fruits and vegetables--does bright colour signal cancer chemopreventive activity? Eur. J. Cancer 41:1931-1940.

[21] Imai M, Yuan B, Kikuchi H, Saito M, Ohyama K, Hirobe C, Oshima T, Hosoya T, Morita H, Toyoda H (2012) Growth inhibition of a human colon carcinoma cell, colo 201, by a natural product, vitex agnus-castus fruits extract, in vivo and in vitro. Adv, Biol. Chem. 2:20-28.

[22] Ohyama K, Akaike T, Imai M, Toyoda H, Hirobe C, Bessho T (2005) Human gastric signet ring carcinoma (kato-iii) cell apoptosis induced by vitex agnus-castus fruit extract through intracellular oxidative stress. Int. J. Biochem. Cell Biol. 37:1496-1510.

[23] Cvorovic J, Tramer F, Granzotto M, Candussio L, Decorti G, Passamonti S (2010) Oxidative stress-based cytotoxicity of delphinidin and cyanidin in colon cancer cells. Arch. Biochem. Biophys. 501:151-157.

[24] Ramos AM, Aller P (2008) Quercetin decreases intracellular gsh content and potentiates the apoptotic action of the antileukemic drug arsenic trioxide in human leukemia cell lines. Biochem. Pharmacol. 75:1912-1923.

[25] Sanchez Y, Simon GP, Calvino E, de Blas E, Aller P (2010) Curcumin stimulates reactive oxygen species production and potentiates apoptosis induction by the antitumor drugs arsenic trioxide and lonidamine in human myeloid leukemia cell lines. J. Pharmacol. Exp. Ther. 335:114-123.

[26] Singh M, Bhatnagar P, Srivastava AK, Kumar P, Shukla Y, Gupta KC (2011) Enhancement of cancer chemosensitization potential of cisplatin by tea polyphenols poly(lactide-co-glycolide) nanoparticles. J. Biomed. Nanotechnol. 7:202.

[27] Hertog MG, Hollman PC, Katan MB, Kromhout D (1993) Intake of potentially anticarcinogenic flavonoids and their determinants in adults in the netherlands. Nutr. Cancer 20:21-29.

[28] Zhang Y, Vareed SK, Nair MG (2005) Human tumor cell growth inhibition by nontoxic anthocyanidins, the pigments in fruits and vegetables. Life Sci. 76:1465-1472.

[29] Taylor PR, Greenwald P (2005) Nutritional interventions in cancer prevention. J. Clin. Oncol. 23:333-345.

[30] Berger D, Schaffner W, Schrader E, Meier B, Brattstrom A (2000) Efficacy of vitex agnus castus l. Extract ze 440 in patients with pre-menstrual syndrome (pms). Arch. Gynecol. Obstet. 264:150-153.

[31] Schellenberg R (2001) Treatment for the premenstrual syndrome with agnus castus fruit extract: Prospective, randomised, placebo controlled study. BMJ 322:134-137.

[32] Hirobe C, Qiao ZS, Takeya K, Itokawa H (1997) Cytotoxic flavonoids from vitex agnus-castus. Phytochemistry 46:521-524.

[33] Itokawa H, Qiao ZS, Hirobe C, Takeya K (1995) Cytotoxic limonoids and tetranortriterpenoids from melia azedarach. Chem. Pharm. Bull. (Tokyo) 43:1171-1175.

[34] Aggarwal BB, Sundaram C, Malani N, Ichikawa H (2007) Curcumin: The indian solid gold. Adv. Exp. Med. Biol. 595:1-75.

[35] Itokawa H, Shi Q, Akiyama T, Morris-Natschke SL, Lee KH (2008) Recent advances in the investigation of curcuminoids. Chin. Med. 3:11.

[36] Seeram NP, Adams LS, Zhang Y, Lee R, Sand D, Scheuller HS, Heber D (2006) Blackberry, black raspberry, blueberry, cranberry, red raspberry, and strawberry extracts inhibit growth and stimulate apoptosis of human cancer cells in vitro. J. Agri. Food Chem. 54:9329-9339.

[37] Afaq F, Zaman N, Khan N, Syed DN, Sarfaraz S, Zaid MA, Mukhtar H (2008) Inhibition of epidermal growth factor receptor signaling pathway by delphinidin, an anthocyanidin in pigmented fruits and vegetables. Int. J. Cancer 123:1508-1515.

[38] Chen PN, Chu SC, Chiou HL, Chiang CL, Yang SF, Hsieh YS (2005) Cyanidin 3-glucoside and peonidin 3-glucoside inhibit tumor cell growth and induce apoptosis in vitro and suppress tumor growth in vivo. Nutr. Cancer 53:232-243.

[39] Hafeez BB, Siddiqui IA, Asim M, Malik A, Afaq F, Adhami VM, Saleem M, Din M, Mukhtar H (2008) A dietary anthocyanidin delphinidin induces apoptosis of human prostate cancer pc3 cells in vitro and in vivo: Involvement of nuclear factor-kappab signaling. Cancer Res. 68:8564-8572.

[40] Reddivari L, Vanamala J, Chintharlapalli S, Safe SH, Miller JC, Jr. (2007) Anthocyanin fraction from potato extracts is cytotoxic to prostate cancer cells through activation of caspase-dependent and caspase-independent pathways. Carcinogenesis 28:2227-2235.

[41] Chang YC, Huang HP, Hsu JD, Yang SF, Wang CJ (2005) Hibiscus anthocyanins rich extract-induced apoptotic cell death in human promyelocytic leukemia cells. Toxicol. Appl. Pharmacol. 205:201-212.

[42] Fanciulli M, Bruno T, Giovannelli A, Gentile FP, Di Padova M, Rubiu O, Floridi A (2000) Energy metabolism of human lovo colon carcinoma cells: Correlation to drug resistance and influence of lonidamine. Clin. Cancer Res. 6:1590-1597.

[43] Warburg O (1956) On respiratory impairment in cancer cells. Science 124:269-270.

[44] Kachadourian R, Day BJ (2006) Flavonoid-induced glutathione depletion: Potential implications for cancer treatment. Free Radic. Biol. Med. 41:65-76.

[45] Elliott AJ, Scheiber SA, Thomas C, Pardini RS (1992) Inhibition of glutathione reductase by flavonoids. A structure-activity study. Biochem. Pharmacol. 44:1603-1608.

[46] Dickancaite E, Nemeikaite A, Kalvelyte A, Cenas N (1998) Prooxidant character of flavonoid cytotoxicity: Structure-activity relationships. Biochem. Mol. Biol. Int. 45:923-930.

[47] Sergediene E, Jonsson K, Szymusiak H, Tyrakowska B, Rietjens IM, Cenas N (1999) Prooxidant toxicity of polyphenolic antioxidants to hl-60 cells: Description of quantitative structure-activity relationships. FEBS Lett. 462:392-396.

[48] Hou DX, Ose T, Lin S, Harazoro K, Imamura I, Kubo M, Uto T, Terahara N, Yoshimoto M, Fujii M (2003) Anthocyanidins induce apoptosis in human promyelocytic leukemia cells: Structure-activity relationship and mechanisms involved. Int. J. Oncol. 23:705-712.

[49] Hou DX, Kai K, Li JJ, Lin S, Terahara N, Wakamatsu M, Fujii M, Young MR, Colburn N (2004) Anthocyanidins inhibit activator protein 1 activity and cell transformation: Structure-activity relationship and molecular mechanisms. Carcinogenesis 25:29-36.

[50] Hou DX, Yanagita T, Uto T, Masuzaki S, Fujii M (2005) Anthocyanidins inhibit cyclooxygenase-2 expression in lps-evoked macrophages: Structure-activity relationship and molecular mechanisms involved. Biochem. Pharmacol. 70:417-425.

[51] Zhang H, Berezov A, Wang Q, Zhang G, Drebin J, Murali R, Greene MI (2007) Erbb receptors: From oncogenes to targeted cancer therapies. J. Clin. Invest. 117:2051-2058.

[52] Ohyama K, Akaike T, Hirobe C, Yamakawa T (2003) Cytotoxicity and apoptotic inducibility of vitex agnus-castus fruit extract in cultured human normal and cancer cells and effect on growth. Biol. Pharm. Bull. 26:10-18.

[53] Imai M, Kikuchi H, Denda T, Ohyama K, Hirobe C, Toyoda H (2009) Cytotoxic effects of flavonoids against a human colon cancer derived cell line, COLO 201: A potential natural anti-cancer substance. Cancer Lett. 276:74-80.

[54] Zhou Y, Liu YE, Cao J, Zeng G, Shen C, Li Y, Zhou M, Chen Y, Pu W, Potters L, Shi YE (2009) Vitexins, nature-derived lignan compounds, induce apoptosis and suppress tumor growth. Clin. Cancer Res. 15:5161-5169.

[55] Epelbaum R, Schaffer M, Vizel B, Badmaev V, Bar-Sela G (2010) Curcumin and gemcitabine in patients with advanced pancreatic cancer. Nutr. Cancer 62:1137-1141.

[56] Kanwar SS, Yu Y, Nautiyal J, Patel BB, Padhye S, Sarkar FH, Majumdar AP (2011) Difluorinated-curcumin (cdf): A novel curcumin analog is a potent inhibitor of colon cancer stem-like cells. Pharm. Res. 28:827-838.

[57] Reuter S, Eifes S, Dicato M, Aggarwal BB, Diederich M (2008) Modulation of anti-apoptotic and survival pathways by curcumin as a strategy to induce apoptosis in cancer cells. Biochem. Pharmacol. 76:1340-1351.

[58] Yoshino M, Haneda M, Naruse M, Htay HH, Tsubouchi R, Qiao SL, Li WH, Murakami K, Yokochi T (2004) Prooxidant activity of curcumin: Copper-dependent formation of 8-hydroxy-2'-deoxyguanosine in DNA and induction of apoptotic cell death. Toxico. in vitro 18:783-789.

[59] Kuo ML, Huang TS, Lin JK (1996) Curcumin, an antioxidant and anti-tumor promoter, induces apoptosis in human leukemia cells. Biochim. Biophys. Acta. 1317:95-100.

[60] Kim MS, Kang HJ, Moon A (2001) Inhibition of invasion and induction of apoptosis by curcumin in h-ras-transformed mcf10a human breast epithelial cells. Arch. Pharm. Res. 24:349-354.

[61] Syng-Ai C, Kumari AL, Khar A (2004) Effect of curcumin on normal and tumor cells: Role of glutathione and bcl-2. Mol. Cancer Ther. 3:1101-1108.

[62] Wahlstrom B, Blennow G (1978) A study on the fate of curcumin in the rat. Acta. Pharmacol. Toxicol. (Copenh) 43:86-92.

[63] Anand P, Kunnumakkara AB, Newman RA, Aggarwal BB (2007) Bioavailability of curcumin: Problems and promises. Mol. Pharm. 4:807-818.

[64] Padhye S, Banerjee S, Chavan D, Pandye S, Swamy KV, Ali S, Li J, Dou QP, Sarkar FH (2009) Fluorocurcumins as cyclooxygenase-2 inhibitor: Molecular docking, pharmacokinetics and tissue distribution in mice. Pharm. Res. 26:2438-2445.

[65] Padhye S, Yang H, Jamadar A, Cui QC, Chavan D, Dominiak K, McKinney J, Banerjee S, Dou QP, Sarkar FH (2009) New difluoro knoevenagel condensates of curcumin, their schiff bases and copper complexes as proteasome inhibitors and apoptosis inducers in cancer cells. Pharm. Res. 26:1874-1880.

[66] Yoshino Y, Yuan B, Miyashita SI, Iriyama N, Horikoshi A, Shikino O, Toyoda H, Kaise T (2009) Speciation of arsenic trioxide metabolites in blood cells and plasma of a patient with acute promyelocytic leukemia. Anal. Bioanal. Chem. 393:689-697.

[67] Kiguchi T, Yoshino Y, Yuan B, Yoshizawa S, Kitahara T, Akahane D, Gotoh M, Kaise T, Toyoda H, Ohyashiki K (2010) Speciation of arsenic trioxide penetrates into cerebrospinal fluid in patients with acute promyelocytic leukemia. Leuk. Res. 34:403-405.

[68] Chen GQ, Shi XG, Tang W, Xiong SM, Zhu J, Cai X, Han ZG, Ni JH, Shi GY, Jia PM, Liu MM, He KL, Niu C, Ma J, Zhang P, Zhang TD, Paul P, Naoe T, Kitamura K, Miller W, Waxman S, Wang ZY, de The H, Chen SJ, Chen Z (1997) Use of arsenic trioxide (As2O3) in the treatment of acute promyelocytic leukemia (APL): I. As2O3 exerts dose-dependent dual effects on apl cells. Blood 89:3345-3353.

[69] Soignet SL, Maslak P, Wang ZG, Jhanwar S, Calleja E, Dardashti LJ, Corso D, DeBlasio A, Gabrilove J, Scheinberg DA, Pandolfi PP, Warrell RP, Jr. (1998) Complete remission after treatment of acute promyelocytic leukemia with arsenic trioxide. N. Engl. J. Med. 339:1341-1348.

[70] Iriyama N, Yoshino Y, Yuan B, Horikoshi A, Hirabayashi Y, Hatta Y, Toyoda H, Takeuchi J (2012) Speciation of arsenic trioxide metabolites in peripheral blood and bone marrow from an acute promyelocytic leukemia patient. J. Hematol. Oncol. 5:1.

[71] Yuan B, Ohyama K, Bessho T, Toyoda H (2006) Contribution of inducible nitric oxide synthase and cyclooxygenase-2 to apoptosis induction in smooth chorion trophoblast cells of human fetal membrane tissues. Biochem. Biophys. Res. Commun. 341:822-827.

[72] Yuan B, Ohyama K, Bessho T, Uchide N, Toyoda H (2008) Imbalance between ros production and elimination results in apoptosis induction in primary smooth chorion trophoblast cells prepared from human fetal membrane tissues. Life Sci. 82:623-630.

[73] Yuan B, Ohyama K, Takeichi M, Toyoda H (2009) Direct contribution of inducible nitric oxide synthase expression to apoptosis induction in primary smooth chorion trophoblast cells of human fetal membrane tissues. Int. J. Biochem. Cell Biol. 41:1062-1069.

[74] Yoshino Y, Yuan B, Kaise T, Takeichi M, Tanaka S, Hirano T, Kroetz DL, Toyoda H (2011) Contribution of aquaporin 9 and multidrug resistance-associated protein 2 to differential sensitivity to arsenite between primary cultured chorion and amnion cells prepared from human fetal membranes. Toxicol. Appl. Pharmacol. 257:198-208.

[75] Dilda PJ, Hogg PJ (2007) Arsenical-based cancer drugs. Cancer Treat Rev. 33:542-564.

[76] Litzow MR (2008) Arsenic trioxide. Expert Opin. Pharmacother. 9:1773-1785.

[77] Ninomiya M, Kajiguchi T, Yamamoto K, Kinoshita T, Emi N, Naoe T (2006) Increased oxidative DNA products in patients with acute promyelocytic leukemia during arsenic therapy. Haematologica 91:1571-1572.

[78] Hou DX, Tong X, Terahara N, Luo D, Fujii M (2005) Delphinidin 3-sambubioside, a hibiscus anthocyanin, induces apoptosis in human leukemia cells through reactive oxygen species-mediated mitochondrial pathway. Arch. Biochem. Biophys. 440:101-109.

[79] Sarkar FH, Adsule S, Padhye S, Kulkarni S, Li Y (2006) The role of genistein and synthetic derivatives of isoflavone in cancer prevention and therapy. Mini. Rev. Med. Chem. 6:401-407.

[80] Sharma H, Sen S, Singh N (2005) Molecular pathways in the chemosensitization of cisplatin by quercetin in human head and neck cancer. Cancer Biol. Ther. 4:949-955.

[81] Akiyama T, Ishida J, Nakagawa S, Ogawara H, Watanabe S, Itoh N, Shibuya M, Fukami Y (1987) Genistein, a specific inhibitor of tyrosine-specific protein kinases. J. Biol. Chem. 262:5592-5595.

[82] Ferry DR, Smith A, Malkhandi J, Fyfe DW, deTakats PG, Anderson D, Baker J, Kerr DJ (1996) Phase I clinical trial of the flavonoid quercetin: Pharmacokinetics and evidence for in vivo tyrosine kinase inhibition. Clin. Cancer Res. 2:659-668.

[83] Sanchez Y, Amran D, Fernandez C, de Blas E, Aller P (2008) Genistein selectively potentiates arsenic trioxide-induced apoptosis in human leukemia cells via reactive oxygen species generation and activation of reactive oxygen species-inducible protein kinases (p38-MAPK, AMPK). Int. J. Cancer 123:1205-1214.

[84] Bode U, Hasan C, Hulsmann B, Fleischhack G (1999) Recancostat compositum therapy does not prevent tumor progression in young cancer patients. Klin. Padiatr. 211:353-355.

[85] Rustin GJ, van der Burg ME, Berek JS (1993) Advanced ovarian cancer. Tumour markers. Ann. Oncol. 4 Suppl 4:71-77.

[86] Kanai M, Yoshimura K, Asada M, Imaizumi A, Suzuki C, Matsumoto S, Nishimura T, Mori Y, Masui T, Kawaguchi Y, Yanagihara K, Yazumi S, Chiba T, Guha S, Aggarwal BB (2011) A phase I/II study of gemcitabine-based chemotherapy plus curcumin for patients with gemcitabine-resistant pancreatic cancer. Cancer Chemother. Pharmacol. 68:157-164.

[87] Thomasset S, Teller N, Cai H, Marko D, Berry DP, Steward WP, Gescher AJ (2009) Do anthocyanins and anthocyanidins, cancer chemopreventive pigments in the diet, merit development as potential drugs? Cancer Chemother. Pharmacol. 64:201-211.

[88] Yuan ZP, Chen LJ, Fan LY, Tang MH, Yang GL, Yang HS, Du XB, Wang GQ, Yao WX, Zhao QM, Ye B, Wang R, Diao P, Zhang W, Wu HB, Zhao X, Wei YQ (2006) Liposomal quercetin efficiently suppresses growth of solid tumors in murine models. Clin. Cancer Res. 12:3193-3199.

[89] Bisht S, Mizuma M, Feldmann G, Ottenhof NA, Hong SM, Pramanik D, Chenna V, Karikari C, Sharma R, Goggins MG, Rudek MA, Ravi R, Maitra A, Maitra A (2010) Systemic administration of polymeric nanoparticle-encapsulated curcumin (nanocurc) blocks tumor growth and metastases in preclinical models of pancreatic cancer. Mol. Cancer Ther. 9:2255-2264.

[90] Kitagawa S (2006) Inhibitory effects of polyphenols on p-glycoprotein-mediated transport. Biol. Pharm. Bull. 29:1-6.

[91] Ishii K, Tanaka S, Kagami K, Henmi K, Toyoda H, Kaise T, Hirano T (2010) Effects of naturally occurring polymethyoxyflavonoids on cell growth, p-glycoprotein function, cell cycle, and apoptosis of daunorubicin-resistant T lymphoblastoid leukemia cells. Cancer Invest. 28:220-229.

[92] Dreiseitel A, Oosterhuis B, Vukman KV, Schreier P, Oehme A, Locher S, Hajak G, Sand PG (2009) Berry anthocyanins and anthocyanidins exhibit distinct affinities for the efflux transporters bcrp and mdr1. Br. J. Pharmacol. 158:1942-1950.

Chapter 9

Pancreatic Islet Beta-Cell Apoptosis in Experimental Diabetes Mellitus

A.V. Smirnov, G.L. Snigur and M.P. Voronkova

Additional information is available at the end of the chapter

1. Introduction

For screening and detailed studying of antidiabetic medications, various genetic and non-genetic experimental models of diabetes mellitus were used (Islam S., Loots D.T. 2009). And though they are not absolutely equivalent to etiopathogenetic mechanisms of human pathological conditions, each of them represents itself as an integral tool for research into genetic, endocrine, metabolic, morphological changes of this disease (Sarvilina I.V., Maclakov Y.S. 2008).

The most commonly used diabetic experimental models are non-genetic models that use hydrophilic β-cell glucose analogues, such as alloxan, streptozotocin, chlorozotocin, cyproheptadine, etc. The common mechanism of action of these substance includes degradation of pancreatic islet β-cells by means of: 1) generation of oxygen free radicals that destroy the integrity of a cell, 2) alkylation of DNA and subsequent activation of poly-ADP-ribose-synthetase - reduction of NAD to β-cell, and 3) inhibition of active transport of calcium and calmodulin-activated protein kinase (Rees D.A., Alcolado J.C. 2005). In this type of experimental models of diabetes mellitus streptozotocin (an N-nitrosourea derivative of glucosamine) is most commonly used (McNeill J.H. 1999). Depending on cytotoxin dosage used in the experiment (45-70 mg/kg) and route of administration (i.p., i.v.), it is possible to model and simulate different states of carbohydrate metabolism based on a specific clinical type of diabetes mellitus (DM mixed (type 1-2) latent or «hidden» diabetes) (Srinivasan K. et al. 2007). Although diabetes mellitus usually has obvious clinical symptoms such as hyperglycemia, glucosuria, polyuria, polydipsia, severe weight loss, it is difficult to measure the contribution of each of the links to the pathogenesis of diabetes and to assess the extent of pancreatic islet β-cell damage and death. The toxic effect of alloxan and streptozotocin on cells in pancreatic islets manifests itself not only by necrosis but also by apoptosis of pancreatic islet β-cells (Daisy Mythili M., et al. 2004). The study of apoptotic mechanisms will enable us to identify specific targets for purposeful creation and development of anti-diabetic medications.

2. Materials and methods

The experiments were performed on 180 adult male Wistar rats weighing from 280.0 to 300.0 g (Table 1). The animals were kept under standard vivarium conditions at pharmacology department of Volgograd Medical University, and were provided with a nutritionally balanced diet that the laboratory animals consume consistently. A pilot study was approved by the Central Regional Independent Ethics Committee (protocol № 1-06; № 43-2006; № 70-2008; № 89-2009), and was performed in accordance with GLP when conducting preclinical studies in Russia. All animal experimentation was carried out in compliance with the International Guidelines of the European Convention for the Protection of Vertebrate Animals used in experimental studies (1997).

Group	Day of experiment	Amount of animals
Intact control	3, 7, 14, 28	40
Alloxan-induced diabetes	3, 7, 14, 28	40
Streptozotocin-induced diabetes	7, 28	20
Immune-dependent diabetes	3, 7, 14, 28	40
Streptozotocin-nicotinamide-induced diabetes	3, 7, 14, 28	40

Table 1. Group distribution of experimental animals

Plasma glucose, insulin and C-peptide concentrations were measured in experimental laboratory animals. The blood glucose was determined in samples obtained from the tail vein of rats by enzymatic method, using "Glucose FKD" assay kits (Russia) on SF-46 spectrophotometer at $\lambda=450$ nm in 10 mm cuvettes. Animals with blood glucose levels >15 mmol/L were enrolled in the experiment (Akbarzadeh A. et al. 2007). Plasma insulin and C-peptide concentrations were determined on an automated enzyme immunoassay «SUNRISE» analyzer (TECAN, Austria), using DRG Insulin ElisaKit and DRG C-peptide ElisaKit.

Alloxan-induced diabetes simulation model was developed by means of intraperitoneal administration of alloxan at a dose of 120 mg/kg. Tissue samples of animals were collected and submitted for routine histopathological investigation at 3, 7, 14 and 28 days of the experiment.

Experimental streptozotocin-induced diabetes simulation model was developed using streptozotocin (Sigma) (45 mg/kg, intravenously once a day) (Baranov V.G. 1983). Tissue samples were collected and submitted for routine histopathological investigation at 7 and 28 days of the experiment.

Experimental immune-dependent diabetes simulation model was developed by giving a subcutaneous injection of 0.2 ml of complete Freund's adjuvant (CFA) (Grand Island Biological Company, USA). Subsequently, daily intravenous injections of streptozotocin (Sigma, USA) (20 mg/kg) were given to the animals for 5 days, resulting in the development of insulin-dependent diabetes (Ziegler B. 1990). Tissue samples were collected and submitted for routine histopathological investigation at 3, 7, 14 and 28 days of the experiment.

Streptozotocin-nicotinamide-induced diabetes model was developed by giving an injection of streptozotocin (Sigma, USA) (intraperitoneally - 65 mg/kg citrate buffer, pH = 4.5) with a preliminary (15 minutes prior to the procedure) administration of nicotinamide (intraperitoneally - 230 mg/kg prepared in 0.9 % solution of sodium chloride) (Islam S., et al. 2009). Tissue samples were collected and submitted for routine histopathological investigation at 3, 7, 14 and 28 days of the experiment.

Pancreatic tissue was divided into three segments including intestinal, gastric and splenic parts, and then they were fixed in 10% solution of neutral buffered formalin (pH 7.4) for 24 hours. 5-6-mm thick slices were obtained on rotary microtomes and were mounted on slides. Tissue sections were stained with hematoxylin and eosin using standard histological stain techniques (Korzhevsky D.E. 2005).

For detection of α-and β- endocrine cells in the islets of Langerhans the primary antibodies against insulin and glucagon were used (Table 2). To study apoptosis, the primary antibodies to proteins, such as caspase 3, TRAIL (TNF-related apoptosis-inducing ligand), MDM2, Bcl 2, p53, Bax, NF-kB, as well as eNOS were used.

№	Antibody	Clone	Manufacturer
1	Insulin	Polyclonal	DakoCytomation, Denmark
		Ab-6 (INS04 + INS05)	LabVision, UK
		2D11-H5	Novocastra, UK
2	Glucagon	Polyclonal	Novocastra, UK
3	Caspase 3	JHM62	Novocastra, UK
		Rb-1197-P0	NeoMarkers, USA
4	TRAIL	27B12	Novocastra, UK
5	MDM2	1B10	Novocastra, UK
6	Bcl 2	sc-7382	Santa Cruz Biotechnology, USA
		124	Dako Cytomation, Denmark
7	p53	Polyclonal	Dako Cytomation, Denmark
		Ab-1 (Pab 240) MS-104-P0	NeoMarkers, USA
8	Bax	Polyclonal	BD Biosciences Pharmingen, USA
9	NOS-3	RN5	Novocastra, UK
10	NF-kB	Polyclonal	Diagnostic BioSystems, USA

Table 2. IHC Primary Antibodies

Immunohistochemistry was performed according to manufacturers' protocols using ABC (Novocastra, UK), «UltraVision» (Lab Vision, UK) and «EnVision» (Dako Cytomation, Denmark) antibody detection systems and a chromogen, diaminobenzidine under the protocol on high temperature antigen unmasking technique using *«Pascal»* mini autoclave (Dako Cytomation, Denmark) (Kumar G.L. et al. 2009). The reliability of the obtained results was defined using both positive and negative control antigens, as well as negative control antibodies.

Immunohistochemical reaction was based on visual evaluation, taking into account the intensity of color, or on determination of the specific amount of positively stained cells (Allred DC, et al. 1998).

Photo images were captured with an «AxioScope» microscope (Carl Zeiss, Germany) and a «PowerShot» digital camera (Canon, Japan). Morphometric analysis was performed using "VideoTestMorfo-4" software (Russia). In the course of the study we determined the relationship between the total α-and β- endocrine cell area and the total area of the islet (S,%), between the volume fraction (ML, %) of islets and exocrine glands, as well as the area of β-cell nuclei (S, μm^2). We also measured apoptotic index (AI), i.e. the relative amount of β-endocrine cells with apoptotic structural and immunohistochemical changes.

The research results were processed using basic statistical analysis techniques as well as "Video TestMorfo-4», Excel Microsoft Office (Microsoft, USA) and STATISTICA 6.0 software (StatSoft Inc., USA). The analysis of the parameters for normally-distributed values was performed using Student's t test. Nonparametric statistics was calculated using Mann-Whitney test. To compare qualitative variables, the chi-square test and the Fisher exact test were used. Differences were considered significant if error probability was p <0.05.

3. Results

All animals in the intact control group showed stable plasma biochemical parameters which did not exceed the physiological norms during an observation period (glucose - 4,02 ± 0,1 mmol/L (Fig. 1), insulin - 1,65 ± 0,04 mU/ml., C-peptide - 4,65 ± 0,16 ng/ml.

Morphological study of the intact control group showed that the exocrine part of the pancreas has an alveolar-tubular structure which exhibited a division into lobules separated by interlobular connective tissue. Exocrinotcytes tended to be highly differentiated and they formed acini. Pancreatic islets in all parts of the pancreas were round or slightly oval in shape. They were single or were arranged in clusters around the intralobular excretory ducts. The total volume fraction of islet tissue in the gastric and splenic segments was almost twice as much as the volume fraction of islets of the intestinal segment (Table 3). The total area of nuclear β-endocrine cells in all segments of the pancreas had no statistically significant differences during all periods of observation. Insulin-positive cells were predominantly found in the central part of the pancreatic islets; however, glucagon-positive cells were more frequently encountered at the periphery. In this case there was a statistically significant increase (p <0.05) in β-cell area of the splenic segment compared with the intestinal and gastric segments of the pancreas (Fig. 2A). The maximum α-endocrine cell area was determined in the intestinal and gastric segments of the pancreas; however, this area was statistically significant lower (p <0.05) in the splenic segment (Table 3). There was a statistically significant (p ≤ 0.05) direct relationship between the total volume fraction of pancreatic islets and the area of β-endocrine cells depending on their location. The amount of islets and β-endocrine cells tended to increase in the following segments of the pancreas: intestinal →gastric → splenic.

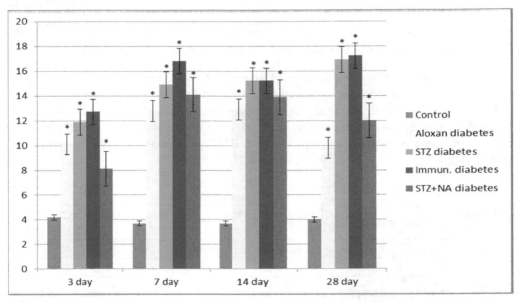

Figure 1. Blood glucose dynamics (mmol/L) in rats with different models of experimental diabetes. * - Significant changes compared with the intact control group.

Indicator	Day of experiment	Segment of the pancreas		
		Intestinal	Gastric	Splenic
Volume fraction of islets,%	3	6,8±2,5	12,6±6,0	13,8±7,3
	7	6,7±2,2	12,0±4,5	12,9±2,1
	14	6,9±3,1	12,1±5,2	14,0±2,9
	28	6,6±4,2	12,8±4,5	14,3±5,2
Total area of β-cells,%	3	51,3±9,2	64,6±9,1	78,2±6,1**
	7	53,8±7,6	63,4±7,0	75,4±8,7**
	14	52,1±5,5	64,5±8,7	79,4±4,6**
	28	54,3±5,4	66,2±3,6	77,3±4,3**
Total area of α-cells,%	3	37,4±5,2	31,2±3,5	22,2±3,4**
	7	39,3±4,3	30,4±4,0	21,3±3,0**
	14	38,5±3,5	32,4±3,2	19,7±2,3**
	28	37,1±1,0	30,2±1,2	17,9±1,3**

Indicator	Day of experiment	Segment of the pancreas		
		Intestinal	Gastric	Splenic
Total area of β-cell nuclei, μm²	3	24,2±1,2	24,4±1,0	23,9±0,2
	7	24,3±1,3	23,8±1,2	24,7±1,0
	14	24,3±1,2	24,8±1,3	24,6±1,3
	28	24,4±1,0	24,3±1,2	24,7±2,0

* - Significant difference in the intestinal segment of the pancreas (p <0.05),
** - Significant difference in the gastric segment of the pancreas (p <0.05).

Table 3. Morphometric indicators of pancreatic islets in the intact control group (M ± m).

Bax protein expression was negative in all islet cells. A weak positive cytoplasmic staining of islet cells in the central part of the islet for Bcl-2 protein was reported. In the nuclei of individual β-endocrine cells ambiguous or weak expression of p53 protein was determined. Most of β-islet cells had a positive nuclear staining for MDM2 protein. In some β-endocrine cells cytoplasmic expression of caspase 3 and TRAIL was poorly defined or ambiguous (Table 4) (Fig.3A, 4A). Apoptotic index was low (Table 5).

Group	Day of experiment	p53	Caspase-3	TRAIL	Bax	Bcl-2	MDM2	NOS-3	NF-kB
				Primary antibodies					
				Intensity of expression					
Intact control	3	-	+	+	-	+	++	-	++
	7	-	+	+	-	+	++	-	++
	14	-	+	+	-	+	++	-	++
	28	-	+	+	-	+	++	-	++
				Amount of cells					
Intact control	3	-	+	+	-	+	+++	-	++
	7	-	+	+	-	+	+++	-	++
	14	-	+	+	-	+	+++	-	++
	28	-	+	+	-	+	+++	-	++

Table 4. Immunohistochemical characteristics of β-cells

Alloxan-induced diabetes

In alloxan-induced diabetes simulation model a significant increase in blood glucose of rats compared with the intact control was reported (Fig. 1).

At day 3 of the experiment a well pronounced interstitial edema associated with hyperemia of blood vessels and capillaries in the acinar tissue of pancreatic islets was determined histologically. Compared with the intact control group, the total volume fraction of the pancreatic islets was not significantly reduced. However, there was a slight decrease in the area of β-cells in all segments of the pancreas (Table 5). The cells with moderate dystrophic and destructive changes were revealed in the central part of the islets of Langerhans. Immunohistochemical reaction in the central part of the pancreatic islets revealed insulin-positive cells with marked expression of insulin. A slight increase in the size of β-cell nuclei was reported. Compared with the intact control group, the total area of islet cells in all segments of the pancreas was slightly reduced. The total glucagon-positive cell area did not change as compared with the control group (Table 5).

At day 7 the islets were swollen, hyperemic and collapsed. There was no statistically significant reduction in the total volume fraction of islets in all segments of the pancreas. In the central part of the islets we observed moderate necrobiotic changes associated with a significant decrease (p <0.05) in the total β-cell area predominantly in the gastric and splenic segments when compared with the intact control group (Table 5). There was a statistically significant increase (p <0.05) in the area of α-endocrine cells in the gastric and splenic segments of the pancreas. β-cell necrosis was accompanied by moderate leukocyte infiltration of the stroma and acinar cells with only few leukocytes in the islets. The area of nuclear β-islet cells in all segments of the pancreas increased by 4.5 μm²; however, no statistically significant changes were identified when compared with the intact control group.

Edema significantly reduced by day 14; however, blood vessels of the acinar tissue and capillaries of the islets of Langerhans were still slightly hyperemic. There were focal sclerotic changes in some islets. Compared with the control group, there were not any statistically significant changes in the islet volume fraction in all segments of the pancreas. Insulin-positive cells tended to be located in a random pattern; however, glucagon-positive cells were predominantly found at the periphery. The total β-endocrine cell area in the gastric and splenic segments, as compared with the intact control group, was significantly decreased (p <0.05), while the total α-cell area was increased (p <0.05) (Table 5) (Fig.2B). There was moderate hypertrophy of β-cell nuclei in all segments of the pancreas.

At 28 day edema, hyperemia of blood vessels and inflammatory infiltration were replaced by focal sclerotic changes of the acinar tissue and pancreatic islets. As before, a slight decrease in the islet volume fraction in all segments of the pancreas was reported. Also, in all segments of the pancreas the total area of β-cells was slightly increased, compared with day 14 of the experiment, but was significantly lower (p <0.05) as compared with the animals in the intact control group. The volume fraction of α-endocrine cells slightly decreased but was statistically greater (p <0.05) than in the intact control group (Table 5). The average area of nuclei was slightly decreased as compared with the intact control group and with day 14 of the experiment.

Indicator	Day of experiment		Segment of the pancreas		
			Intestinal	Gastric	Splenic
Volume fraction of islets,%	3	intact	6,8±2,5	12,6±6,0	13,8±7,3
		diabetes	6,7±2,1	11,8±5,4	13,9±10,2
	7	intact	6,7±2,2	12,0±4,5	12,9±2,1
		diabetes	4,5±3,1	11,2±3,2	11,1±5,2

Indicator	Day of experiment		Segment of the pancreas		
			Intestinal	Gastric	Splenic
	14	intact	6,9±3,1	12,1±5,2	14,0±2,9
		diabetes	4,2±1,4	8,4±1,6	9,3±2,1
	28	intact	6,8±2,1	11,9±3,7	13,7±3,2
		diabetes	3,1±1,1	7,6±2,3	9,5±2,4
Total area of β-cells, %	3	intact	51,3±9,2	64,6±9,1	78,2±6,1
		diabetes	50,4±8,5	65,9±8,2	69,8±4,9
	7	intact	53,8±7,6	63,4±7,0	75,4±7,7
		diabetes	42,2±2,1	45,5±1,4*	58,1±1,4*
	14	intact	52,1±5,5	64,5±8,7	79,4±4,6
		diabetes	40,4±2,4	39,7±2,2*	57,4±2,0*
	28	intact	51,1±7,5	65,2±6,0	77,4±4,4
		diabetes	32,3±1,3*	40,2±1,4*	49,7±2,1*
Total area fraction of α-cells,%	3	intact	37,4±5,2	31,2±3,5	22,2±3,4
		diabetes	36,5±4,8	30,8±2,5	25,7±2,4
	7	intact	39,3±4,3	30,4±4,0	21,3±3,0
		diabetes	41,1±4,1	49,3±2,1*	35,4±2,2*
	14	intact	38,5±3,5	32,4±3,2	19,7±2,3
		diabetes	32,2±1,5	47,4±2,0*	32,2±2,1*
	28	intact	30,0±4,4	31,4±3,9	18,9±3,2
		diabetes	30,1±2,1*	42,3±3,1*	30,1±4,1*
Total area of β-cell nuclei, μm²	3	intact	24,2±1,2	24,4±1,0	23,9±0,2
		diabetes	27,8±5,9	27,5±5,5	26,6±4,6
	7	intact	24,3±1,3	23,8±1,2	24,7±1,0
		diabetes	29,5±4,3	28,4±3,2	29,6±4,3
	14	intact	24,3±1,2	24,8±1,3	24,6±1,3
		diabetes	30,3±3,2	29,6±2,3	30,4±3,4
	28	intact	23,9±1,2	24,2±1,2	24,6±1,2
		diabetes	28,5±5,1	27,4±3,4	28,5±4,0

* - Significant difference compared with the intact control group (p <0.05).

Table 5. Morphometric indicators of pancreatic islets in rats with alloxan-induced diabetes (M ± m)

Immunohistochemical study which, was performed at day 3 and day 28, showed negative staining of β-endocrine cells for p53, TRAIL, endothelial NO-synthase and Bcl-2 proteins (Fig. 3B) in all segments of the pancreas. At day 3 we observed mild cytoplasmic staining for anti-caspase-3 and anti-Bax-antibody in some endocrine cells. Compared with the intact control group, apoptotic index was significantly increased (p <0.05) during all observation periods. The expression of NF-kB and MDM2 proteins tended to decrease. At 7-14 day moderate cytoplasmic staining for caspase-3 and Bax proteins accompanied by decreased expression [59]. However, the antioxidants, NAC, L-ascorbic acid, alpha-tocopherol, catalase and superoxide dismutase, all effectively prevented curcumin-induced apoptosis, suggesting that curcumin-mediated apoptosis was closely related to the increase in intracellular ROS [59]. Besides hematopoietic cancer cells, curcumin-mediated apoptosis in human breast epithelial cells (H-ras MCF10A) involved generation of ROS as well as down-regulation of Bcl-2 and up-regulation of Bax, suggesting redox signaling as a mechanism responsible for curcumin-induced apoptosis in these cells [60]. Syng-Ai et al. also demonstrated that curcumin-induced apoptosis in human breast tumor cell lines (MCF-7, MDAMB) and HepG2 cells is also mediated through the generation of ROS, and that depletion of glutathione by buthionine sulfoximine (BSO) promoted the increased generation of ROS, thereby further sensitizing the cells to curcumin [61]. Interestingly, curcumin had no cytocidal effect on normal rat hepatocytes, because of no superoxide generation [61]. These observations suggest that curcumin with broad biological actions could be developed into an effective chemopreventive and chemotherapeutic agent based on its ability to modulate intracellular redox status. However, the use of curcumin as a therapeutic agent has met with considerable skepticism, since as much as 75% of curcumin is excreted in the feces [62] and also undergoes repaid inactivation by glucuronidation [63], similar to metabolisms of flavonoids [2]. Recently, in order to increase its metabolic stability, numerous approaches have been undertaken, such as generating the fluoro-analog of curcumin termed Diflourinated-Curcumin (referred to as CDF) that exhibits increased metabolic stability [64, 65]. Furthermore, the CDF has been found to exhibit superior growth inhibitory properties against cancer cells to the parental compound curcumin [56, 64, 65].

3. Potential future application of polyphenolic compounds, alone or in combination with, anticancer drugs

As mentioned in the previous section, the deregulation and sustained activation of multiple tumorigenic pathways are typically implicated in cancer development and progression with locally advanced and aggressive nature. Consequently, the use of therapeutic agents acting on different deregulated gene products, alone or in combination therapy, may represent a potentially better strategy than the targeting on one specific oncogenic product to overcome treatment resistance and prevent cancer development and disease recurrence [2-4, 7]. So far, one of the most successful models for combinatory cancer therapies is all-*trans* retinoic acid (ATRA)/arsenic trioxide (ATO, arsenite) combination as a synergistic therapy for acute promyelocytic leukemia (APL) patients, in which ATRA synergizes ATO activity to provide a superior efficacy of combination therapy in patient through promoting the effects of ATO on several signaling pathway, such as apoptosis induction, differentiation as well as the degradation of PML-RARα [a fusion gene between promyelocytic leukemia (PML) gene and retinoic acid receptor (RAR) α], a causative gene for APL [4, 13]. In order to understand the mode of action of ATO and provide an effective treatment protocol for individual APL patients, we recently conducted studies on the pharmacokinetics of ATO in APL patients using biological samples such as peripheral blood (PB) and cerebrospinal fluid, and demonstrated that not only inorganic arsenic but also methylated arsenic metabolites amount of cells with NF-kB-and MDM2-positive stained nuclei was reported. At day 7 there was a significant ($p < 0.05$) increase in the amount of apoptotic β-cells, compared with day 3 of the experiment, which was followed by a significant ($p < 0.05$) decrease by day 14 (Table 12). At day 28 the amount of cells and the intensity of expression of apoptotic markers significantly decreased and anti-apoptotic proteins were expressed in large quantities in the endocrine cells of pancreatic islets. Apoptotic index was significantly decreased as compared with day 7; however, it was still statistically more significant than in the intact control group (Table 6).

Group	Day of experiment	Primary antibodies							
		p53	Caspase-3	TRAIL	Bax	Bcl-2	MDM2	NOS-3	NF-kB
Intensity of expression									
Intact control	3	-	+	+	-	+	++	-	++
	7	-	+	+	-	+	++	-	++
	14	-	+	+	-	+	++	-	++
	28	-	+	+	-	+	++	-	++
Diabetes	3	-	+	-	+	-	++	-	++
	7	-	++	-	++	-	+	-	+
	14	-	++	-	++	-	+	-	+
	28	-	+	-	+	-	++	-	++
Amount of cells									
Intact control	3	-	+	+	-	+	+++	-	++
	7	-	+	+	-	+	+++	-	++
	14	-	+	+	-	+	+++	-	++
	28	-	+	+	-	+	+++	-	++
Diabetes	3	-	+	-	+	-	+	-	+
	7	-	++	-	++	-	++	-	+
	14	-	++	-	++	-	++	-	+
	28	-	+	-	+	-	++	-	+

Table 6. Immunohistochemical characteristics of β-endocrine cells in alloxan-induced diabetes

Within the period from day 7 to day 14 we observed cytoplasmic staining for caspase-3 and Bax proteins accompanied by decreased amount of cells with NF-kB-and MDM2-positive stained nuclei (Fig. 4B). At day 7 there was a significant ($p < 0.05$) increase in the amount of apoptotic β-cells compared with day 3 of the experiment, which was followed by a significant ($p < 0.05$) decrease by day 14 (Table 12).

Thus, in alloxan-induced diabetes we observe a decrease in the total β-cell area from day 7 to day 28 of observation which is accompanied by a simultaneous increase in the total α-cell area. These changes occur in all segments of the pancreas; however, they are predominantly obvious and are statistically significant in the gastric and splenic segments. Along with pronounced necrotic changes in the cells of the insular apparatus of the pancreas, there is also an increase in apoptogenic activity of β-endocrine cells due to the damage to mitochondrial membranes (increased expression of Bax and inhibition of Bcl-2 proteins), which might be caused by impaired intracellular calcium homeostasis and activation of the mitochondrial apoptotic pathway with subsequent activation of caspase cascade without the involvement of p53 protein (Table 12).

Streptozotocin-induced diabetes

In streptozotocin-induced diabetes a statistically significant elevation of plasma glucose levels, as compared with those in the intact control animals, was reported (Fig. 1).

Histological investigation which was performed at day 7 showed pronounced hyperemia of blood vessels associated with lymphocytic infiltration and marked edema of the interlobular connective tissue. There were not any statistically significant differences in the total volume fraction of islets either with regard to the intact control group, or to different segments of the pancreas. We revealed marked necrotic changes in endocrine cells both in the central part and at the periphery of the islets (Fig. 2C). We also stated a statistically significant decrease (p <0.05) in the area of β-cells in all segments of the pancreas. However, a significant increase (p <0.05) in the volume fraction of α-cells exclusively in the gastric and splenic segments of the pancreas was reported. There was marked hypertrophy of β-islet cell nuclei (p <0.05) predominantly in the splenic segment of the pancreas (Table 7).

By day 28 moderate hyperemia of blood capillaries of pancreatic islets and accumulation of lymphocytes at some sites were observed. Focal necrotic changes were reported in some islets, whereas other islets underwent sclerotic changes. No significant changes in the islet volume fraction were observed. The total β-islet cell area was still significantly lower (p <0.05) compared with the intact control group during the same period of observation. The total volume fraction of α-endocrine cells was significantly increased (p <0.05) exclusively in the splenic segment of the pancreas. The nuclei of β-cells were hypertrophic and hypertrophy was statistically significant (p <0.05) in the cells of the splenic segment of the pancreas (Table 7).

Indicator	Day of experiment		Segment of the pancreas		
			Intestinal	Gastric	Splenic
Volume fraction of islets,%	7	intact	6,7±2,2	12,0±4,5	12,9±2,1
		diabetes	4,7±3,1	10,2±6,7	11,8±1,5
	28	intact	6,8±2,1	11,9±3,7	13,7±3,2
		diabetes	3,2±2,1	8,2±3,1	9,7±2,3
Total area of β-cells,%	7	intact	53,8±7,6	63,4±7,0	75,4±8,7
		diabetes	24,3±2,1*	39,2±3,2*	40,1±2,0*
	28	intact	51,1±7,5	65,2±6,0	77,4±4,4
		diabetes	29,8±5,5*	43,2±5,4*	47,7±8,2*
Total area of α-cells,%	7	intact	39,3±4,3	30,4±4,0	21,3±3,0
		diabetes	44,2±4,5	42,3±2,2*	29,8±2,1*

Indicator	Day of experiment		Segment of the pancreas		
			Intestinal	Gastric	Splenic
Total area of β-cell nuclei, μm²	28	intact	40,0±4,4	31,4±3,9	18,9±3,2
		diabetes	40,9±4,8	42,7±4,8	**30,3±4,7***
	7	intact	24,3±1,3	23,8±1,2	24,7±1,0
		diabetes	23,8±3,1	25,7±2,5	**29,5±1,8***
	28	intact	23,9±1,2	24,2±1,2	24,6±1,2
		diabetes	25,3±1,2	25,9±1,0	**29,3±1,1***

* - Significant difference compared with the intact control group (p <0.05).

Table 7. Morphometric indicators of pancreatic islets in rats with streptozotocin-induced diabetes (M ± m)

Immunohistochemical reaction, which was performed at day 7, revealed that most endocrine cells exhibited marked expression of proapoptotic proteins, such as caspase-3, TRAIL and Bax (Fig. 3C, 4C). Activation of apoptosis was accompanied by inhibited expression of anti-apoptotic Bcl-2, MDM2, nuclear factor NF-kB proteins (Table 8). There was a statistically significant increase in the amount of cells with morphological features of apoptosis when compared with the intact control group (Table 8).

By day 28 there was a decrease in the amount of β-cells as well as in the intensity of expression of caspase 3 in the surviving cells (Table 12). We did not observe any significant changes in expression of either mitochondrial proapoptotic Bax protein or membrane TRAIL (Table 8). Compared with the intact control group, there was a statistically significant increase in apoptotic index. With regard to day 7 of observation, a statistically significant decrease in apoptotic index was stated (Table 8). Moderate expression of endothelial NO-synthase in the capillaries of pancreatic islets was found during all observation periods (Table 8).

Characteristic	Day of experiment	Primary antibodies							
		p53	Каспаза-3	TRAIL	Bax	Bcl-2	MDM2	NOS-3	NF-kB
Intensity of expression									
Intact control	7	-	+	+	-	+	++	-	++
	28	-	+	+	-	+	++	-	++
Diabetes	7	-	+++	+	++	-	+	+	+
	28	-	++	+	++	-	+	+	+
Amount of cells									
Intact control	7	-	+	+	-	+	+++	-	++
	28	-	+	+	-	+	+++	-	++
Diabetes	7	-	++	+	+	-	+	+	+
	28	-	+	+	+	-	++	+	++

Table 8. Immunohistochemical characteristics of β-endocrine cells in streptozotocin-induced diabetes

Thus, the development of streptozotocin-induced diabetes is associated not only with hyperglycemia, but also with marked necrotic changes in endocrine cells of pancreatic islets, decreased β-islet cell area accompanied by hyperplasia of their nuclei, increased α-endocrine cell area, as well as increased apoptotic index (Table 12). High expression levels of endothelial NO-synthase in the capillaries of pancreatic islets are indicative of endothelial dysfunction.

Immune-dependent diabetes

Persistent hyperglycemia and a reduction in plasma insulin by 67.8% were reported in immune-dependent diabetes (Fig. 1), as compared with the intact control group.

Microscopic examination, which was performed at day 7, found that the total volume fraction of the islets decreased when compared with the intact control group. The islets had an irregular shape due to marked necrotic changes in the endocrine cells, hyperemia of the capillaries, as well as mild lymphocytic infiltration. There were single insulin-positive cells or they were arranged in small clusters in the central part of the islets around the capillaries. We observed a significant decrease ($p < 0.05$) in the total β-cell area in all segments of the pancreas as compared with the intact controls (Table 9). The total α-endocrine cell area was significantly increased exclusively in the splenic segment of the pancreas.

At day 14 some symptoms of insulitis, i.e. inflammation of insular cells, persisted. Compared with the intact control group, the area covered by β-endocrine cells in all segments of the pancreas tended to become smaller ($p < 0.05$) due to marked necrotic changes in β-endocrine cells (Fig. 2D). However, we observed moderate hypertrophy of the nuclei of β-islet cells as well as an increase in the area covered by α-endocrine cells in the splenic segment of the pancreas (Table 9). Some pancreatic islets were reported to undergo sclerotic changes.

By day 28, as before, a statistically significant reduction ($p < 0.05$) in the area covered by β-islet cells in all segments of the pancreas, as compared with the intact control group, was observed. Inflammatory infiltration, swelling and hyperemia of blood vessels tended to reduce, too. However, the amount of islets which developed sclerotic changes increased and β-endocrine cells had moderately hypertrophic nuclei. There was an increase in the area covered by α-endocrine cells in the gastric and splenic segments of the pancreas (Table 9).

Indicator	Day of experiment		Section of the pancreas		
			Intestinal	Gastric	Splenic
Volume fraction of islets,%	7	intact	6,7±2,2	12,0±4,5	12,9±2,1
		diabetes	4,7±3,1	10,2±6,7	11,8±1,5
	14	intact	6,9±3,1	12,1±5,2	14,0±2,9
		diabetes	4,2±1,2	8,9±2,0	9,2±2,2
	28	intact	6,8±2,1	11,9±3,7	13,7±3,2
		diabetes	4,4±1,0	7,4±1,3	8,7±3,4

Indicator	Day of experiment	Section of the pancreas		
		Intestinal	Gastric	Splenic
Total area of β-cells,%	7 intact	53,8±7,6	63,4±7,0	75,4±8,7
	diabetes	29,8±5,5*	43,2±5,4*	47,7±8,2*
	14 intact	52,1±5,5	64,5±8,7	79,4±4,6
	diabetes	25,4±3,2*	40,1±3,2*	42,3±1,2*
	28 intact	51,1±7,5	65,2±6,0	77,4±4,4
	diabetes	30,2±4,5*	44,3±2,3*	40,1±3,2*
Total area of α-cells,%	7 intact	39,3±4,3	30,4±4,0	21,3±3,0
	diabetes	45,2±1,2	40,4±5,1	56,7±4,6*
	14 intact	38,5±3,5	32,4±3,2	19,7±2,3
	diabetes	44,1±2,1	42,3±4,2	56,4±3,4*
	28 intact	40,0±4,4	31,4±3,9	18,9±3,2
	diabetes	33,9±3,2	49,5±4,4*	50,9±3,2*
Total area of β-cell nuclei, μm^2	7 intact	24,3±1,3	23,8±1,2	24,7±1,0
	diabetes	25,8±1,1	25,2±2,3	27,5±2,2
	14 intact	24,3±1,2	24,8±1,3	24,6±1,3
	diabetes	25,6±2,3	24,6±2,3	26,4±2,1
	28 intact	23,9±1,2	24,2±1,2	24,6±1,2
	diabetes	24,9±2,0	25,1±1,2	25,8±2,2

* - Significant difference compared with the intact control group (p <0.05).

Table 9. Morphometric indicators of pancreatic islets in rats with insulin-dependent diabetes (M ± m)

Immunohistochemical study during all observation periods showed absence or doubtful expression of p53, Bax, MDM2 as well as Bcl-2 proteins in certain β-islet cells. Within the period from day 7 to day 14 we observed increased expression of TRAIL proapoptotic protein and caspase 3 protein (compared with the intact control group and at day 14 compared with day 7 of the experiment) (Table 10, 12, Fig. 3D, 4D) in most β-cells accompanied by a slight reduction by day 28. Simultaneously, apoptotic index during all observation periods was significantly increased compared with the intact control group (Table 5). The expression of NO-synthase in the endothelium of the pancreatic islet capillaries during all observation periods enhanced.

Group	Day of experiment, days	Primary antibodies							
		p53	Caspase-3	TRAIL	Bax	Bcl-2	MDM2	NOS-3	NF-kB
		Intensity of expression							
Intact control	7	-	+	+	-	+	++	-	++
	14	-	+	+	-	+	++	-	++
	28	-	+	+	-	+	++	-	++

Group	Day of experiment, days	Primary antibodies							
		p53	Caspase-3	TRAIL	Bax	Bcl-2	MDM2	NOS-3	NF-kB
Diabetes	7	-	+++	+++	++	-	+	++	+
	14	-	+++	++	++	-	+	++	+
	28	-	++	+	+	-	++	++	+
Amount of cells									
Intact control	7	-	+	+	-	+	+++	-	++
	14	-	+	+	-	+	+++	-	++
	28	-	+	+	-	+	+++	-	++
Diabetes	7	-	++	+	++	-	++	+	+
	14	-	++	+	++	-	++	+	+
	28	-	++	+	+	-	++	+	+

Table 10. Expression of pro-and β-anti-apoptotic proteins in endocrine cells of insulin-dependent diabetic rats

Thus, it was found that immuno-dependent diabetes is characterized by the following changes: hyperglycemia, reduced area of β-islet cells, inflammatory infiltration of islets, marked β-necrotic changes in endocrine cells, focal sclerosis of the islets of Langerhans.

Streptozotocin-nicotinamide-induced diabetes

The results of the 28-day streptozotocin-nicotinamide-induced diabetes study indicated that plasma C-peptide concentration in the control animals with diabetes increased by 36.8% (p<0,05), as compared with intact rats. Blood glucose levels in rats with strep-tozototsin-nicotinamide-induced diabetes during the entire experiment was significantly superior to glycemia in intact animals (Fig. 1).

Histologically, the maximum reduction in the volume fraction of islets was found at day 3 with a tendency to increase by day 28 in all segments of the pancreas.

At day 3 the pancreatic tissue appeared unevenly swollen and hyperemic. There were minor destructive changes of endocrine cells in pancreatic islets. The proportion of β-islet cells in all segments of the pancreas markedly decreased. At the same time no significant changes in the area of β-cells in islets were observed. The islets showed uneven accumulation of insulin-positive cells. Along with the cells which exhibited a marked accumulation of insulin, endocrine cells showing uneven distribution of immunopositive material in the cytoplasm with moderate or weak staining were singled out. Endocrine cells in the splenic segment of the pancreas were characterized by moderate hypertrophy of the nuclei and it was statistically significant, as compared to the intact control (Table 11).

By day 7 swelling hyperemia decreased and a microscopic mosaic image was observed. Both histologically intact and impaired islets of Langerhans were found. In a number of islets we could observe focal necrotic changes of endocrine cells, while others exhibited focal proliferation of connective tissue, and sometimes poorly defined intralobular and periductal sclerotic changes. Compared with the intact control group, there was a statistically significant reduction in the area of β-cells in the splenic segment. Immunopositive material was unevenly distributed both in the insular cells and within the cytoplasm of an insular cell. Hypertrophy of the nuclei of islet cells was reported in all segments of the pancreas.

At day 14 a mosaic pattern of cell damage was still noted. Both histologically intact islets and islets revealing sclerotic changes were defined. Their amount tended to decrease in all segments of the pancreas. Simultaneously, the total β-cell area decreased in the gastric and splenic segments of the pancreas. Hypertrophy of the nuclei of β-cells persisted in all segments of the pancreas (Fig. 2E).

By day 28 a statistically significant reduction in β-cell area in the gastric segment of the pancreas was reported, as compared with the control group. Hypertrophy of the nuclei of β-cells persisted (Table 11).

Indicator	Day of experiment		Segment of the pancreas		
			Intestinal	Gastric	Splenic
Total volume fraction of islets,%	3	intact	6,8±2,5	12,6±6,0	13,8±7,3
		diabetes	3,8±1,0	8,3±1,0	5,4±3,2
	7	intact	6,7±2,2	12,0±4,5	12,9±2,1
		diabetes	4,2±1,2	7,8±1,4	6,7±2,5
	14	intact	6,8±2,1	11,9±3,7	13,7±3,2
		diabetes	5,4±1,3	6,9±1,2	8,9±4,2
	28	intact	6,6±4,2	12,8±4,5	14,3±5,2
		diabetes	5,3±1,1	7,9±1,0	11,3±2,1
Total area of β-cells,%	3	intact	51,3±9,2	64,6±9,1	78,2±6,1
		diabetes	43,4±4,9	52,4±3,4	65,7±3,4
	7	intact	53,8±7,6	63,4±7,0	75,4±7,7
		diabetes	44,6±4,2	53,2±6,1	57,7±3,3*
	14	intact	51,1±7,5	65,2±6,0	77,4±4,4
		diabetes	37,4±5,6	47,6±4,3*	56,2±4,2*
	28	intact	54,3±5,4	66,2±3,6	77,3±4,3
		diabetes	41,2±2,3	50,2±3,1*	67,8±2,6
Total area of β-cell nuclei, μm²	3	intact	24,2±1,2	24,4±1,0	23,9±0,2
		diabetes	26,4±1,1	26,8±1,0	26,5±0,9*
	7	intact	24,3±1,3	23,8±1,2	24,7±1,0
		diabetes	29,8±1,0*	29,6±1,2*	30,1±0,6*
	14	intact	23,9±1,2	24,2±1,2	24,6±1,2
		diabetes	29,1±1,3*	29,6±1,0*	30,0±1,0*
	28	intact	24,4±1,0	24,3±1,2	24,7±2,0
		diabetes	28,7±1,2*	28,7±1,1*	29,7±1,2*

* - Significant difference compared with the intact control (p <0.05).

Table 11. Morphometric indicators of pancreatic islets in rats with experimental streptozotocin-nicotinamide-induced diabetes (M ± m)

Day experiment	Intact control	Alloxan-induced diabetes	Streptozotocin-induced diabetes	Immune-dependent diabetes	Streptozotocin-nicotinamide-induced diabetes
3	4,1±0,2	10,2±1,2*	3,9±0,1	-	5,1±1,8
7	3,9±0,1	17,4±0,9* ^	4,0±1,1	28,6±1,5*	5,4±2,1
14	4,2±1,0	14,9±1,0* ^	32,3±4,2*	35,2±2,9*#	4,8±1,9
28	4,0±1,1	12,5±0,6* #	21,3±2,2* #	30,3±2,7*	5,0±4,1

* - Significant difference compared with the intact control (p <0.05) ^ - significant difference compared with day 3 of experiment (p <0.05); # - significant difference compared with day 7 of experiment (p <0.05).

Table 12. Apoptotic index of β-cells (%)

Thus, hyperglycemia as well as reduced plasma C-peptide concentrations are common to experimental streptozotocin-nicotinamide-induced diabetes. Along with biochemical changes, it displays some characteristic morphological changes, including a statistically significant decrease in insular cell area associated with hypertrophy of β-cell nuclei, mosaic islet damage resulting in the destruction of endocrine cells as well as sclerotic changes of pancreatic islets during long-term observation period.

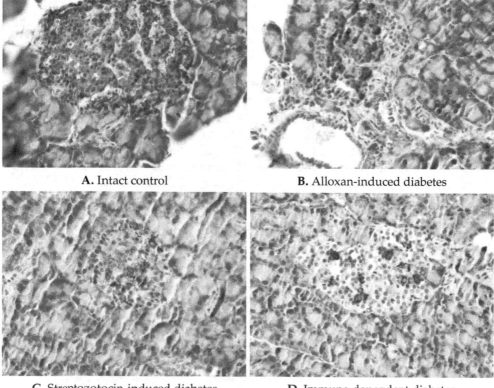

A. Intact control   B. Alloxan-induced diabetes

C. Streptozotocin-induced diabetes   D. Immune-dependent diabetes

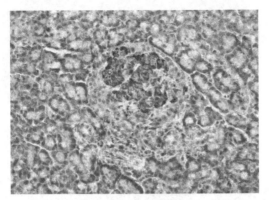

E. Streptozotocin-nicotinamide-induced diabetes

Figure 2. Reduction in the amount of cells under different experimental diabetes conditions, day 14 (Streptozotocin-induced diabetes, day 7), splenic segment of the pancreas. Primary antibody against insulin, DAB staining. Magnification: x400

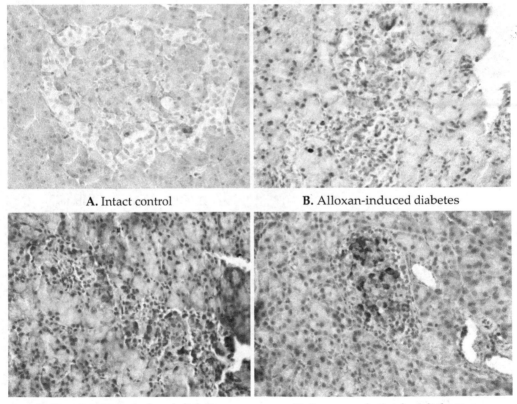

A. Intact control   B. Alloxan-induced diabetes

C. Streptozotocin-induced diabetes   D. Immune-dependent diabetes

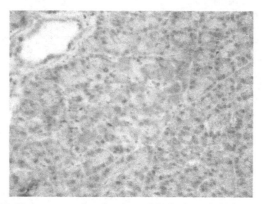

E. Streptozotocin-nicotinamide-induced diabetes

Figure 3. TRAIL expression in pancreatic islet cells under different experimental diabetes conditions, day 14 (Streptozotocin-induced diabetes, day 7), splenic segment of the pancreas. TRAIL primary antibody, DAB staining. Magnification: x400.

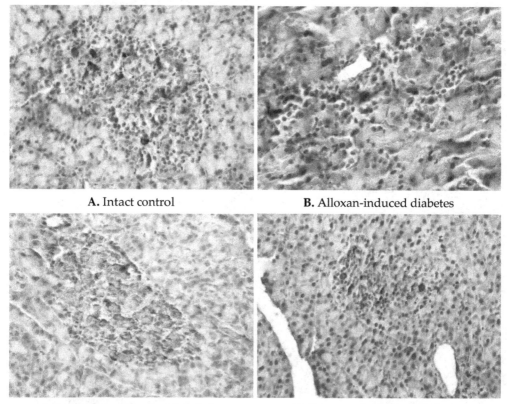

A. Intact control   **B.** Alloxan-induced diabetes

C. Streptozotocin-induced diabetes   **D.** Immune-dependent diabetes

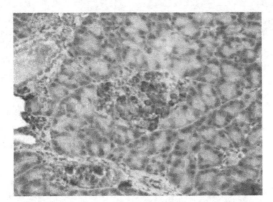

E. Streptozotocin-nicotinamide-induced diabetes

Figure 4. Apoptotic cells under different experimental diabetes conditions, day 14 (Streptozotocin-induced diabetes, day 7), splenic segment of the pancreas. Caspase 3 primary antibody, DAB staining. Magnification: x400.

4. Discussion

It is known that type I diabetes develops when pancreatic β-cells are damaged due to certain inflammatory, autoimmune and other pathological processes. Selective organ-specific tissue destruction of the insulin-producing pancreatic β-cells is associated with insulin deficiency resulting in impairment of glucose homeostasis. A large body of experimental evidence emphasizes the key role of apoptosis in the pathogenesis of diabetes mellitus (Kim B.M. et al. 2001; Severgina E.S. 2002; Butler A.E. et al. 2003; Bertalli E. et al. 2005; Rees D.A., et al. 2005; Srinivasan K., et al. 2007; Islam S., et al. 2009; Pisarev V.B. et al. 2009). Toxic effects of certain chemicals (e.g. alloxan, streptozotocin, etc.) used to induce diabetes specifically in pancreatic β-cells, manifest themselves by alkylation of DNA and formation of toxic compounds, such as superoxide anion, peroxynitrite, and nitric oxide. Damage to DNA and intracellular structures causes necrosis and activates apoptosis of pancreatic β-cells (Daisy Mythili M., et al. 2004).

Selective toxicity of streptozotocin can be explained by destruction of antiradical protective system and pancreatic β-cell DNA fragmentation. Numerous experiments show that the principal cause of streptozotocin-induced β-cell death is alkylation of DNA. Exposure of cells to streptozotocin results in the formation of toxic compounds, such as superoxide anion, peroxynitrite, and nitric oxide (NO). However, the contribution of NO to the cytotoxic activity remains controversial, because low concentrations of NO in the cells inhibit the inducible forms of NO-synthase, thus reducing DNA fragmentation (Szkudelski T. 2001; Lenzen S. 2008).

Receptor- or mitochondrial-mediated pathway may trigger programmed pancreatic β-cell death (Szkudelski T. 2001). Induction of β-cell apoptosis can occur through a TRAIL-mediated mechanism. DNA damage can activate p53 gene which regulates the expression of

DR5 and/or DR4 death receptors, thus enhancing TRAIL-induced apoptosis (Haupt S., et al. 2003; Kelley R.F., et al. 2005). Our studies provide evidence for activation of TRAIL-induced apoptosis in pancreatic β-cells by administration of streptozotocin on models of streptozotocin-induced and immune-dependent diabetes.

Mitochondrial-mediated apoptotic pathway plays an important role in the development of pancreatic islet damage in diabetes mellitus (Sakurai K., et al. 2001). In alloxan-induced diabetes intracellular calcium homeostasis is typically impaired. *In vitro* and *in vivo* experiments have shown that cytoplasmic calcium concentrations increase in the cells of pancreatic islets. This effect is due to depolarization of cell and mitochondrial membranes of β-endocrine cells which is associated with excessive entry of calcium from the extracellular fluid and intracellular calcium mobilization (Crow M.T., et al. 2004; Jung J.Y., et al. 2004). Thus, considerable gross damage to the intracellular structures by free radicals, oxidation of SH-proteins and impairment of intracellular calcium homeostasis typically result in necrosis and apoptosis, with necrotic processes being more common in alloxan-induced diabetes.

Comparison of morphological data and the dynamics of impairment of β-cell functional activity in experimental models (alloxan-, streptozotocin-, nicotinamide-streptozotocin-induced and immune-dependent diabetes) gives insight into the pathogenesis of experimental diabetes. A high apoptotic index and the prevalence of destructive processes, which enable us to assess not only the extent of pancreatic tissue damage but also to evaluate biochemical parameters, such as plasma glucose, insulin, C-peptide, etc., can be used as markers of severity of experimental diabetes (Snigur G.L., Smirnov A.V. 2010).

Thus, apoptosis is a complex, multi-step, poorly regulated process. Any regulatory impairment can lead to the development of pathological changes in the endocrine apparatus of pancreatic islets including the development of experimental diabetes (Reed J.C., Green D.R. 2011). Each level and element of this system is a potential drug target. The significance of apoptosis determines the need for the refocusing of biomedical research from theoretical biology to clinical medicine. Investigation of apoptotic mechanisms in various pathological conditions enable us to make more accurate diagnoses and prognoses as well as to adjust therapy. The possibility of deliberate control of apoptotic processes is considered to be the basis for developing new medications used to treat many socially important diseases, including the development of anti-diabetic drugs.

Experimental diabetic rats exhibited significant pathohistological changes in the pancreatic islets caused by cytotoxine (streptozotocin+CFA) at all levels of tissue structure. The development of diabetes mellitus was associated with decreased volume fraction of pancreatic islets and β-cells due to inflammation (insulitis) and pronounced destructive changes (e.g. necrosis and apoptosis).

Author details

A.V. Smirnov and G.L. Snigur

Department of Pathologic Anatomy, Volgograd State Medical University, Russia

M.P. Voronkova
Department of Pharmacology, Volgograd State Medical University, Russia

Acknowledgement

We express our gratitude to the Head of pharmacology department of Volgograd State Medical University, Academician of Russian Academy for Medical Sciences, Honored Scientist of Russia, MD, full professor, Spasov Alexander A., assistant professor of pharmacology department of Volgograd State Medical University, PhD, Cheplyaeva Natalia I., assistant professor of pharmacology department of Volgograd State Medical University, PhD, Chepurnova Mariya V. for their assistance in conducting the present study.

5. References

[1] Akbarzadeh A., et al. (2007) Induction of diabetes by streptozotocin in rats. Indian Journal of Clinical Biochemistry Vol. 22. №2. pp.60-64

[2] Allred D.C., Harvey J.M., Berardo M., Clark G.M. (1998) Prognostic and predictive factors in breast cancer by immunohistochemical analysis. Mod. Pathol.11. pp.155-68

[3] Baranov V.G. (1983) Experimental diabetes mellitus. Leningrad

[4] Bertalli E., Bendayan M. (2005) Association between endocrine pancreas and ductal system. More than an epiphenomenon of endocrine differentiation and development? J. Histochem. & Cytochem. Vol.53. N3. pp.1071-1086

[5] Butler A.E., Janson J., Bonner-Weir S., et al. (2003) β-cell deficit and increased β-cell apoptosis in humans with type 2 diabetes. Diabetes Vol. 52. N1.pp.102-110

[6] Crow M.T., Mani K., Nam Y.J., et. al. (2004) The mitochondrial death pathway and cardiac myocyte apoptosis. Circ Res. Vol. 95. N 10. pp.957-70

[7] Daisy M., Rashmi V., Akila G., Gunasekaran S. (2004) Effect of streptozotocin on the ultrastructure of rat pancreatic islets. Microsc. Res. Tech. Vol.63. N 5. pp.274 – 281

[8] Haupt S., Berger M., Goldberg Z., Haupt Y. (2003) Apoptosis - the p53 network. J. Cell Sci. Vol. 116. N 20. pp.4077-85

[9] Islam S., Loots D.T. (2009) Experimental rodent models of type 2 diabetes: a review. Methods Find Exp. Clin. Pharmacol. Vol. 4. N31. pp.249-261

[10] Jung J.Y., Kim W.J. (2004) Involvement of mitochondrial- and Fas-mediated dual mechanism in CoCl(2)-induced apoptosis of rat PC12 cells. Neurosci Lett. Vol. 371. N 2-3. pp.85-90

[11] Kelley R.F., Totpal K., Lindstrom S.H., et al. (2005) Receptor-selective mutants of Apo2L/TRAIL reveal a greater contribution of DR5 than DR4 to apoptosis signaling. J Biol Chem. Vol. 280. N 3. pp.2205-12

[12] Kim B.M., Ham Y.M., Shin Y.J., et al. (2001) Clusterin expression during regeneration of pancreatic islet β-cell in streptozotocin-induced diabetic rats. Diabetologia. Vol. 44. N 12. pp.2192-2202

[13] Korzhevsky A. (2005) Summary of the foundations of histological techniques for doctors, laboratory technicians as well as histologists. St. Petersburg

[14] Kumar G.L., et al. (2009) Education Guide: Immunohistochemical Staining Methods. 5th Edition. Dako North America, Carpinteria, California

[15] Lenzen S. (2008) The mechanisms of alloxan- and streptozotocin-induced diabetes. Diabetologia. Vol. 51. pp.216-226

[16] McNeill J.H. (1999) Experimental models of diabetes. Boca Raton., Fla., CRC Press

[17] Pisarev V.B., Snigur G.L., Spasov A.A., et al. (2009) Mechanisms of toxic action of streptozotocin on pancreatic β-cell islets. Bull. Exper. Biol. and Med. Vol. 148. N. 12. pp. 700-702

[18] Reed J.C., Green D.R. (2011) Apoptosis: physiology and pathology. Cambridge University Press

[19] Rees D.A., Alcolado J.C. (2005) Animal models of diabetes mellitus. Diabet. Med. Vol.22. N 4. pp.359-370

[20] Sakurai K., Katoh M., Someno K., et al. (2001) Apoptosis and mitochondrial damage in INS-1 cells treated with alloxan. Biol. Pharm. Bull. Vol. 24. N 8. pp.876–882

[21] Sarvilina I.V., Maklyakov Y.S., Krishtop A.V., et. al. (2008) The search for new targets for the development of antidiabetic drugs on the basis of biomodeling type 2 diabetes and proteomic technologies. Biomedicine. N1. pp.5-13

[22] Severgina E.S. (2002) Insulin-dependent diabetes mellitus - a view of morphology. Moscow: Publishing House Vidar-M.

[23] Snigur G.L., Smirnov A.V. (2010) To the problem of standardization of pathohystological diagnostics of diabetes mellitus. Bull. Volgograd State Medical University. Vol. 35. N. 3. pp.112-115

[24] Srinivasan K., Ramarao P. (2007) Animal models in type 2 diabetes research: an overview. Indian J. Med. Res. Vol. 125. N 3. pp.451-472

[25] Szkudelski T. (2001) The Mechanism of Alloxan and Streptozotocin Action in β-cells of the rat. Pancreas Physiol. Res. V. 50. pp.536-546

[26] Ziegler B., Kohler E., Kloting I., Besch W. (1990) Survival of islet isografts despite cytotoxicity against pancreatic islets measured in vitro. Exp Clin Endocrinol. V. 95(1). pp.31-38

Immunohistochemistry of Neuronal Apoptosis in Fatal Traumas: The Contribution of Forensic Molecular Pathology in Medical Science

Qi Wang, Tomomi Michiue and Hitoshi Maeda

Additional information is available at the end of the chapter

1. Introduction

The most important part of forensic pathology is investigation of the cause and process of death, especially in violent and unexpected sudden deaths, which involve social and medicolegal issues of ultimate, personal and public concern. Forensic pathologists are expected to respond to social requests by reliable interpretation of these issues in routine casework on the basis of research activities to develop, improve and sophisticate the procedures as well as to establish an autopsy database within the framework of social and legal systems. Systematic investigations are needed for comprehensive assessment of pathological findings, making full use of the available procedures; while classical morphology remains a core procedure to investigate deaths in forensic pathology, a spectrum of ancillary procedures has been developed and incorporated to detail the pathology. In addition to postmortem biochemistry, experimental and practical investigations using molecular biological procedures in the context of forensic pathology (molecular forensic pathology) have suggested the usefulness of detecting dynamic functional changes involved in the dying process that cannot be detected by morphology (pathophysiological vital reactions) (Maeda et al., 2010; Maeda et al., 2011). These procedures may effectively be included in routine casework as part of forensic laboratory investigations (forensic molecular pathology). The purpose of forensic molecular pathology is to provide a general explanation of the process or pathophysiology of human death caused by insults involving forensic issues as well as the assessment of individual deaths on the basis of biological molecular evidence; in forensic investigation of death, the genetic background, dynamics of gene expression (up-/down-regulation), and vital phenomena, involving biological mediators and degenerative products, are detected by DNA analysis, relative quantification of mRNA transcripts using real-time reverse transcription-PCR (RT-

PCR), and immunohisto-/immunocytochemistry combined with biochemistry, respectively. These observations will also contribute to understanding life-threatening events after traumas in the clinical management of critical patients.

In forensic and clinical medicine, head injury is a major trauma, and primary or secondary brain damage, e.g. due to ischemic, hypoxic and toxic insults, is involved in most fatal traumas and diseases; thus, the investigation of brain damage after such insults is essential to assess the etiology and evaluate the severity of brain impairment relevant to central nervous system (CNS) dysfunction (Oehmichen et al., 2006). Necrosis and apoptosis are involved in morphological deterioration of the brain, involving cell and tissue decay (Fawthrop et al., 1991). Neuronal apoptosis is involved in both early and delayed responses after insults; however, this type of neuronal degeneration and cell death is of greater importance in connection with delayed or intermittent CNS dysfunction (Martin et al., 1998). This chapter reviews neuronal apoptosis and related pathologies in the brain after fatal traumas and diseases as demonstrated in forensic autopsy casework, summarizing previous observations (Michiue et al., 2008; Wang et al., 2011a; Wang et al., 2012a; Wang et al., 2012b).

2. Brain neuronal apoptosis in human death

Apoptosis is programmed cell death, regulated by specific 'death genes.' The process involves active protein synthesis, initiated by changes in the microenvironment and impaired metabolic and tropic supply (Alison & Sarraf, 1992), with the participation of immediate early gene transcription factors (e.g. c-jun, jun-B, jun-D, c-fos, AP-1, ATF and nuclear factor (NF)-κB), proteases (e.g. calpains and caspases), and glutamate-mediate toxicity, including free radicals, protein kinases, Ca^{2+} homeostasis, and second messenger systems (Vaux & Strasser, 1996). It is known that microglial cells have an anti-apoptotic function to protect neurons from apoptotic death (Polazzi et al., 2001). Mechanical brain injury is accompanied by the apoptosis of neurons and glial cells surrounding the site of contusion and hemorrhage, which undergoes cell degeneration and necrosis (Oehmichen et al., 2006). Apoptosis begins hours after a traumatic event, and remains demonstrable for about 3 days (Yakovlev & Faden, 2001). These survival time-dependent changes are useful for timing brain contusions and hemorrhages in forensic pathology (Hausmann et al., 2004); however, apoptosis has been detected in the white matter as long as 1 year after injury (Williams et al., 2001). Apoptosis is also induced by other insults, including cerebral ischemia and hypoxia/asphyxia (Rosenblum, 1997), carbon monoxide (CO) intoxication (Piantadosi et al., 1997) and drug toxicity (Cadet & Krasnova, 2009). It is of particular importance that apoptosis may be involved in delayed neuronal loss (Becker & Bonni, 2004), which may contribute to delayed death or posttraumatic neurological disorders and sequelae.

Neuronal apoptosis is usually detected by *in situ* labeling of DNA fragments, e.g. terminal deoxynucleotidyl-transferase-mediated dUDP nick end-labeling (TUNEL) or *in situ* nick translation (ISNT) (Clark et al., 2001; Gavrieli et al., 1992; Rink et al., 1995). However, experimental studies have shown that single-stranded DNA (ssDNA) degradation precedes

Immunohistochemistry of Neuronal Apoptosis in Fatal Traumas: The Contribution of Forensic Molecular
Pathology in Medical Science

249

DNA double-strand breaks (DNAdsb) during a delayed neuronal death process caused by reperfusion after transient brain ischemia or intracerebral hemorrhage, possibly due to oxidative stress (Chen et al., 1997; Gong et al., 2001; Love, 1999; Nakamura et al., 2005). Thus, ssDNA can be used as an earlier marker of apoptosis and programmed cell death, which causes neuronal loss (Chen et al., 1997; Frankfurt et al., 1996; Michiue, 2008). This marker may contribute to the investigation of neuronal damage in acute death and also the timing of brain injury in the early phase (Chen et al., 1997; Hausmann et al., 2004).

Animal experimentation has shown that ssDNA positivity could be detected after as little as 1 min of reperfusion following transient brain ischemia, showing a progressive increase, and exclusively in neurons exhibiting normal nuclear morphology within the first hour of reperfusion before the appearance of DNAdsb, whereas DNAdsb was first detected after 1 h of reperfusion. Thereafter, at 16–72 h of reperfusion, both ssDNA and DNAdsb positivity were found in many neurons and astrocytes, showing morphological changes consistent with apoptosis (Chen et al., 1997). Alternatively, ssDNA-positive neurons may be decreased after several hours of reperfusion, possibly due to active DNA repair. These findings suggest that damage to nuclear DNA is an early event after neuronal ischemia and that the accumulation of unrepaired DNA single-strand breaks due to oxidative stress may contribute to delayed ischemic neuronal death by triggering apoptosis. Other experimental studies have suggested that oxidative stress contributes to DNA damage and brain injury after intracerebral hemorrhage (Gong et al., 2001; Nakamura et al., 2005). These observations indicate that neuronal ssDNA positivity can be a marker of early brain damage, possibly within the first hour after an insult involving oxidative stress, including reperfusion and hemorrhage (Michiue et al., 2008). The detection of neuronal ssDNA may depend on the cause of death and survival time after a fatal insult. Brain reperfusion during cardiopulmonary resuscitation (CPR) may also contribute to positivity.

Astrocytes are essential for the structural integrity of neurons and also for maintaining their physiological environment, involving electrolyte and water homeostasis, pH and osmotic regulation, and elimination of transmitter amino acids and plasma proteins, as well as the control of vascular tone and intercellular transport of molecules from the vessel to the neuron, supporting the blood-brain barrier (BBB) (Nag, 2011). In forensic neuropathology, glial fibrillary acidic protein (GFAP) and S100β, as specific markers of differentiated astrocytes in the brain, are used to detect their morphological and functional alterations involved in brain damage (Liedtke et al., 1996; Stroick et al., 2006). GFAP is normally detected in fibrous astrocytes in the white matter and molecular layer of the cerebral cortex, but is usually not detectable in protoplasmatic astrocytes in the cerebral cortex by a routine immunohistochemical procedure (Li et al., 2009b; Oehmichen et al., 2006). GFAP is essential for fibrous astrocyte functions, including maintenance of the integrity of CNS white matter and the blood-brain barrier (Liedtke et al., 1996), and can therefore be used to detect the morphological and functional alterations of astrocytes due to brain damage; the decrease of white matter GFAP immunopositivity indicates the disruption of astrocytes, while reactive astrogliosis involves an increase in the gray matter (Wang et al., 2011a; Wang et al., 2012a). S100β is a calcium-binding peptide and is used as a clinical parameter of glial activation and/or death in a spectrum of CNS disorders (Stroick et al., 2006); S100β levels in serum and cerebrospinal fluid (CSF) can be used as a marker of brain damage in clinical and postmortem investigations (Korfias et al., 2006; Li et al., 2006a; Li et al., 2009a). Basic fibroblast growth factor (bFGF) is closely involved in neuronal protection and repair after ischemic, metabolic or traumatic brain injury, and has emerged as a central player in acute brain damage (Bikfalvi et al., 1997); the increase of glial bFGF positivity indicates a self-protective response (Wang et al., 2011a; Wang et al., 2012a; Wang et al., 2012b). Thus, bFGF can be used to monitor the self-protective capacity of the brain after injury.

Previous studies of neuronal apoptosis in forensic pathology have mostly focused on the healing process at the site of brain injury for wound timing in the forensic context (Hausmann et al., 2004; Tao et al., 2006); however, it is of great forensic and clinical importance to investigate overall brain damage to evaluate the severity of insults. Immunohistochemistry of neuronal apoptosis and related molecular pathology using biological markers, including ssDNA, bFGF, GFAP and S100β, demonstrated various type of brain damage due to head injury, ischemia/hypoxia or asphyxia, intoxication, burns, and extreme ambient temperatures (hyperthermia and hypothermia) (Wang et al., 2011a; Wang et al., 2012a; Wang et al., 2012b). Details are described below.

3. Brain injury

3.1. General considerations

Mechanical brain injury is a major trauma in both forensic and clinical medicine and is caused by various insults, resulting in various types of brain damage, often accompanied by secondary brain dysfunction, involving brain edema, swelling and compression; these are subdivided into focal and diffuse brain injury (Greenfield & Ellison, 2008; Knight & Saukko, 2004; Oehmichen et al., 2006). Classic concepts of CNS dysfunction due to mechanical brain injury comprise the disruption of brain structures by laceration and contusion, subarachnoid hemorrhage (SAH), compression by space-occupying intracranial hematoma or increased intracranial pressure due to edema, axonal injury, ischemic brain damage and primary acute brain swelling, especially in infancy. Brain compression or swelling accompanied by increased intracranial pressure is critical for survival in the early phase after brain injury in most cases. Previous studies have made great strides in investigating the morphology and causal mechanism of brain injury and dysfunction; forensic neuropathological case studies have demonstrated findings useful for establishing practical investigation procedures. Estimation of the age of brain injury and hematomas at the site or in the area adjacent to the injury has especially important criminological implications (Bratzke, 2004; Dressler et al., 2007; Hausmann & Betz, 2001; Hausmann et al., 1999; Hausmann et al., 2000; Oehmichen et al., 2003; Takamiya et al., 2007); however, some patients may survive for months or years after severe brain injury, while it may be difficult to explain the causal relationship between a focal brain injury and death in some fatalities. Cerebral edema/swelling alone may be a distinct finding of brain injury at autopsy, with mild or even no other structural lesions, to explain the cause of death, involving increased intracranial pressure affecting vital centers in

Immunohistochemistry of Neuronal Apoptosis in Fatal Traumas: The Contribution of Forensic Molecular
Pathology in Medical Science

251

the brainstem. Animal experiments have demonstrated the rapid onset of brain edema following injury (Byard et al., 2009). Considering the anatomical and metabolic species differences, however, it is necessary to investigate human materials. Moreover, human brain injury is rarely as simple as in experimental models; thus, the changes to the whole human brain after injury should be clarified to establish the relationship to death. Brain damage to a part distant from primary lesions may provide more significant information about the whole brain condition. In particular, the evaluation of human brain damage with regard to parahippocampal herniation or secondary brainstem hemorrhage of Duret as a macroscopic sign of brain swelling and compression is important since they are believed to be closely related to a fatal outcome, causing brainstem dysfunction.

Immunohistochemical investigation of the expressions of bFGF and GFAP in glial cells as well as ssDNA positivity in the neurons as a sign of neuronal apoptosis at sites distant from the primary injury to detect survival time-dependent changes in forensic autopsy cases of fatal mechanical brain injury demonstrated characteristic posttraumatic glial and neuronal changes in regions that were not involved in the primary injury, with regard to the influence of brain swelling and compression (Fig. 1 and Table 1). These changes involved early glial changes in peracute to subacute death with survival time within 12 h and neuronal loss in prolonged death after 3 days, which depended on brain swelling and compression, irrespective of the type of primary brain injury, as follows.

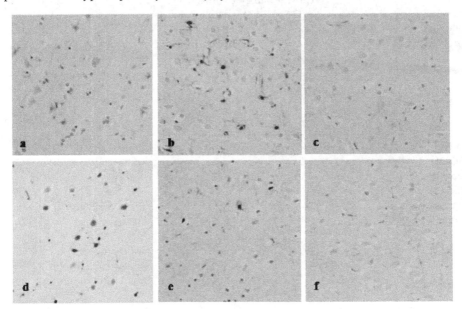

Figure 1. Immunohistochemistry of single-stranded DNA (ssDNA), basic fibroblast growth factor (bFGF) and glial fibrillary acidic protein (GFAP) in the parietal cortex of mechanical brain injury cases: 1) early death without Duret hemorrhage (2-day survival), showing low ssDNA (a) and high bFGF (b) positivity with unaffected GFAP positivity (c); 2) prolonged death with Duret hemorrhage (9 days survival), showing high ssDNA (d), and low bFGF (e) and GFAP (f) positivity

Macropathology	Peracute death ST, minutes	Acute death ST <0.5 h	Subacute death ST, 0.5–12 h	Early death ST, 12 h–3 days	Prolonged death ST >3 days
Open skull fractures with brain lacerations	Decreased white matter astrocyte GFAP positivity				
Brain contusions/ SAH/SDH without brain swelling or compression sign		Decreased white matter astrocyte GFAP positivity	→-----→-----→	Increased cortical and white matter bFGF positivity with hippocampal involvement	→-----→------ + partial cortical neuronal loss
Brain contusions/ SAH/SDH with brain swelling and compression sign without craniotomy		Increased cortical astrocyte bFGF positivity	Increased cortical and white matter astrocyte bFGF positivity	→-----→-----→ + hippocampal involvement and partly low GFAP positivity	Diffuse astrocyte loss with decreased GFAP positivity and neuronal loss with increased ssDNA positivity
Brain contusions/ SAH/SDH with brain swelling and compression sign with craniotomy				Cortical and white matter astrocyte loss with decreased GFAP positivity and neuronal loss with increased ssDNA positivity	

ST, survival time; SAH, subarachnoid hemorrhage; SDH, subdural hemorrhage or hematoma; ssDNA, single-stranded DNA, bFGF, basic fibroblast growth factor; GFAP, glial fibrillary acidic protein

Table 1. Immunohistochemical findings of apoptosis-related biomarkers in mechanical brain injury with regard to the survival time

Immunohistochemistry of Neuronal Apoptosis in Fatal Traumas: The Contribution of Forensic Molecular
Pathology in Medical Science

253

3.2. Diffuse brain injury

Diffuse mechanical brain injury clinically involves primary and secondary CNS dysfunction, which may result in permanent disability or fatal outcome. This type of mechanical brain injury is morphologically associated with specific white matter injury, usually termed diffuse axonal injury (DAI); however, other non-specific factors, including disrupted BBB, ischemia and vascular injury, also contribute to posttraumatic CNS dysfunction (Oehmichen et al., 2006). The macropathology may present with brain swelling and hemorrhages in the deep part of white matter, for which histology often involves focal edema and demyelination, accompanied by axonal injury, along the junction of gray and white matter, but these findings cannot be detected in very short survival cases. In such cases of peracute or instantaneous death within minutes, involving severe open head injury and apparently fatal structural brain damage, immunohistochemistry detected decreased glial GFAP positivity in the parietal white matter without glial or neuronal loss; however, this finding was not evident in the cerebral cortex (Wang et al., 2012b). GFAP as a marker of fibrous astrocytes in the white matter and molecular layer of the cerebral cortex is usually not detectable in protoplasmatic astrocytes or neurons in the cerebral cortex by routine immunohistochemistry (Li et al., 2009b). An increase in GFAP immunoreactivity in the cerebral cortex may be detected in classic astrocytic activation or astrogliosis; however, GFAP immunopositivity in the cerebral cortex showed no difference among all mechanical brain injury and control groups, irrespective of survival times, indicating a morphologically intact cerebral cortex. In peracute death, however, a significant decrease of white matter GFAP immunopositivity indicated the immediate, diffuse disruption of brain white matter; such findings were not detected in the hippocampus. Similar findings were detected in acute and subacute deaths (survival time <12 h) due to closed head injury without parahippocampal hernia as a brain compression sign, irrespective of the type of brain injury. This glial change in the parietal white matter may represent damage to the whole brain white matter immediately due to mechanical brain injury, suggesting fatal CNS dysfunction without brain swelling (Graham et al., 1988).

3.3. Brain swelling and compression

Brain swelling and compression, which cause brainstem dysfunction, are critically life-threatening events in clinical trauma care. In patients with a brain compression sign, accompanied by increased brain weight, glial bFGF positivity in the parietal cerebral cortex was increased in acute death (survival time <0.5 h), followed by an increase of glial bFGF positivity in the parietal white matter in subacute death (survival time of 6–8 h). Such a finding was not detected in the hippocampus in acute–subacute deaths. The bFGF has been well documented as a neuroprotective and neurotrophic factor, both *in vitro* and *in vivo* (Bikfalvi et al., 1997; Dietrich et al., 1996; Louis et al., 1993); thus, the increase of glial bFGF positivity in these cases suggests a self-protective response to maintain BBB function in the early phase of brain swelling after trauma (Deguchi et al., 2002), which may start in the cerebral cortex and spread into the white matter, despite the fatal brain compression, as suggested by parahippocampal herniation. However, in some cases of acute death with or without the brain compression sign, the above-mentioned findings were not detected, suggesting other mechanisms of acute death, including rapid cardiorespiratory failure induced by SAH around the brainstem (Macmillan et al., 2002).

In early and prolonged death cases (survival time >12 h), parahippocampal hernia may not be identified because of brain softening (encephalomalacia) around the hippocampus, but Duret hemorrhage in the brainstem as a sign of advanced descending transtentorial herniation can be used as an indicator of fatally severe high intracranial pressure (Graham et al., 1987; Parizel et al., 2002). In early deaths (survival time of 12 h–3 days), cases without Duret hemorrhage, irrespective of craniotomy, as well as those with Duret hemorrhage without depression craniotomy had increased glial bFGF positivity in the parietal cortex and white matter as well as the hippocampus, without significant glial loss; however, GFAP positivity in parietal white matter began to decrease in cases with the sign of brain compression (Fig. 1) (Wang et al., 2012b). The up-regulation of bFGF in these cases may reflect the self-protective responses of the brain after brain injury. Furthermore, the bFGF may be involved in the anti-apoptosis pathways; exogenous application of bFGF could prevent apoptosis (Ay et al., 2001; Tamatani et al., 1998). In these early death cases, mostly involving subdural hemorrhage/hematoma (SDH), high glial bFGF positivity accompanied by low neuronal ssDNA expression is consistent with the function of endogenous bFGF as an anti-apoptosis factor in traumatized brains (Wang et al., 2011a); self-protective activity in the cerebrum is maintained despite a fatal outcome, even in patients with Duret hemorrhage as a sign of fatal brainstem compression. In patients without such a brain compression sign, death may be attributed to overall brain damage without brain swelling, accompanied by SDH (Graham et al., 1988).

Early deaths with Duret hemorrhage and decompressive craniectomy (survival time of 12–60 h), often involving massive contusions, presented quite different findings, involving glial and neuronal losses in the parietal cortex and/or hippocampus, accompanied by decreased glial GFAP positivity in the parietal white matter and hippocampus, with overall low glial bFGF positivity and high neuronal ssDNA positivity (Fig. 1) (Wang et al., 2012b). These findings suggest that the brain failed to generate sufficient bFGF to prevent apoptosis as a consequence of serious brain damage involving uncontrollable progressive brain edema and swelling, which developed fatal brainstem compression and Duret hemorrhage.

In prolonged deaths (survival time >3 days), patients without Duret hemorrhage as a brain compression sign, irrespective of craniotomy, had increased glial bFGF positivity in the parietal cortex and white matter as well as the hippocampus without glial loss; however, neuronal loss without a significant increase of neuronal ssDNA positivity was detected in the parietal cortex, showing no neuronal loss in the hippocampus. This suggests gradual cortical neurodegeneration after trauma despite anti-apoptotic neuroprotective activity, as indicated by increased glial bFGF positivity, and different mechanisms of cell death involved in mechanical brain injury besides apoptosis (Castejon & Arismendi, 2006; Stoica & Faden, 2010). The fatal complication of secondary pneumonia was more frequent in these cases than in deaths with Duret hemorrhage; secondary complications may play an important role in patients without a brain compression sign. Prolonged deaths with Duret hemorrhage, however, showed advanced glial and neuronal losses in the parietal cortex and hippocampus, accompanied by decreased GFAP positivity in the parietal white matter and hippocampus, overall low glial bFGF positivity, and high neuronal ssDNA positivity in the parietal cortex and hippocampus, which were more evident than in the early deaths with Duret hemorrhage and decompressive craniectomy described above. These findings suggested fatal CNS dysfunction due to posttraumatic progressive deterioration of whole brain involving the hippocampus, lacking glial bFGF and GFAP activation for neuroprotection and repair, as a consequence of unimproved brain swelling.

Of note, there were significantly different findings depending on the survival time in cases of brain compression, as mentioned above. Acute and subacute death cases (survival time <12 h) as well as early death cases without decompressive craniectomy had higher glial bFGF and GFAP, and lower neuronal ssDNA positivity without glial and neuronal losses, whereas glial and neuronal losses, accompanied by lower glial bFGF and GFAP, and higher neuronal ssDNA positivity, were evident in early death despite decompressive craniectomy and prolonged death, suggesting different pathologies and mechanisms of brain edema/swelling, depending on the time after brain injury (Wang et al., 2012b). In a classic concept, brain edema is divided into two types based on its pathogenesis (Klatzo, 1994; Unterberg et al., 2004): a) 'vasogenic (extracellular)' edema due to BBB disruption, resulting in extracellular water accumulation, and b) 'cytotoxic (intracellular)' edema due to sustained intracellular water collection. However, brain edema after mechanical brain injury is considered to be a mixed form; vasogenic edema may be predominant in the acute phase, followed by prolonged cytotoxic edema (Barzo et al., 1997). In acute and subacute phases, the mechanical/physical impact on the brain may injure blood vessels with subsequent disruption of endothelial membranes (Hellal et al., 2004) and minor damage to astrocytes, leading to BBB opening. Thereafter, in longer survival cases, increased intracranial pressure may be involved with diffuse cytotoxic brain edema (Marmarou et al., 2000; Unterberg et al., 2004), in which glial swelling is a major mediator (Kimelberg, 1995). The activation of glial bFGF in acute and subacute death cases and decreased glial GFAP positivity in longer survival cases with brain swelling suggest a self-protective response to maintain BBB function in the early phase after mechanical brain injury and the structural damage of astrocytes caused by cytotoxic edema, respectively. Furthermore, astrocyte damage can in turn deteriorate the extracellular microenvironment (e.g. persistent increase of extracellular glutamate levels), which causes both glial and neuronal damage (Barbeito et al., 2004; Matute et al., 2006).

To summarize, characteristic immunohistochemical findings were detected with regard to the influence of cerebral compression and survival time in mechanical brain injury (Table 1) (Wang et al., 2012b). Peracute deaths with severe open head injury without brain swelling presented with glial injury in the parietal white matter. Other fatalities without a brain compression sign did not show a significant loss of glial cells; however, glial injury in the parietal white matter was seen during a survival time of <12 h, while glial responses involving bFGF positivity were detected overall after 12 h–3-day survival, and delayed neuronal loss without an increase of neuronal ssDNA positivity was seen after 3 days at the time of death, mostly due to complications. Fatalities with signs of brain swelling and compression showed gradual losses of glial cells and neurons with an early increase of glial bFGF positivity in the parietal cerebral cortex, which was followed by an increase of glial bFGF positivity in the parietal white matter and hippocampus, and final decreases of glial bFGF and GFAP positivity with increased neuronal ssDNA positivity in the parietal lobe and hippocampus, suggesting the involvement of neuronal apoptosis in progressive brain damage after injury. Such findings were detected earlier in death despite decompression craniotomy. These observations suggested different mechanisms of whole brain damage in the death process, depending on the severity of brain compression. ssDNA, bFGF and GFAP immunohistochemistry is useful to investigate such different death processes after brain injury with regard to the survival time. These findings may also contribute to wound timing when the pathology of the primary injury involving brain contusion and hematoma is considered.

4. Cerebral ischemia and hypoxia/asphyxia

4.1. General considerations

Ischemia implies a local loss of blood supply due to arterial occlusion/disruption or vasoconstriction, or as part of systemic circulatory insufficiency or blood loss, resulting in a lack of oxygen (ischemic hypoxia), while other causes are also involved in hypoxia (oxygen deficiency), for which ischemic hypoxia is the simplest model (Table 2). Brain ischemia and hypoxia are common consequences of trauma or disease involving severe cardiac and peripheral vascular injury; the brain is more susceptible to ischemia/hypoxia than other viscera. Asphyxia in the forensic context implies systemic hypoxia associated with carbon dioxide retention due to a mechanical insult, causing acidosis, which aggravates tissue damage involving the brain; however, a lack of atmospheric oxygen (suffocation) is also included. Susceptibility of neurons in the brain to oxygen deficiency depends on the vasculature and the vulnerability of individual neurons; ischemic hypoxia first affects the watershed/arterial border zone of the frontal gyri, the globus pallidus, the Ammon horn (hippocampus), and the cerebral cortex (Oehmichen et al., 2006). These site-dependent susceptibilities of the brain to ischemia/hypoxia present with various pathologies of neurons and glial cells following cardiac arrest and asphyxia, depending on the survival time.

4.2. Cerebral ischemia

It is known that transient cerebral ischemia induces neuronal apoptosis (Chan, 2004); however, the usual feature of global ischemia involves neuronal necrosis in the cerebral cortex (watershed/arterial border zone of the frontal gyri), the globus pallidus, the hippocampus and the celebellar Purkinje cells. Immunohistochemistry detected no evident changes of the brain in sudden death due to acute heart attack (simple cerebral ischemia); however, prolonged deaths under intensive medical care, possibly involving reperfusion,

Immunohistochemistry of Neuronal Apoptosis in Fatal Traumas: The Contribution of Forensic Molecular
Pathology in Medical Science

257

showed higher parietal glial bFGF positivity and neuronal loss with low ssDNA positivity, indicating incomplete necrosis or selective neuronal necrosis without positive evidence of apoptosis (Table 3) (Wang et al., 2011a).

I	Hypoxia:	1. Ischemic hypoxia – diminished blood supply
		2. Hypoxic hypoxia – reduced blood oxygenation in the lung
		3. Others: e.g. anemic, stagnant, oxygen affinity and histotoxic hypoxia
II	Asphyxia:	1. Neck compression – hanging and ligature/manual strangulation
		2. Smothering – obstruction of the airway orifices (nose and mouth)
		3. Choking – foreign body in the airway
		4. Suffocation – lack of atmospheric oxygen

Table 2. Major causes of hypoxia and asphyxia in the forensic context

4.3. Cerebral hypoxia – Asphyxia

The classification of asphyxia in the forensic context is not uniform (Byard, 2011; Sauvageau & Boghossian, 2010). From a practical point of view, however, the causes of mechanical asphyxia can grossly be divided into types with and without neck compression; the former (strangulation) involves lethal factors including brain ischemia/congestion due to closure of the blood vessels and/or air passages of the neck, whereas the latter (choking and smothering) causes hypoxia due to obstruction of the air passages. In addition, neurogenic cardiac suppression may be involved in both types (Oehmichen et al., 2006). The diagnosis of mechanical asphyxia as a cause of death is one of the most difficult tasks in forensic pathology, especially in cases lacking significant pathological evidence, even when 'classic signs of asphyxia' are apparent; for example, a very careful examination is needed to discriminate between smothering and sudden cardiac attack in cases without bruises or abrasions around the nose and mouth. The diagnosis of choking may also be obstructed when a foreign body has been removed in resuscitation measures. Furthermore, it is difficult to determine whether a food bolus in the air passages was the cause of death or a result of agonal or postmortem spillage; therefore, various procedures have been developed to detect and explain the pathophysiology of asphyxial death (Ishida et al., 2002; Zhu et al., 2000). In prolonged death cases, however, it is difficult to differentiate asphyxia from heart attack. With respect to this, immunohistochemistry of the brain detected no specific findings in acute asphyxial death, compared with sudden cardiac death; however, prolonged asphyxial death showed lower parietal glial GFAP positivity and neuronal loss with increased ssDNA positivity as a sign of apoptosis following advanced brain hypoxia, which was usually not detected in cardiac death (Table 3) (Wang et al., 2011a).

Traumatic insult	Acute/subacute death	Prolonged death
Asphyxia	Poor glial and neuronal changes in acute death	Low parietal glial bFGF positivity and neuronal loss with high ssDNA positivity
Fire fatality Burns	High parietal neuronal ssDNA Positivity	Parietal neuronal loss and increased glial cells with increased cortical and white matter bFGF positivity, and higher cortical GFAP and lower white matter GFAP positivity; overall low neuronal ssDNA positivity
CO intoxication	High parietal neuronal ssDNA Positivity	Parietal neuronal loss without glial activation; overall low glial bFGF and GFAP positivities; high neuronal ssDNA positivity
CO intoxication	High neuronal ssDNA positivity in the pallidum	
Drug abuse	High neuronal ssDNA positivity in the cerebral cortex, pallidum and midbrain substantia nigra	
Hypothermia (cold exposure)	Increased glial bFGF positivity in the cerebral cortex and white matter, and high S100β positivity in the cerebral cortex	
Hyperthermia (heatstroke)	Low glial GFAP and S100β positivities in the white matter, and high neuronal ssDNA positivity in the cerebral cortex and hippocampus, with high glial bFGF and S100β positivities in the cerebral cortex	
Cardiac attack	Poor glial and neuronal changes in acute death	Increased parietal glial bFGF positivity and neuronal loss with low ssDNA positivity

CO, carbon monoxide; ssDNA, single-stranded DNA, bFGF, basic fibroblast growth factor; GFAP, glial fibrillary acidic protein

Table 3. Immunohistochemical findings of apoptosis-related biomarkers in non-injury traumas with regard to the survival time

Immunohistochemistry of Neuronal Apoptosis in Fatal Traumas: The Contribution of Forensic Molecular
Pathology in Medical Science

259

5. Intoxication

5.1. General considerations

Numerous chemical substances are involved in accidental, suicidal and even homicidal intoxication; it is a very difficult task to screen and identify individual intoxication in forensic and clinical routine work, especially in cases where anamnesis or circumstantial evidence is obscure or absent, since intoxication often presents with non-specific signs and symptoms, or poor morphological findings. It is also important for forensic pathologists to discriminate other insults as the cause of death or contributory factor even when drugs or poisons are detected. Previous studies showed systemic deterioration in fatal intoxication, involving CNS, using biochemical markers (Maeda et al., 2011). Some drugs and poisons primarily affect the nervous system, and secondary brain damage is almost inevitable in any kind of intoxication; however, CO intoxication and drug abuse are most frequent in forensic routine work.

5.2. Carbon monoxide

The histo-/cytotoxicity of CO is due to its high affinity to iron-containing structures such as hemoglobin and myoglobin, as well as to specific sites of the brain, including the globus pallidus and the midbrain substantia nigra, which is different from cyanide, sulfide and azide (Knight & Saukko, 2004; Oehmichen et al., 2006). CO also depresses myocardial function, resulting in severe hypotension and subsequent global cerebral ischemia and hypoxia. Bilateral necrosis of the globus pallidus and the pars reticulata of the midbrain substatia nigra are known as non-specific alterations (Oehmichen et al., 2006); however, neuronal apoptosis in the pallidum has been suggested as an early change due to CO intoxication in an animal experiment (Piantadosi et al., 1997). With respect to this, immunohistochemistry of ssDNA demonstrated high positivity as a sign of apoptosis in the pallidum of the human brain in fatal CO intoxication (Table 3) (Michiue et al., 2008; Wang et al., 2011a). Similar findings were detected in acute and delayed fire fatalities having a fatal level of blood carboxyhemoglobin (COHb) saturation, different from those with lower COHb level, as described below. These findings indicate the specific neurotoxicity of CO to the pallidum.

5.3. Drug abuse

A variety of psychostimulants, narcotics and hallucinogens are involved in drug abuse, which results in brain damage and functional impairment. Among these drugs, animal experiments have shown that amphetamine and its derivatives, such as methamphetamine and ecstasy, induce apoptosis of cortical and striatal neurons, and cerebellar granular cells via various pathways (Cunha-Oliveira et al., 2008). Neuronal apoptosis has also been suggested with cocaine and opiates (Cunha-Oliveira et al., 2008). In forensic autopsy materials, immunohistochemistry also detected high neuronal ssDNA positivity as a sign of apoptosis in the cerebral cortex, pallidum and substantia nigra in fatal abuse of sedative hypnotics as well as methamphetamine, suggesting selective neuronal damage (Table 3) (Michiue et al., 2008).

5.4. Others

Very little knowledge is available with regard to the contribution of apoptosis to the neurotoxicity of other drugs and poisons at present; however, drug-related hyperthermia may induce neuronal apoptosis in a similar manner to that in heatstroke, described below. Various chemicals that trigger oxidative stress can induce neuronal apoptosis. Animal experiments showed that organophosphorus compounds caused acute necrosis of neurons in the brain at toxic doses, but induced apoptotic neuronal death at sublethal doses (Abou-Donia, 2003).

6. Fire fatality

6.1. General considerations

Fire fatality involves complex causes of death; major lethal factors involved in fire death are burns and inhalation of toxic gases, including CO and cyanide, which are produced by combustion, accompanied by smoke and ambient oxygen depletion (Stefanidou et al., 2008). Despite recent advances in clinical burn and CO intoxication care measures (Ipaktchi & Arbabi, 2006; Prockop & Chichkova, 2007), most fire victims are found dead, and in those found alive, severe burns or brain damage from CO intoxication can cause death despite intensive clinical care. In such cases, it is necessary to clarify the cause and process of death in a fire. In forensic casework, however, it may be difficult to determine the predominant cause of death due to fire or to exclude other causes of death, for which acute heart attack and asphyxiation are of particular interest, especially when clinical toxicological data are not available in cases of prolonged death without severe burns. In this respect, previous studies showed pulmonary pathology, and systemic hematological and biochemical disorders due to burns (Zhu et al., 2001a; Zhu et al., 2001b), while brain immunohistochemistry suggested specific findings of CO intoxication (Michiue et al., 2008). Thus, immunohistochemical markers in the brain that are involved in neuronal damage, apoptosis, degeneration and repair, including ssDNA, GFAP and bFGF, are useful to detect the specific neuropathology of CO intoxication for differentiation from fatal burns as well as other fatal insults.

6.2. Burns

Severe burns in a fire cause systemic disorders involving hypovolemic shock accompanied by hypoalbuminemia (burn shock), and hemolysis and skeletal muscle injury due to deep burns, followed by systemic inflammatory responses and hypoxia (Jeschke et al., 2008), which are usually detected by pathomorphology and biochemistry in postmortem investigation (Bohnert et al., 2010; Quan et al., 2009; Zhu et al., 2001b). Macro- and microscopic signs of vitality in fire death include soot deposits and thermal injury in the upper airways, but these findings may partly be sparse or even absent, especially in peracute deaths, making the diagnosis difficult (Bohnert et al., 2003). Recent immunohistochemical studies of the respiratory tract and lungs demonstrated intravital reactions in fatal burns (Boehm et al., 2010; Bohnert et al., 2010; Marschall et al., 2006);

Immunohistochemistry of Neuronal Apoptosis in Fatal Traumas: The Contribution of Forensic Molecular
Pathology in Medical Science

261

however, the pathophysiological process leading to death is still unclear. In particular, the influences of toxic gases usually do not leave significant pathology that is detectable after death, except that bilateral pallidum necrosis is occasionally seen in CO intoxication. With respect to this, previous studies detected specific neuronal damage in the pallidum due to CO intoxication by immunohistochemistry (Michiue et al., 2008; Piantadosi et al., 1997); immunohistochemical markers in the brain may be used to differentiate pathological conditions of the neurons and glial cells due to ischemic, metabolic, toxic and traumatic brain injury (Chen et al., 1997; Piantadosi et al., 1997; Zhang et al., 2010).

In acute fire fatality, immunohistochemistry demonstrated higher neuronal ssDNA immunopositivity in the parietal cortex than in acute cardiac and asphyxial deaths, suggesting the induction of neuronal apoptosis, irrespective of the blood COHb level; however, such findings were not seen in cases of postmortem burns (Wang et al., 2011a). These suggest that brain damage due to a fire is not simply caused by ischemia or hypoxia, but also involves cytotoxic factors, including massive thermal tissue injury and hemolysis, which can induce systemic oxidative stress involving the brain (Gatson et al., 2009). However, neuronal ssDNA immunopositivity in the pallidum was lower in cases of a low level of blood COHb saturation than in those with a fatal level of blood COHb saturation (>60%). Therefore, increased neuronal ssDNA immunopositivity in the parietal cortex and pallidum can be used as a vitality finding in acute fire deaths, with consideration of other pathological findings; these findings can be used to interpret death due to burns or CO intoxication in a fire. Furthermore, the topographical distribution of neuronal ssDNA immunopositivity in the brain may be helpful for determining the immediate cause of death in cases of other potentially fatal traumas or diseases, e.g. strangulation, drug abuse and acute cardiac attack. However, higher neuronal ssDNA immunopositivity was sporadically detected in other cases, suggesting the partial contribution of unspecific neuronal damage due to reperfusion, possibly involved with cardiopulmonary resuscitation measures (Li et al., 2010); this should be carefully considered when determining the cause of death, especially in cases where the vitality findings are sparse.

In prolonged deaths, the macro- and microscopic signs of vitality in fire death, described above, may become obscure, making the pathological diagnosis quite difficult. In immunohistochemical study of the brain, however, there were significant differences between fatal burns and CO intoxication in prolonged fire deaths under critical clinical care (Fig. 2). Neuronal loss was seen in those with burns and CO intoxication as well as in patients with a fatal ischemic heart attack and prolonged asphyxial deaths, while glial cells were increased in burns and heart attack; the glial cell number was larger in fatality due to burns than in CO intoxication and asphyxiation, regardless of temporary cardiopulmonary arrest (CPA) after insult, suggesting glial activation. The increase in glial cells in cases of fatal burns was accompanied by higher glial bFGF immunopositivity in the parietal cortex and white matter, and higher and lower glial GFAP immunopositivity in the cortex and white matter, respectively, showing low neuronal ssDNA immunopositivity. The above-mentioned findings differed from those in prolonged death due to heart attack or mechanical asphyxiation involving simple cerebral ischemia or hypoxia, regardless of temporary CPA after insult. These observations suggest neuronal loss accompanied by active glial responses after severe burns regardless of CPA after insult. Lower glial GFAP immunopositivity in the white matter in prolonged deaths due to burns may be related to BBB damage, as discussed below.

Both *in vitro* and *in vivo*, bFGF has been well established as a neuroprotective and neurotrophic factor (Dietrich et al., 1996; Louis et al., 1993). Severe trauma, including burn injury, can result in whole body tissue damage, for which an important early sign is systemic inflammatory response syndrome (SIRS), which may lead to multiple organ dysfunction syndrome (MODS). The early appearance of inflammatory cytokines in the systemic circulation has been demonstrated following thermal injury both in humans (Cannon et al., 1992) and animals (Kataranovski et al., 1999). Systemic inflammatory responses also develop in the brain (Reyes et al., 2006), which may induce, enhance or accompany astrogliosis (Balasingam et al., 1994). Thus, the up-regulation of bFGF and GFAP as well as increased numbers of glial cells in the parietal cerebral cortex in prolonged death due to burns may reflect self-protective responses of the brain. In acute death, such glial responses may not be apparent due to the shorter survival time. Furthermore, bFGF may be involved in anti-apoptotic pathways; the exogenous application of bFGF prevented apoptosis in both *in vitro* and *in vivo* studies (Ay et al., 2001; Tamatani et al., 1998). High cerebral cortex glial bFGF immunopositivity accompanied by low neuronal ssDNA expression suggests that endogenous bFGF is an anti-apoptosis factor in the brain.

The BBB between systemic circulation and the cerebral parenchyma is composed of interendothelial tight junctions, basal lamina and perivascular astrocytes, and may also be damaged by severe burns; thus, BBB permeability can be increased, causing advanced brain edema (Reyes et al., 2009). Low glial GFAP immunopositivity in the white matter in prolonged deaths due to severe burns suggests astrocyte damage related to BBB dysfunction. In addition, systemic inflammatory responses followed by hyperthermia may also induce BBB dysfunction, which is characterized by vasogenic brain edema (Sharma, 2006). Further investigation is needed to clarify the mechanism of BBB dysfunction in prolonged deaths due to severe burns. Meanwhile, high bFGF positivity in the white matter, which was detected in prolonged deaths due to severe burns, suggests that the self-protective system involving bFGF is activated to maintain BBB function (Deguchi et al., 2002); the damaged brain does not lose its self-protective capacity after severe burns.

To summarize, typical pathologies in the brain after fatal burns are: 1) in acute deaths, increased neuronal ssDNA immunopositivity in the cerebral cortex, irrespective of the severity of the burns and CO intoxication; 2) neuronal loss in prolonged death; 3) increase in glial cells in prolonged death, accompanied by higher glial bFGF immunopositivity in the cerebral cortex and white matter, higher and lower glial GFAP immunopositivity in the cortex and white matter, respectively, with low neuronal ssDNA immunopositivity (Table 3) (Wang et al., 2011a). These findings suggest that: 1) increased neuronal ssDNA positivity, together with other pathological findings, can be used as a vitality finding in acute fire death; 2) the brain retained self-protective response capacity in fire victims who died due to severe burns. However, progressive systemic deterioration after severe burns, accompanied by preexisting disorders or physical predispositions, can cause fatality due to respiratory failure, hypoxic brain damage, hypovolemic shock and secondary infection involving sepsis, even under critical life support care (Barber et al., 2007; Williams et al., 2009). These death processes should be assessed based on individual evidence.

6.3. Toxic gases

A spectrum of toxic or asphyxiating gases produced by combustion, including CO, cyanide and carbon dioxide, can contribute to death in a fire. CO has histo-/cytotoxicity due to its high affinity to specific sites of the brain, including the globus pallidus and the midbrain substantia nigra, but cyanide did not have such neurotoxicity (Oehmichen et al., 2006). In fire fatality with a fatal level of blood COHb saturation (>60%), immunohistochemistry detected higher neuronal ssDNA immunopositivity in the pallidum than in cases of a lower COHb level (<60%), as described above, suggesting CO-specific neuronal damage (Michiue et al., 2008; Tofighi et al., 2006; Wang et al., 2011a). Cyanide did not appear to contribute to neuronal ssDNA immunopositivity.

In prolonged deaths, neuron and glial cell number was decreased in CO intoxication as well as asphyxiation, regardless of temporary CPA after insult, suggesting reduced glial reactivity due to CO intoxication and asphyxiation. Glial bFGF and GFAP immunopositivity was low at each site, but neuronal ssDNA immunopositivity was high in prolonged deaths due to CO intoxication (Fig. 2) (Wang et al., 2011a). These findings differed from those in prolonged death due to heart attack or mechanical asphyxiation involving simple cerebral ischemia or hypoxia regardless of temporary CPA after insult. These observations suggest neuronal loss and progressive apoptosis without glial responses after CO intoxication. When the crucial functions of glial bFGF and GFAP in the self-protective responses of the brain are considered, high neuronal ssDNA immunopositivity accompanied by low glial bFGF and GFAP expressions in prolonged deaths due to CO intoxication, as indicated above, suggests that the brain has failed to generate sufficient bFGF to prevent apoptosis, which may indicate serious damage to the brain due to CO intoxication; CO can exert direct damage on cells by inducing apoptosis (Tofighi et al., 2006). Low bFGF and GFAP positivity in the white matter in prolonged death due to CO intoxication suggests delayed effects of CO, characterized by bilateral, confluent lesions that reflect diffuse demyelination (Lo et al., 2007). Such injury may also be caused by slowly progressive cytotoxic edema related to the direct toxic effect of CO. These findings suggest persistent and irreversible damage to the brain white matter due to CO intoxication. Similar findings suggesting damage to the BBB in the white matter were partly seen in prolonged asphyxial deaths, but were milder in ischemic heart attack. Such white matter damage may be responsible for delayed CNS deterioration due to CO intoxication and asphyxiation (Lo et al., 2007; Strackx et al., 2008).

To summarize, typical pathologies in the brain of fire fatality with a fatal level of blood COHb saturation are: 1) in acute deaths, increased neuronal ssDNA immunopositivity in the cerebral cortex, irrespective of the severity of CO intoxication, and higher neuronal ssDNA positivity in the pallidum due to CO intoxication; 2) neuronal loss in prolonged death; 3) overall low glial bFGF and GFAP immunopositivities with high neuronal ssDNA immunopositivity in prolonged death due to CO intoxication (Table 3) (Wang et al., 2011a). These findings suggest neuronal loss and progressive apoptosis without glial responses after CO intoxication; the brain sustained serious damage involving the loss of self-protective capacity in CO intoxication, thus causing delayed death.

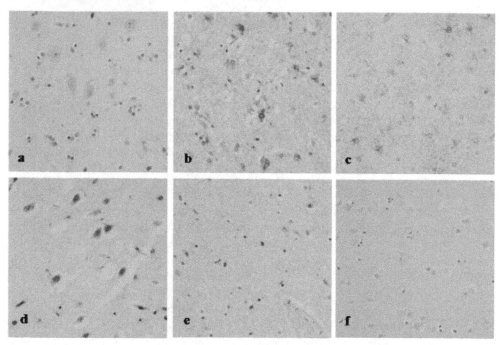

Figure 2. Immunohistochemistry of single-stranded DNA (ssDNA), basic fibroblast growth factor (bFGF) and glial fibrillary acidic protein (GFAP) in the parietal cortex of prolonged fire death cases: 1) a case of fatal burns and low blood cardoxyhemoglobin (COHb) saturation (72 h survival), showing low ssDNA (a), and high bFGF (b) and GFAP (c) positivity; 2) a case of a fatal level of blood COHb saturation (48 h survival), showing high ssDNA (d), and low bFGF (e) and GFAP (f) positivity

7. Extreme ambient temperature

7.1. General considerations

In forensic practice, the diagnosis of death due to extreme environmental temperatures involving hypothermia (cold exposure) and hyperthermia (heat stroke) is often difficult because of poor or nonspecific gross and microscopic findings, although hypothermia may present with typical pathologies, including frost erythema and hemorrhagic gastric erosions (Wischnewski spots) (Green et al., 2001; Nixdorf-Miller et al., 2006; Schuliar et al., 2001; Turk, 2010). Besides diagnosis by exclusion, histology, immunohistochemistry, biochemistry and molecular biology can be used for detailed investigation of functional deaths (Madea &

Saukko, 2010; Madea et al., 2010); previous studies have suggested that postmortem biochemistry, immunohistochemistry and molecular biology can detect systemic functional alterations in these fatalities (Fineschi et al., 2005; Ishikawa et al., 2008; Jakubeniene et al., 2009; Maeda et al., 2011; Yoshida et al., 2011). Immunohistochemistry of the brain using ssDNA, bFGF, GFAP and S100 can also demonstrate functional alterations in fatalities due to extreme ambient temperature, involving glial responses and neuronal apoptosis (Wang et al., 2012a)

7.2. Hypothermia (Cold exposure)

When the human body cannot compensate for heat loss in an extremely cold environment, the body temperature decreases progressively, resulting in cerebral and cardiorespiratory dysfunction, and finally fatal arrhythmia and asystole. Metabolic deterioration involves dehydration, acidosis, azotemia and enhanced fat metabolism with ketonemia/ketouria, but myocardial and brain tissue damage are usually mild (Maeda et al., 2011).

In immunohistochemcal investigation of the brain, hypothermia cases showed higher glial bFGF immunopositivity in the cerebral cortex and white matter, and higher S100β immunopositivity in the cerebral cortex with a lower CSF S100β concentration, without glial or neuronal loss (Fig. 3 and Table 3) (Wang et al., 2012a). The up-regulation of glial bFGF and S100β in the cerebral cortex suggests the self-protective responses of the brain and possible neurotrophic properties, respectively (Gomide & Chadi, 1999). Furthermore, since bFGF may be involved in the anti-apoptotic pathways (Ay et al., 2001; Tamatani et al., 1998), high glial bFGF immunopositivity accompanied by low neuronal ssDNA expression in hypothermia cases can indicate the activation of endogenous bFGF as an anti-apoptosis factor in the brain, which is similar to previous findings in prolonged fire fatality due to burns (Wang et al., 2011a). As above, the brain may retain self-protective response capacity without marked glial or neuronal damage in fatal hypothermia. The mechanism of death may mainly involve cardiac dysfunction, including ventricular fibrillation or asystole, resulting from myocardial ischemia, hypoxia, electrolyte abnormalities and elevated catecholamine levels (Turk, 2010), although there have been few postmortem investigations (Ishikawa et al., 2010; Wang et al., 2011b). To summarize, fatal hypothermia cases showed neuroprotective glial responses without marked neuronal or glial damage, which can serve as a condition for possible recovery and survival by means of adequate resuscitation and life-supporting measures.

7.3. Hyperthermia (Heatstroke)

A high ambient temperature in combination with predisposing factors and individual susceptibility ultimately impairs thermoregulation, and the body temperature rises precipitously; the main pathophysiology of heatstroke consists of hyperpyrexia involving impaired thermoregulation, accompanied by dehydration and profound systemic hypoxia, which is followed by further complications of pulmonary edema, renal tubular necrosis, adrenal hemorrhage, hepatic necrosis, myocardial necrosis, rhabdomyolysis, systemic inflammatory response syndrome (SIRS), disseminated intravascular coagulation (DIC), and ultimately MODS. Clinical diagnosis of heatstroke and related syndromes is usually not difficult, considering hyperpyrexia and laboratory findings, and excluding other causes of hyperpyrexia; however, postmortem diagnosis is obstructed by a lack of specific findings. The diagnosis should be established by collecting pathological findings compatible with heatstroke, related to the predisposition, drug abuse, and physical abuse or neglect, and to differentiate other insults, in combination with toxicology and biochemistry (Maeda et al., 2011). Circumstantial evidence may also be considered when available.

In immunohistochemcal investigation of the brain, characteristic findings in hyperthermia cases were lower glial GFAP and S100β immunopositivity in the white matter, and higher neuronal ssDNA immunopositivity in the cerebral cortex and hippocampus, accompanied by high glial bFGF and S100β immunopositivity in the cerebral cortex, without glial or neuronal loss (Fig. 3 and Table 3) (Wang et al., 2012a). Survival in hospital for days under a clinical diagnosis of heatstroke showed similar findings. Increased cortical glial bFGF and S100β may indicate self-protective responses of the brain, as described above for hypothermia; however, these findings were milder in hyperthermia than in hypothermia, involving neuronal and glial damage described below, and may also be related to the initiation of inflammatory processes involved in the systemic inflammatory response leading to MODS, in which encephalopathy predominates (Bouchama & Knochel, 2002).

Hyperthermia can exert direct damage on tissue cells by inducing apoptosis (Basile et al., 2008; Vogel et al., 1997); increased neuronal ssDNA expression can be used as evidence of brain dysfunction involving apoptosis as part of MODS from hyperthermia. These observations suggest diffuse neuronal apoptosis despite initiation of neuroprotective cortical astrocyte reactions in hyperthermia. Furthermore, the BBB, composed of endothelial tight junctions, basal lamina and perivascular astrocytes, may be damaged by hyperthermia, characterized by vasogenic brain edema (Sharma, 2006; Sharma & Hoopes, 2003). Low glial GFAP and S100β immunopositivity in the white matter in hyperthermia cases suggests that astrocyte damage may be involved in BBB dysfunction. In addition, low bFGF immunopositivity in the white matter in hyperthermia cases indicates that white matter loses the capacity for a compensatory response.

These observations suggest characteristic brain responses in the death process due to an extreme environmental temperature; hyperthermia as well as hypothermia involved higher glial bFGF positivity in the cerebral cortex, indicating activation of neuroprotective processes. To summarize, fatal hyperthermia cases showed diffuse neuronal apoptosis despite the initiation of neuroprotective cortical astrocyte responses, accompanied by glial damage in the white matter; diffuse neuronal and glial deterioration in the brain may lead to a fatal outcome even under critical medical care. Further investigation is needed to clarify the underlying mechanisms.

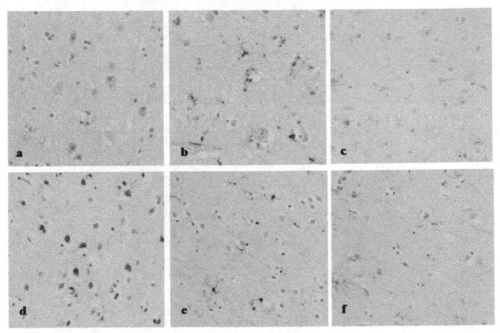

Figure 3. Immunohistochemistry of single-stranded DNA (ssDNA), basic fibroblast growth factor (bFGF) and glial fibrillary acidic protein (GFAP) in the parietal cerebral cortex in fatalities due to extreme ambient temperatures: 1) hypothermia (cold exposure), showing low ssDNA (a) and high bFGF (b) positivity with unaffected GFAP positivity (c); 2) hyperthermia (heatstroke), showing high ssDNA (d) and bFGF (e) positivity with unaffected GFAP positivity (f)

8. Limitations and outlook

Different from animal experimentation, forensic and clinical materials are not homogenous owing to the complexity of insults and the consequent brain damage, varied susceptibility of subjects, and intensive clinical intervention. In addition, forensic autopsy materials partly include cases where the estimated survival time and/or postmortem interval depend on obscure circumstantial evidence. Therefore, it is difficult to elucidate the time course of cellular responses after individual specified insults in detail. It is important, however, to collect postmortem human data involving the whole brain pathology, which are not clinically or experimentally available. Further investigation is needed, including other markers involved in apoptotic pathways as well as in water homeostasis, BBB integrity and inflammatory responses, combined with the systematic analysis of related gene expressions.

9. Conclusion

A serial study of forensic autopsy cases suggested the involvement of neuronal apoptosis at specific sites of the brain, possibly contributing to CNS damage and dysfunction, which was characteristic of traumatic insults, including progressive or delayed brain damage due to mechanical head injury, involving brain swelling and compression, as well as due to asphyxia, CO intoxication, fire fatality, and hyperthermia (heatstroke). Molecular pathological investigation of neuronal apoptosis and related biological responses in forensic materials can provide specific information in medical science for understanding the death process after traumatic insults. These studies will contribute not only in forensic casework but also to the clinical management of critically traumatized patients.

Author details

Qi Wang, Tomomi Michiue and Hitoshi Maeda
Department of Legal Medicine, Osaka City University Medical School, Abeno, Osaka, Japan

Acknowledgement

Serial studies by the authors and coworkers were supported in part by Grants-in-Aid for Scientific Research from the Japan Society for the Promotion of Science and the Ministry of Education, Culture, Sports, Science and Technology, Japan (Grant Nos. 11670425, 12470109, 08307006, 15390217, 15590585, 21790612 and 22590642).

10. References

Abou-Donia, M. B. (2003). Organophosphorus ester-induced chronic neurotoxicity. *Archives of Environmental Health*, 58(8), pp. 484-497.

Alison, M. R. & Sarraf, C. E. (1992). Apoptosis: a gene-directed programme of cell death. *Journal of the Royal College of Physicians of London*, 26(1), pp. 25-35.

Ay, I., Sugimori, H. & Finklestein, S. P. (2001). Intravenous basic fibroblast growth factor (bFGF) decreases DNA fragmentation and prevents downregulation of Bcl-2 expression in the ischemic brain following middle cerebral artery occlusion in rats. *Brain Research. Molecular Brain Research*, 87(1), pp. 71-80.

Balasingam, V., Tejada-Berges, T., Wright, E., Bouckova, R. & Yong, V. W. (1994). Reactive astrogliosis in the neonatal mouse brain and its modulation by cytokines. *Journal of Neuroscience*, 14(2), pp. 846-856.

Barbeito, L. H., Pehar, M., Cassina, P., Vargas, M. R., Peluffo, H., Viera, L., Estevez, A. G. & Beckman, J. S. (2004). A role for astrocytes in motor neuron loss in amyotrophic lateral sclerosis. *Brain Research. Brain Research Reviews*, 47(1-3), pp. 263-274.

Barber, R. C., Aragaki, C. C., Chang, L. Y., Purdue, G. F., Hunt, J. L., Arnoldo, B. D. & Horton, J. W. (2007). CD14-159 C allele is associated with increased risk of mortality after burn injury. *Shock*, 27(3), pp. 232-237.

Barzo, P., Marmarou, A., Fatouros, P., Hayasaki, K. & Corwin, F. (1997). Contribution of vasogenic and cellular edema to traumatic brain swelling measured by diffusion-weighted imaging. *Journal of Neurosurgery*, 87(6), pp. 900-907.

Basile, A., Biziato, D., Sherbet, G. V., Comi, P. & Cajone, F. (2008). Hyperthermia inhibits cell proliferation and induces apoptosis: relative signaling status of P53, S100A4, and Notch

Immunohistochemistry of Neuronal Apoptosis in Fatal Traumas: The Contribution of Forensic Molecular
Pathology in Medical Science

269

in heat sensitive and resistant cell lines. *Journal of Cellular Biochemistry,* 103(1), pp. 212-220.

Becker, E. B. & Bonni, A. (2004). Cell cycle regulation of neuronal apoptosis in development and disease. *Progress in Neurobiology,* 72(1), pp. 1-25.

Bikfalvi, A., Klein, S., Pintucci, G. & Rifkin, D. B. (1997). Biological roles of fibroblast growth factor-2. *Endocrine Reviews,* 18(1), pp. 26-45.

Boehm, J., Fischer, K. & Bohnert, M. (2010). Putative role of TNF-alpha, interleukin-8 and ICAM-1 as indicators of an early inflammatory reaction after burn: a morphological and immunohistochemical study of lung tissue of fire victims. *Journal of Clinical Pathology,* 63(11), pp. 967-971.

Bohnert, M., Werner, C. R. & Pollak, S. (2003). Problems associated with the diagnosis of vitality in burned bodies. *Forensic Science International,* 135(3), pp. 197-205.

Bohnert, M., Anderson, J., Rothschild, M. A. & Bohm, J. (2010). Immunohistochemical expression of fibronectin in the lungs of fire victims proves intravital reaction in fatal burns. *International Journal of Legal Medicine,* 124(6), pp. 583-588.

Bouchama, A. & Knochel, J. P. (2002). Heat stroke. *New England Journal of Medicine,* 346(25), pp. 1978-1988.

Bratzke, H. (2004). Research in forensic neurotraumatology. *Forensic Science International,* 144(2-3), pp. 157-165.

Byard, R. W., Bhatia, K. D., Reilly, P. L. & Vink, R. (2009). How rapidly does cerebral swelling follow trauma? Observations using an animal model and possible implications in infancy. *Legal Medicine (Tokyo),* 11 Suppl 1, pp. S128-131.

Byard, R. W. (2011). Commentary on: Sauvageau A, Boghossian E. Classification of asphyxia: the need for standardization. *Journal of Forensic Sciences,* 56(1), pp. 264.

Cadet, J. L. & Krasnova, I. N. (2009). Molecular bases of methamphetamine-induced neurodegeneration. *International Review of Neurobiology,* 88, pp. 101-119.

Cannon, J. G., Friedberg, J. S., Gelfand, J. A., Tompkins, R. G., Burke, J. F. & Dinarello, C. A. (1992). Circulating interleukin-1 beta and tumor necrosis factor-alpha concentrations after burn injury in humans. *Critical Care Medicine,* 20(10), pp. 1414-1419.

Castejon, O. J. & Arismendi, G. J. (2006). Nerve cell death types in the edematous human cerebral cortex. *Journal of Submicroscopic Cytology and Pathology,* 38(1), pp. 21-36.

Chan, P. H. (2004). Mitochondria and neuronal death/survival signaling pathways in cerebral ischemia. *Neurochemical Research,* 29(11), pp. 1943-1949.

Chen, J., Jin, K., Chen, M., Pei, W., Kawaguchi, K., Greenberg, D. A. & Simon, R. P. (1997). Early detection of DNA strand breaks in the brain after transient focal ischemia: implications for the role of DNA damage in apoptosis and neuronal cell death. *Journal of Neurochemistry,* 69(1), pp. 232-245.

Clark, R. S. B., Chen, M., Kochanek, P. M., Watkins, S. C., Jin, K. L., Draviam, R., Nathaniel, P. D., Pinto, R., Marion, D. W. & Graham, S. H. (2001). Detection of single- and double-strand DNA breaks after traumatic brain injury in rats: comparison of in situ labeling techniques using DNA polymerase I, the Klenow fragment of DNA polymerase I, and terminal deoxynucleotidyl transferase. *Journal of Neurotrauma,* 18(7), pp. 675-689.

Cunha-Oliveira, T., Rego, A. C. & Oliveira, C. R. (2008). Cellular and molecular mechanisms involved in the neurotoxicity of opioid and psychostimulant drugs. *Brain Research Reviews*, 58(1), pp. 192-208.

Deguchi, Y., Okutsu, H., Okura, T., Yamada, S., Kimura, R., Yuge, T., Furukawa, A., Morimoto, K., Tachikawa, M., Ohtsuki, S., Hosoya, K. & Terasaki, T. (2002). Internalization of basic fibroblast growth factor at the mouse blood-brain barrier involves perlecan, a heparan sulfate proteoglycan. *Journal of Neurochemistry*, 83(2), pp. 381-389.

Dietrich, W. D., Alonso, O., Busto, R. & Finklestein, S. P. (1996). Posttreatment with intravenous basic fibroblast growth factor reduces histopathological damage following fluid-percussion brain injury in rats. *Journal of Neurotrauma*, 13(6), pp. 309-316.

Dressler, J., Hanisch, U., Kuhlisch, E. & Geiger, K. D. (2007). Neuronal and glial apoptosis in human traumatic brain injury. *International Journal of Legal Medicine*, 121(5), pp. 365-375.

Fawthrop, D. J., Boobis, A. R. & Davies, D. S. (1991). Mechanisms of cell death. *Archives of Toxicology*, 65(6), pp. 437-444.

Fineschi, V., D'Errico, S., Neri, M., Panarese, F., Ricci, P. A. & Turillazzi, E. (2005). Heat stroke in an incubator: an immunohistochemical study in a fatal case. *International Journal of Legal Medicine*, 119(2), pp. 94-97.

Frankfurt, O. S., Robb, J. A., Sugarbaker, E. V. & Villa, L. (1996). Monoclonal antibody to single-stranded DNA is a specific and sensitive cellular marker of apoptosis. *Experimental Cell Research*, 226(2), pp. 387-397.

Gatson, J. W., Maass, D. L., Simpkins, J. W., Idris, A. H., Minei, J. P. & Wigginton, J. G. (2009). Estrogen treatment following severe burn injury reduces brain inflammation and apoptotic signaling. *Journal of Neuroinflammation*, 22(6), pp. 30.

Gavrieli, Y., Sherman, Y. & Ben-Sasson, S. A. (1992). Identification of programmed cell death in situ via specific labeling of nuclear DNA fragmentation. *Journal of Cell Biology*, 119(3), pp. 493-501.

Gomide, V. C. & Chadi, G. (1999). The trophic factors S-100beta and basic fibroblast growth factor are increased in the forebrain reactive astrocytes of adult callosotomized rat. *Brain Research*, 835(2), pp. 162-174.

Gong, C., Boulis, N., Qian, J., Turner, D. E., Hoff, J. T. & Keep, R. F. (2001). Intracerebral hemorrhage-induced neuronal death. *Neurosurgery*, 48(4), pp. 875-883.

Graham, D. I., Lawrence, A. E., Adams, J. H., Doyle, D. & McLellan, D. R. (1987). Brain damage in non-missile head injury secondary to high intracranial pressure. *Neuropathology and Applied Neurobiology*, 13(3), pp. 209-217.

Graham, D. I., Lawrence, A. E., Adams, J. H., Doyle, D. & McLellan, D. R. (1988). Brain damage in fatal non-missile head injury without high intracranial pressure. *Journal of Clinical Pathology*, 41(1), pp. 34-37.

Green, H., Gilbert, J., James, R. & Byard, R. W. (2001). An analysis of factors contributing to a series of deaths caused by exposure to high environmental temperatures. *American Journal of Forensic Medicine and Pathology*, 22(2), pp. 196-199.

Greenfield, J. G. & Ellison, D. (2008). *Greenfield's neuropathology*. Hodder Arnold.

Hausmann, R., Kaiser, A., Lang, C., Bohnert, M. & Betz, P. (1999). A quantitative immunohistochemical study on the time-dependent course of acute inflammatory cellular response to human brain injury. *International Journal of Legal Medicine*, 112(4), pp. 227-232.

Hausmann, R., Riess, R., Fieguth, A. & Betz, P. (2000). Immunohistochemical investigations on the course of astroglial GFAP expression following human brain injury. *International Journal of Legal Medicine*, 113(2), pp. 70-75.

Hausmann, R. & Betz, P. (2001). Course of glial immunoreactivity for vimentin, tenascin and alpha1-antichymotrypsin after traumatic injury to human brain. *International Journal of Legal Medicine*, 114(6), pp. 338-342.

Hausmann, R., Biermann, T., Wiest, I., Tubel, J. & Betz, P. (2004). Neuronal apoptosis following human brain injury. *International Journal of Legal Medicine*, 118(1), pp. 32-36.

Hellal, F., Bonnefont-Rousselot, D., Croci, N., Palmier, B., Plotkine, M. & Marchand-Verrecchia, C. (2004). Pattern of cerebral edema and hemorrhage in a mice model of diffuse brain injury. *Neuroscience Letters*, 357(1), pp. 21-24.

Ipaktchi, K. & Arbabi, S. (2006). Advances in burn critical care. *Critical Care Medicine*, 34(9 Suppl), pp. S239-244.

Ishida, K., Zhu, B. L. & Maeda, H. (2002). A quantitative RT-PCR assay of surfactant-associated protein A1 and A2 mRNA transcripts as a diagnostic tool for acute asphyxial death. *Legal Medicine (Tokyo)*, 4(1), pp. 7-12.

Ishikawa, T., Quan, L., Li, D. R., Zhao, D., Michiue, T., Hamel, M. & Maeda, H. (2008). Postmortem biochemistry and immunohistochemistry of adrenocorticotropic hormone with special regard to fatal hypothermia. *Forensic Science International*, 179(2-3), pp. 147-151.

Ishikawa, T., Yoshida, C., Michiue, T., Perdekamp, M. G., Pollak, S. & Maeda, H. (2010). Immunohistochemistry of catecholamines in the hypothalamic-pituitary-adrenal system with special regard to fatal hypothermia and hyperthermia. *Legal Medicine (Tokyo)*, 12(3), pp. 121-127.

Jakubeniene, M., Irnius, A., Chaker, G. A., Paliulis, J. M. & Bechelis, A. (2009). Post-mortem investigation of calcium content in liver, heart, and skeletal muscle in accidental hypothermia cases. *Forensic Science International*, 190(1-3), pp. 87-90.

Jeschke, M. G., Chinkes, D. L., Finnerty, C. C., Kulp, G., Suman, O. E., Norbury, W. B., Branski, L. K., Gauglitz, G. G., Mlcak, R. P. & Herndon, D. N. (2008). Pathophysiologic response to severe burn injury. *Annals of Surgery*, 248(3), pp. 387-401.

Kataranovski, M., Magic, Z. & Pejnovic, N. (1999). Early inflammatory cytokine and acute phase protein response under the stress of thermal injury in rats. *Physiological Research*, 48(6), pp. 473-482.

Kimelberg, H. K. (1995). Current concepts of brain edema. Review of laboratory investigations. *Journal of Neurosurgery*, 83(6), pp. 1051-1059.

Klatzo, I. (1994). Evolution of brain edema concepts. *Acta Neurochirurgica. Supplementum*, 60, pp. 3-6.

Knight, B. & Saukko, P. J. (2004). *Knight's Forensic pathology* (3rd ed.). Arnold.

Korfias, S., Stranjalis, G., Papadimitriou, A., Psachoulia, C., Daskalakis, G., Antsaklis, A. & Sakas, D. E. (2006). Serum S-100B protein as a biochemical marker of brain injury: a review of current concepts. *Current Medicinal Chemistry*, 13(30), pp. 3719-3731.

Li, D. R., Zhu, B. L., Ishikawa, T., Zhao, D., Michiue, T. & Maeda, H. (2006a). Postmortem serum protein S100B levels with regard to the cause of death involving brain damage in medicolegal autopsy cases. *Legal Medicine (Tokyo)*, 8(2), pp. 71-77.

Li, D. R., Zhu, B. L., Ishikawa, T., Zhao, D., Michiue, T. & Maeda, H. (2006b). Immunohistochemical distribution of S-100 protein in the cerebral cortex with regard to the cause of death in forensic autopsy. *Legal Medicine (Tokyo)*, 8(2), pp. 78-85.

Li, D. R., Michiue, T., Zhu, B. L., Ishikawa, T., Quan, L., Zhao, D., Yoshida, C., Chen, J. H., Wang, Q., Komatsu, A., Azuma, Y. & Maeda, H. (2009a). Evaluation of postmortem S100B levels in the cerebrospinal fluid with regard to the cause of death in medicolegal autopsy. Legal Medicine (Tokyo), 11 Suppl 1, pp. S273-275.

Li, D. R., Ishikawa, T., Zhao, D., Michiue, T., Quan, L., Zhu, B. L. & Maeda, H. (2009b). Histopathological changes of the hippocampus neurons in brain injury. *Histology and Histopathology*, 24(9), pp. 1113-1120.

Li, J., Han, B., Ma, X. & Qi, S. (2010). The effects of propofol on hippocampal caspase-3 and Bcl-2 expression following forebrain ischemia-reperfusion in rats. *Brain Research*, 1356, pp. 11-23.

Liedtke, W., Edelmann, W., Bieri, P. L., Chiu, F. C., Cowan, N. J., Kucherlapati, R. & Raine, C. S. (1996). GFAP is necessary for the integrity of CNS white matter architecture and long-term maintenance of myelination. *Neuron*, 17(4), pp. 607-615.

Lo, C. P., Chen, S. Y., Lee, K. W., Chen, W. L., Chen, C. Y., Hsueh, C. J. & Huang, G. S. (2007). Brain injury after acute carbon monoxide poisoning: early and late complications. *American Journal of Roentgenology*, 189(4), pp. W205-211.

Louis, J. C., Magal, E., Gerdes, W. & Seifert, W. (1993). Survival-promoting and protein kinase C-regulating roles of basic FGF for hippocampal neurons exposed to phorbol ester, glutamate and ischaemia-like conditions. *European Journal of Neuroscience*, 5(12), pp. 1610-1621.

Love, S. (1999). Oxidative stress in brain ischemia. *Brain Pathology*, 9(1), pp. 119-131.

Macmillan, C. S., Grant, I. S. & Andrews, P. J. (2002). Pulmonary and cardiac sequelae of subarachnoid haemorrhage: time for active management? *Intensive Care Medicine*, 28(8), pp. 1012-1023.

Madea, B. & Saukko, P. (2010). Molecular pathology in forensic medicine. Preface. *Forensic Science International*, 203(1-3), pp. 1-2.

Madea, B., Saukko, P., Oliva, A. & Musshoff, F. (2010). Molecular pathology in forensic medicine--Introduction. *Forensic Science International*, 203(1-3), pp. 3-14.

Maeda, H., Zhu, B. L., Ishikawa, T. & Michiue, T. (2010). Forensic molecular pathology of violent deaths. *Forensic Science International*, 203(1-3), pp. 83-92.

Maeda, H., Ishikawa, T. & Michiue, T. (2011). Forensic biochemistry for functional investigation of death: concept and practical application. *Legal Medicine (Tokyo)*, 13(2), pp. 55-67.

Immunohistochemistry of Neuronal Apoptosis in Fatal Traumas: The Contribution of Forensic Molecular
Pathology in Medical Science

273

Marmarou, A., Fatouros, P. P., Barzo, P., Portella, G., Yoshihara, M., Tsuji, O., Yamamoto, T., Laine, F., Signoretti, S., Ward, J. D., Bullock, M. R. & Young, H. F. (2000). Contribution of edema and cerebral blood volume to traumatic brain swelling in head-injured patients. *Journal of Neurosurgery*, 93(2), pp. 183-193.

Marschall, S., Rothschild, M. A. & Bohnert, M. (2006). Expression of heat-shock protein 70 (Hsp70) in the respiratory tract and lungs of fire victims. *International Journal of Legal Medicine*, 120(6), pp. 355-359.

Martin, L. J., Al-Abdulla, N. A., Brambrink, A. M., Kirsch, J. R., Sieber, F. E. & Portera-Cailliau, C. (1998). Neurodegeneration in excitotoxicity, global cerebral ischemia, and target deprivation: A perspective on the contributions of apoptosis and necrosis. *Brain Research Bulletin*, 46(4), pp. 281-309.

Matute, C., Domercq, M. & Sanchez-Gomez, M. V. (2006). Glutamate-mediated glial injury: mechanisms and clinical importance. *Glia*, 53(2), pp. 212-224.

Michiue, T., Ishikawa, T., Quan, L., Li, D. R., Zhao, D., Komatsu, A., Zhu, B. L. & Maeda, H. (2008). Single-stranded DNA as an immunohistochemical marker of neuronal damage in human brain: an analysis of autopsy material with regard to the cause of death. *Forensic Science International*, 178(2-3), pp. 185-191.

Nag, S. (2011). Morphology and properties of astrocytes. *Methods in Molecular Biology*, 686, pp. 69-100.

Nakamura, T., Keep, R. F., Hua, Y., Hoff, J. T. & Xi, G. (2005). Oxidative DNA injury after experimental intracerebral hemorrhage. *Brain Research*, 1039(1-2), pp. 30-36.

Nixdorf-Miller, A., Hunsaker, D. M. & Hunsaker, J. C., 3rd. (2006). Hypothermia and hyperthermia medicolegal investigation of morbidity and mortality from exposure to environmental temperature extremes. *Archives of Pathology and Laboratory Medicine*, 130(9), pp. 1297-1304.

Oehmichen, M., Walter, T., Meissner, C. & Friedrich, H. J. (2003). Time course of cortical hemorrhages after closed traumatic brain injury: statistical analysis of posttraumatic histomorphological alterations. *Journal of Neurotrauma*, 20(1), pp. 87-103.

Oehmichen, M., Auer, R. N. & König, H. G. (2006). *Forensic neuropathology and associated neurology*. Springer Verlag.

Parizel, P. M., Makkat, S., Jorens, P. G., Ozsarlak, O., Cras, P., Van Goethem, J. W., van den Hauwe, L., Verlooy, J. & De Schepper, A. M. (2002). Brainstem hemorrhage in descending transtentorial herniation (Duret hemorrhage). *Intensive Care Medicine*, 28(1), pp. 85-88.

Piantadosi, C. A., Zhang, J., Levin, E. D., Folz, R. J. & Schmechel, D. E. (1997). Apoptosis and delayed neuronal damage after carbon monoxide poisoning in the rat. *Experimental Neurology*, 147(1), pp. 103-114.

Polazzi, E., Gianni, T. & Contestabile, A. (2001). Microglial cells protect cerebellar granule neurons from apoptosis: evidence for reciprocal signaling. *Glia*, 36(3), pp. 271-280.

Prockop, L. D. & Chichkova, R. I. (2007). Carbon monoxide intoxication: an updated review. *Journal of the Neurological Sciences*, 262(1-2), pp. 122-130.

Quan, L., Ishikawa, T., Michiue, T., Li, D. R., Zhao, D., Yoshida, C., Chen, J. H., Komatsu, A., Azuma, Y., Sakoda, S., Zhu, B. L. & Maeda, H. (2009). Analyses of cardiac blood cells and serum proteins with regard to cause of death in forensic autopsy cases. *Legal Medicine (Tokyo)*, 11 Suppl 1, pp. S297-300.

Reyes, R., Guo, M., Swann, K., Shetgeri, S. U., Sprague, S. M., Jimenez, D. F., Barone, C. M. & Ding, Y. (2009). Role of tumor necrosis factor-alpha and matrix metalloproteinase-9 in blood-brain barrier disruption after peripheral thermal injury in rats. *Journal of Neurosurgery*, 110(6), pp. 1218-1226.

Reyes, R., Jr., Wu, Y., Lai, Q., Mrizek, M., Berger, J., Jimenez, D. F., Barone, C. M. & Ding, Y. (2006). Early inflammatory response in rat brain after peripheral thermal injury. *Neuroscience Letters*, 407(1), pp. 11-15.

Rink, A., Fung, K. M., Trojanowski, J. Q., Lee, V. M., Neugebauer, E. & McIntosh, T. K. (1995). Evidence of apoptotic cell death after experimental traumatic brain injury in the rat. *American Journal of Pathology*, 147(6), pp. 1575-1583.

Rosenblum, W. I. (1997). Histopathologic clues to the pathways of neuronal death following ischemia/hypoxia. *Journal of Neurotrauma*, 14(5), pp. 313-326.

Sauvageau, A. & Boghossian, E. (2010). Classification of asphyxia: the need for standardization. *Journal of Forensic Sciences*, 55(5), pp. 1259-1267.

Schuliar, Y., Savourey, G., Besnard, Y. & Launey, J. C. (2001). Diagnosis of heat stroke in forensic medicine. Contribution of thermophysiology. *Forensic Science International*, 124(2-3), pp. 205-208.

Sharma, H. S. & Hoopes, P. J. (2003). Hyperthermia induced pathophysiology of the central nervous system. *International Journal of Hyperthermia*, 19(3), pp. 325-354.

Sharma, H. S. (2006). Hyperthermia induced brain oedema: current status and future perspectives. *Indian Journal of Medical Research*, 123(5), pp. 629-652.

Stefanidou, M., Athanaselis, S. & Spiliopoulou, C. (2008). Health impacts of fire smoke inhalation. *Inhalation Toxicology*, 20(8), pp. 761-766.

Stoica, B. A. & Faden, A. I. (2010). Cell death mechanisms and modulation in traumatic brain injury. *Neurotherapeutics*, 7(1), pp. 3-12.

Strackx, E., Van den Hove, D. L., Steinbusch, H. P., Steinbusch, H. W., Vles, J. S., Blanco, C. E. & Gavilanes, A. W. (2008). A combined behavioral and morphological study on the effects of fetal asphyxia on the nigrostriatal dopaminergic system in adult rats. *Experimental Neurology*, 211(2), pp. 413-422.

Stroick, M., Fatar, M., Ragoschke-Schumm, A., Fassbender, K., Bertsch, T. & Hennerici, M. G. (2006). Protein S-100B--a prognostic marker for cerebral damage. *Current Medicinal Chemistry*, 13(25), pp. 3053-3060.

Takamiya, M., Fujita, S., Saigusa, K. & Aoki, Y. (2007). Simultaneous detections of 27 cytokines during cerebral wound healing by multiplexed bead-based immunoassay for wound age estimation. *Journal of Neurotrauma*, 24(12), pp. 1833-1844.

Tamatani, M., Ogawa, S., Nunez, G. & Tohyama, M. (1998). Growth factors prevent changes in Bcl-2 and Bax expression and neuronal apoptosis induced by nitric oxide. *Cell Death and Differentiation*, 5(10), pp. 911-919.

Tao, L., Chen, X., Qin, Z. & Bian, S. (2006). Could NF-kappaB and caspase-3 be markers for estimation of post-interval of human traumatic brain injury? *Forensic Science International*, 162(1-3), pp. 174-177.

Tofighi, R., Tillmark, N., Dare, E., Aberg, A. M., Larsson, J. E. & Ceccatelli, S. (2 Hypoxia-independent apoptosis in neural cells exposed to carbon monoxide in v *Brain Research*, 1098(1), pp. 1-8.

Turk, E. E. (2010). Hypothermia. *Forensic Sci Med Pathol*, 6(2), pp. 106-115.

Unterberg, A. W., Stover, J., Kress, B. & Kiening, K. L. (2004). Edema and brain trauma. *Neuroscience*, 129(4), pp. 1021-1029.

Vaux, D. L. & Strasser, A. (1996). The molecular biology of apoptosis. *Proceedings of the National Academy of Sciences of the United States of America*, 93(6), pp. 2239-2244.

Vogel, P., Dux, E. & Wiessner, C. (1997). Evidence of apoptosis in primary neuronal cultures after heat shock. *Brain Research*, 764(1-2), pp. 205-213.

Wang, Q., Ishikawa, T., Michiue, T., Zhu, B. L. & Maeda, H. (2011a). Evaluation of human brain damage in fire fatality by quantification of basic fibroblast growth factor (bFGF), glial fibrillary acidic protein (GFAP) and single-stranded DNA (ssDNA) immunoreactivities. *Forensic Science International*, 211(1-3), pp. 19-26.

Wang, Q., Michiue, T., Ishikawa, T., Zhu, B. L. & Maeda, H. (2011b). Combined analyses of creatine kinase MB, cardiac troponin I and myoglobin in pericardial and cerebrospinal fluids to investigate myocardial and skeletal muscle injury in medicolegal autopsy cases. *Legal Medicine (Tokyo)*, 13(5), pp. 226-232.

Wang, Q., Ishikawa, T., Michiue, T., Zhu, B. L., Guan, D. W. & Maeda, H. (2012a). Evaluation of human brain damage in fatalities due to extreme environmental temperature by quantification of basic fibroblast growth factor (bFGF), glial fibrillary acidic protein (GFAP), S100beta and single-stranded DNA (ssDNA) immunoreactivities. Forensic Science International, doi:10.1016/j.forsciint.2012.01.015

Wang, Q., Ishikawa, T., Michiue, T., Zhu, B. L., Guan, D. W. & Maeda, H. (2012b). Quantitative immunohistochemical analysis of human brain basic fibroblast growth factor, glial fibrillary acidic protein and single-stranded DNA expressions following traumatic brain injury. Forensic Science International, doi:10.1016/j.forsciint.2012.04.025.

Williams, F. N., Herndon, D. N., Hawkins, H. K., Lee, J. O., Cox, R. A., Kulp, G. A., Finnerty, C. C., Chinkes, D. L. & Jeschke, M. G. (2009). The leading causes of death after burn injury in a single pediatric burn center. *Critical Care*, 13(6), pp. R183.

Williams, S., Raghupathi, R., MacKinnon, M. A., McIntosh, T. K., Saatman, K. E. & Graham, D. I. (2001). In situ DNA fragmentation occurs in white matter up to 12 months after head injury in man. *Acta Neuropathological*, 102(6), pp. 581-590.

Yakovlev, A. G. & Faden, A. I. (2001). Caspase-dependent apoptotic pathways in CNS injury. *Molecular Neurobiology*, 24(1-3), pp. 131-144.

Yoshida, C., Ishikawa, T., Michiue, T., Quan, L. & Maeda, H. (2011). Postmortem biochemistry and immunohistochemistry of chromogranin A as a stress marker with special regard to fatal hypothermia and hyperthermia. *International Journal of Legal Medicine*, 125(1), pp. 11-20.

Zhang, D., Hu, X., Qian, L., O'Callaghan, J. P. & Hong, J. S. (2010). Astrogliosis in CNS pathologies: is there a role for microglia? *Molecular Neurobiology*, 41(2-3), pp. 232-241.

ıu, B. L., Ishida, K., Fujita, M. Q. & Maeda, H. (2000). Immunohistochemical investigation of a pulmonary surfactant in fatal mechanical asphyxia. *International Journal of Legal Medicine*, 113(5), pp. 268-271.

Zhu, B. L., Ishida, K., Oritani, S., Quan, L., Taniguchi, M., Li, D. R., Fujita, M. Q. & Maeda, H. (2001a). Immunohistochemical investigation of pulmonary surfactant-associated protein A in fire victims. *Legal Medicine (Tokyo)*, 3(1), pp. 23-28.

Zhu, B. L., Ishida, K., Quan, L., Taniguchi, M., Oritani, S., Kamikodai, Y., Fujita, M. Q. & Maeda, H. (2001b). Post-mortem urinary myoglobin levels with reference to the causes of death. *Forensic Science International*, 115(3), pp. 183-188.

Permissions

The contributors of this book come from diverse backgrounds, making this book a truly international effort. This book will bring forth new frontiers with its revolutionizing research information and detailed analysis of the nascent developments around the world.

We would like to thank Doctor Tobias M. Ntuli, BScHons, Msc, PhD, for lending his expertise to make the book truly unique. He has played a crucial role in the development of this book. Without his invaluable contribution this book wouldn't have been possible. He has made vital efforts to compile up to date information on the varied aspects of this subject to make this book a valuable addition to the collection of many professionals and students.

This book was conceptualized with the vision of imparting up-to-date information and advanced data in this field. To ensure the same, a matchless editorial board was set up. Every individual on the board went through rigorous rounds of assessment to prove their worth. After which they invested a large part of their time researching and compiling the most relevant data for our readers. Conferences and sessions were held from time to time between the editorial board and the contributing authors to present the data in the most comprehensible form. The editorial team has worked tirelessly to provide valuable and valid information to help people across the globe.

Every chapter published in this book has been scrutinized by our experts. Their significance has been extensively debated. The topics covered herein carry significant findings which will fuel the growth of the discipline. They may even be implemented as practical applications or may be referred to as a beginning point for another development. Chapters in this book were first published by InTech; hereby published with permission under the Creative Commons Attribution License or equivalent.

The editorial board has been involved in producing this book since its inception. They have spent rigorous hours researching and exploring the diverse topics which have resulted in the successful publishing of this book. They have passed on their knowledge of decades through this book. To expedite this challenging task, the publisher supported the team at every step. A small team of assistant editors was also appointed to further simplify the editing procedure and attain best results for the readers.

Our editorial team has been hand-picked from every corner of the world. Their multi-ethnicity adds dynamic inputs to the discussions which result in innovative outcomes. These outcomes are then further discussed with the researchers and contributors who give their valuable feedback and opinion regarding the same. The feedback is then collaborated with the researches and they are edited in a comprehensive manner to aid the understanding of the subject.

Apart from the editorial board, the designing team has also invested a significant amount of their time in understanding the subject and creating the most relevant covers. They scrutinized every image to scout for the most suitable representation of the subject and create an appropriate cover for the book.

The publishing team has been involved in this book since its early stages. They were actively engaged in every process, be it collecting the data, connecting with the contributors or procuring relevant information. The team has been an ardent support to the editorial, designing and production team. Their endless efforts to recruit the best for this project, has resulted in the accomplishment of this book. They are a veteran in the field of academics and their pool of knowledge is as vast as their experience in printing. Their expertise and guidance has proved useful at every step. Their uncompromising quality standards have made this book an exceptional effort. Their encouragement from time to time has been an inspiration for everyone.

The publisher and the editorial board hope that this book will prove to be a valuable piece of knowledge for researchers, students, practitioners and scholars across the globe.

List of Contributors

Zhao Hongmei
Tianjin Key Laboratory of Food Biotechnology, College of Biotechnology and Food Science, Tianjin University of Commerce, Tianjin, China Chinese Academy of Medical Science & Peking Union Medical School, China

Ji Yean Kwon, Hisashi Naito, Takeshi Matsumoto and Masao Tanaka
Graduate School of Engineering Science, Osaka University, Japan

María A. García
Research Unit, Hospital Universitario Virgen de las Nieves, Granada, Spain

Esther Carrasco, Gema Jiménez, Manuel Picón, Houria Boulaiz and Juan Antonio Marchal
Biopathology and Regenerative Medicine Institute (IBIMER), Department of Human Anatomy and Embryology, Universidad de Granada, Granada, Spain

Alberto Ramírez, Elena López-Ruiz and Macarena Perán
Department of Health Sciences, Universidad de Jaén, Jaén, Spain

Joaquín Campos
Department of Pharmaceutical and Organic Chemistry, Universidad de Granada, Granada, Spain

Yasuhiko Hirabayashi
Department of Rheumatology, Hikarigaoka Spellman Hospital, Japan Department of Hematology & Rheumatology, Tohoku University Hospital, Japan

Silvina Grasso, Estefanía Carrasco-García, Lourdes Rocamora-Reverte, Ángeles Gómez-Martínez and José A. Ferragut
Instituto de Biología Molecular y Celular, Universidad Miguel Hernández, Elche (Alicante), Spain

M. Piedad Menéndez-Gutiérrez
Instituto de Biología Molecular y Celular, Universidad Miguel Hernández, Elche (Alicante), Spain Centro Nacional de Investigaciones Cardiovasculares, Departamento de Desarrollo y Reparación Cardiovascular, Madrid, Spain

Leticia Mayor-López, Elena Tristante and Isabel Martínez-Lacaci
Instituto de Biología Molecular y Celular, Universidad Miguel Hernández, Elche (Alicante), Spain Unidad AECC de Investigación Traslacional en Cáncer, Hospital Universitario Virgen de la Arrixaca, Instituto Murciano de Investigación Biosanitaria, Murcia, Spain

Pilar García-Morales and Miguel Saceda
Instituto de Biología Molecular y Celular, Universidad Miguel Hernández, Elche (Alicante), Spain Unidad de Investigación, Hospital General Universitario de Elche, Elche (Alicante), Spain

Tatyana O. Volkova
Department of Biochemistry, School of Medicine, Petrozavodsk State University, Petrozavodsk, Russia

Alexander N. Poltorak
Department of Pathology, School of Medicine, Tufts University, Boston, USA

Mayumi Tsuji and Katsuji Oguchi
Department of Pharmacology, Showa University, School of Medicine, Tokyo, Japan

Bo Yuan, Masahiko Imai, Hidetomo Kikuchi, Shin Fukushima, Shingo Hazama, Takenori Akaike, Yuta Yoshino, Kunio Ohyama and Hiroo Toyoda
Department of Clinical Molecular Genetics, School of Pharmacy, Tokyo University of Pharmacy & Life Sciences, Hachioji, Tokyo, Japan

Xiaomei Hu
National Therapeutic Center of Hematology of Traditional Chinese Medicine, XiYuan Hospital, China Academy of Traditional Chinese Medicine, Beijing, P.R. China

Xiaohua Pei
The Third Affiliated Hospital of Beijing University of Traditional Chinese Medicine, Beijing, P.R. China

A.V. Smirnov and G.L. Snigur
Department of Pathologic Anatomy, Volgograd State Medical University, Russia

M.P. Voronkova
Department of Pharmacology, Volgograd State Medical University, Russia

Qi Wang, Tomomi Michiue and Hitoshi Maeda
Department of Legal Medicine, Osaka City University Medical School, Abeno, Osaka, Japan